ADVANCES IN
ORGAN BIOLOGY

Volume 6 • 1998

MYOCARDIAL PRESERVATION
AND CELLULAR ADAPTATION

ADVANCES IN ORGAN BIOLOGY

MYOCARDIAL PRESERVATION AND CELLULAR ADAPTATION

Series Editor: E. EDWARD BITTAR
Department of Physiology
University of Wisconsin Medical School
Madison, Wisconsin

Guest Editor: DIPAK K. DAS
Department of Surgery
University of Connecticut Health Center
Farmington, Connecticut

VOLUME 6 • *1998*

JAI PRESS INC.

Stamford, Connecticut *London, England*

100 Prospect Street
Stamford, Connecticut 06901

JAI PRESS LTD.
38 Tavistock Street
Covent Garden
London WC2E 7PB
England

ISBN: 0-7623-0391-3

Transferred to Digital Printing 2009

CONTENTS

LIST OF CONTRIBUTORS

John A. Auchampach	Division of Cardiology University of Louisville Louisville, Kentucky
Anirban Banerjee	Department of Surgery University of Colorado Health Sciences Center Denver, Colorado
Roberto Bolli	Division of Cardiology University of Louisville Louisville, Kentucky
Mario Cagnoni	Instituto di Clinica Medica Generalle University Degli Studi Italy
Brian C. Cain	Department of Surgery University of Colorado Health Sciences Center Denver, Colorado
Charles B. Cairns	Department of Surgery University of Colorado Health Sciences Center Denver, Colorado
Marco Matucci Cerinic	Instituto di Clinica Medica Generalle University Degli Studi Italy
Naranjan S. Dhalla	Department of Physiology Faculty of Medicine University of Manitoba Winnipeg, Manitoba, Canada

Dipak K. Das	Department of Surgery University of Connecticut School of Medicine Farmington, Connecticut
Tanya Das	Bose Institute Calcutta, India
Gary T. Elliott	Department of Pharmaceutical Development Ribi ImmunoChem Research, Inc. Hamilton, Montana
Sergio Generini	Instituto di Clinica Medica Generalle University Degli Studi Italy
Satish Ghatpande	Department of Anatomy and Cell Biology SUNY Health Center Brooklyn, New York
Ben C.G. Gho	Department of Experimental Cardiology Erasmus University Rotterdam, The Netherlands
Garrett J. Gross	Department of Pharmacology and Toxicology Medical College of Wisconsin Milwaukee, Wisconsin
Alden H. Harken	Department of Surgery University of Colorado Health Sciences Center Denver, Colorado
Kyong Joo	Department of Surgery University of Colorado Health Sciences Center Denver, Colorado
Frantisek Kolar	Institute of Physiology Academy of Sciences of the Czech Republic Prague, Czech Republic
Nilanjana Maulik	Molecular Cardiology Laboratory University of Connecticut School of Medicine Farmington, Connecticut

David A. Mei	Department of Pharmacology and Toxicology Medical College of Wisconsin Milwaukee, Wisconsin
Daniel R. Meldrum	Department of Surgery University of Colorado Health Sciences Center Denver. Colorado
Xianzhong Meng	Department of Surgery University of Colorado Health Sciences Center Denver, Colorado
Tsuneo Mizumura	Department of Pharmacology and Toxicology Medical College of Wisconsin Milwaukee, Wisconsin
Ernes E. Moore	Department of Surgery University of Colorado Health Sciences Center Denver, Colorado
Kasem Nithipatikom	Department of Pharmacology and Toxicology Medical College of Wisconsin Milwaukee, Wisconsin
Bohuslav Ostadal	Institute of Physiology Academy of Sciences of the Czech Republic Prague, Czech Republic
Ivana Ostadalova	Institute of Physiology Academy of Sciences of the Czech Republic Prague, Czech Republic
James R. Parratt	Department of Physiology and Pharmacology University Strathclyde, Royal College Glasgow Scotland
Vaclav Pelouch	Institute of Physiology Academy of Sciences of the Czech Republic Prague, Czech Republic
Alberto Pignone	Instituto di Clinica Medica Generalle University Degli Studi Italy

Peipei Ping — Division of Cardiology
University of Louisville
Louisville, Kentucky

Yumin Qiu — Division of Cardiology
University of Louisville
Louisville, Kentucky

T. Ramasarma — Department of Biochemistry
Indian Institute of Science
Bangalore, India

Prasanta K. Ray — Bose Institute
Calcutta, India

Fabia Gamboni Robertson — Department of Surgery
University of Colorado Health Sciences Center
Denver, Colorado

M.A.Q. Siddiqui — Department of Anatomy and Cell Biology
SUNY Health Center
Brooklyn, New York

Nabuakira Takeda — School of Medicine
Jikei University
Tokyo, Japan

Xian-Liang Tang — Division of Cardiology
University of Louisville
Louisville, Kentucky

Arpad Tosaki — Department of Surgery
University of Connecticut School of Medicine
Farmington, Connecticut

Mirella A. van den Doel — Department of Experimental Cardiology
Erasmus University
Rotterdam, The Netherlands

Agnes Vegh — Department of Pharmacology
Albert Szent-Gyogyi Medical University
Szeged, Hungary

Pieter D. Verdouw
Department of Experimental Cardiology
Erasmus University
Rotterdam, The Netherlands

Michael Wagner
Department of Anatomy and Cell Biology
SUNY Health Center
Brooklyn, New York

Patricia A. Weber
Department of Pharmaceutical Development
Ribi ImmunoChem Research, Inc.
Hamilton, Montana

PREFACE

Living organisms exhibit specific responses when confronted with sudden changes in their environmental conditions. The ability of the cells to acclimate to their new environment is the integral driving force for adaptive modification of the cells. Such adaptation involves a number of cellular and biochemical alterations including metabolic homeostasis and reprogramming of gene expression. Changes in metabolic pathways are generally short-lived and reversible, while the consequences of gene expression are a longterm process and may lead to permanent alternation in the pattern of adaptive responses.

The heart possesses remarkable ability to adapt itself against any stressful situation by increasing resistance to the adverse consequences. Stress composes the foundation of many degenerative heart diseases including atherosclerosis, spasm, thrombosis, cardiomyopathy, and congestive heart failure. Based on the concept that excessive stress may play a crucial role in the pathogenesis of ischemic heart disease, attempts were made to design methods for prevention of myocardial injury. Creation of stress reactions by repeated ischemia and reperfusion or subjecting the hearts to heat or oxidative stress enable them to meet the future stress challenge. Repeated stress exposures adapt the heart to withstand more severe stress reactions probably by upregulating the cellular defense and direct accumulation of intracellular mediators, which presumably constitute the material basis of increased adaptation to stress. Thus, the powerful cardioprotective effect of adaptation is likely to originate at the cellular and molecular levels that compose fundamental processes in the prophylaxis of such diseases.

Volume six of the *Advances in Organ Biology* series contains state-of-the-art reviews on myocardial preservation and cellular adaptation from the leading authorities in this subject. The editor hopes that this volume serves as an up-to-date source of information for scientists as well as clinicians interested in applying the concept of Stress Adaptation to cure heart diseases. The editor would like to thank the contributing authors for their excellent contributions and cooperation.

Dipak K. Das
Guest Editor

PRECONDITIONING INDUCES BOTH IMMEDIATE AND DELAYED PROTECTION AGAINST ARRHYTHMIAS RESULTING FROM ISCHEMIA AND REPERFUSION

James R. Parratt and Agnes Vegh

Advances in Organ Biology
Volume 6, pages 1-20.

ISBN: 0-7623-0391-3

I. INTRODUCTION

It has now been over 10 years since the germinal paper by Murry et al. (1986) demonstrating that short periods of ischemia can protect the myocardium against the effects of a subsequent, more prolonged period of ischemia. This finding was reflected in a reduction in myocardial infarct size, an aspect of the protection afforded by preconditioning that has already been discussed in another chapter in the present volume. Brief periods of ischemia also protect against arrhythmias that result from coronary artery occlusion and reperfusion and, in many ways, this must be perhaps the most important aspect of this particular form of adaptation. The ability to reduce myocardial cell death by a combination of coronary thrombolysis and drug therapy (e.g., low-dose aspirin, β-adrenoceptor blockade, and angiotensin-converting enzyme [ACE] inhibitors) is now well established and is a major advance in the treatment of patients with acute myocardial infarction.

In contrast, the problem of sudden cardiac death from ventricular fibrillation (or reperfusion) outside the hospital setting remains the biggest problem facing clinical cardiology, despite the introduction of implantable defibrillators, the limited success of mobile coronary care units, and the training of nonmedical personnel in the use of defibrillators. In fact, drug therapy has proved to be largely ineffective except in the hospital setting. Thus, any phenomenon that has been shown to reduce ventricular fibrillation in the experimental setting is to be welcomed and, if we understood the mechanisms involved, there is the potential for clinical exploitation. Indeed, there are situations in which previous anginal attacks have reduced the severity of a subsequent acute myocardial infarction (Kloner et al., 1995). What follows is a summary of the current understanding of the powerful antiarrhythmic effect of ischemic preconditioning and cardiac pacing, with particular emphasis on the possible mechanisms involved.

II. REDUCTION IN ARRHYTHMIA SEVERITY DURING MYOCARDIAL ISCHEMIA BY BRIEF CORONARY ARTERY OCCLUSIONS AND BY CARDIAC PACING

There were already several references to protective (antiarrhythmic) effects of brief periods of ischemia prior to the commencement of the preconditioning era in 1986. For example, Harris (1950) described the effects of occluding a coronary artery in two stages—partial occlusion followed by complete closure—as a means of reducing arrhythmia severity; ventricular fibrillation is common when a coronary artery is acutely and completely occluded but is reduced when this is done in two stages. Harris used this technique to study the arrhythmias that occurred hours (or days) after coronary occlusion; arrhythmias that are similar in mechanism to those seen in clinical coronary care units.

This particular experimental model has been used extensively to examine the effects of potential antiarrhythmic drugs against these late ventricular arrhythmias. Other evidence came from the studies of Gulker and colleagues (1977), who showed that the reduction in the ventricular fibrillation threshold, which occurs during coronary artery occlusion, was less marked with subsequent occlusions, while Barber (1983) also showed that short (5-minute) serial occlusions of the left anterior descending coronary artery in anesthetized dogs resulted in fewer ectopic beats during the second, and subsequent, occlusions provided the reperfusion time was less than 40 minutes.

In our own experiments (e.g., Marshall and Parratt [1980]), where the primary purpose was to examine changes in blood flow in developing infarcts, we often used the technique of occluding the coronary artery for brief periods and then reperfusing in order to reduce the arrhythmias that would normally occur when a coronary artery was occluded for a prolonged period of time. A similar protection against those arrhythmias that result following reperfusion of the ischemic myocardium was first described by Shiki and Hearse in 1987. This particular study examined in great detail the effect of varying the time between two coronary artery occlusions of the same (short) duration by increasing the recovery period to hours and even days. This almost certainly represented the first attempt to examine whether brief periods of ischemia are able to protect the myocardium long after the initial stimulus, a phenomenon that has since aroused considerable interest. The fact that their own results were negative, and that no protection was seen several hours after the initial preconditioning stimulus, might well be because, in order to achieve delayed protection, several brief periods of coronary artery occlusion or cardiac pacing are required.

The stimulus for our own studies on the antiarrhythmic effects of ischemic preconditioning were the experiments of Podzuweit and colleagues (1989), who examined the arrhythmic effects of locally infused noradrenaline directly into the myocardium. They demonstrated that the pacemaker activity that occurred during these infusions was abolished if the coronary artery supplying the infusion area was occluded. When the artery was reopened, ventricular arrhythmias resumed within seconds of the release of the occlusion. They termed this phenomenon "the antiarrhythmic effect of ischaemia" and suggested that "the ischaemic myocardium might have previously unrecognised antiarrhythmic properties." They wondered whether reperfusion arrhythmias might result from vanishing ischemic protection. A similar concept is that a variety of potentially protective substances are released from the ischemic myocardium, and particularly from endothelial cells, and that these might modify the effects of subsequent occlusions (Parratt, 1987, 1993).

It was Sadayoshi Komori, working in the Glasgow department. who was the first to demonstrate the marked antiarrhythmic effects of brief periods of coronary artery occlusion. He was interested in the question of whether survival from a prolonged period ischemic insult could be modified if the myocardium had been subjected to short (preconditioning) coronary artery occlusions. He showed. in anesthetized rats, that a brief period of coronary artery occlusion (the optimum period was 3 minutes) led to a marked reduction in arrhythmia severity when that artery was reoccluded several minutes later (Komori et al.,1990a, 1990b). Later, similar marked antiarrhythmic effects of brief periods of coronary artery occlusion were demonstrated in rat isolated hearts (Lawson et al., 1993a, 1993b; Piacentini et al., 1993; and recently reviewed by Connaughton et al., 1996).

Clearly, it is easier to examine potential mechanisms in larger animals, and we repeated these studies in anesthetized dogs (Vegh et al., 1990, 1992a). The original protocol was to occlude the anterior descending coronary artery, in dogs anesthetized with urethane and chloralose, for one or two 5-minute periods (with a 20-minute reperfusion period between); then, at various times later, to occlude the same artery for a prolonged period of time (usually 25 minutes); and at the end of that period to rapidly reperfuse the ischemic myocardium. More recently, we have examined the effects of four brief coronary artery occlusions to determine whether this modifies the time course of the protection. This arrhythmia model is a particularly severe one. In control (nonpreconditioned) dogs, ventricular fibrillation is common (usually around 50% of the animals fibrillate at some time during the occlusion period), ventricular tachycardia (VT) is the norm, with many such periods of VT during the occlusion period, and the number of single or coupled ventricular premature (ectopic) beats is large (i.e., around 500 during the 25-minute period, which means that approximately 1 in 20 beats is ectopic). At the end of the 25-minute period, rapid reperfusion invariably results in ventricular fibrillation; thus there are very few survivors from the combined ischemia–reperfusion insult. This model is therefore a particularly good one for examining potential antiarrhythmic effects of brief periods of ischemia. In contrast, in dogs anesthetized with pentobarbitone there are rather few ventricular premature beats and very few episodes of VT (Przyklenk and Kloner, 1995), making it a somewhat inappropriate model to use in examining the antiarrhythmic effects of ischemic preconditioning. In our hands, brief periods of ischemia markedly reduce the severity of arrhythmias that occur during a subsequent coronary artery occlusion. This is illustrated in Figure 1. Ventricular fibrillation is rare in preconditioned dogs and the incidence and number of episodes of ventricular tachycardia, and the number of ventricular premature beats, is markedly reduced.

Similar protection can be achieved by brief periods of rapid right ventricular pacing (Vegh et al., 1991b). This is illustrated in Figure 2. Pacing to such high rates (in this case 300 beats-minute^{-1}) presumably results in some degree of myocardial ischemia, particularly in the subendocardial regions of the left ventricular wall. Perfusion (arterial) pressure is markedly reduced during pacing and left ventricular filling pressures are elevated, resulting in a marked reduction in subendocardial driving pressure (Marshall and Parratt, 1974). The main difference between preconditioning by short periods of coronary artery occlusion and by periods of rapid cardiac pacing is that the duration of the protection by pacing is somewhat less, presumably indicating a less powerful preconditioning stimulus.

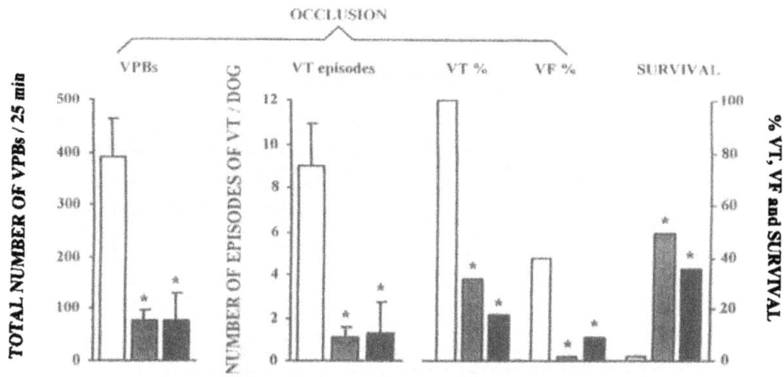

Figure 1. The incidence and severity of ventricular arrhythmias during a 25-minute occlusion of the anterior descending branch of the left coronary artery, and survival following reperfusion at the end of the occlusion period, in control dogs (open columns) and in dogs subjected to preconditioning, either by two (striped columns) or four 5-minute (solid columns) coronary artery occlusions. The severity of these ventricular arrhythmias during such a prolonged occlusion is markedly reduced when the dogs had been previously preconditioned, either by two or four brief periods of occlusion of that same artery. **P* < .05 cp. control dogs. VF, ventricular fibrillation; VPBs, ventricular premature beats; VT, ventricular tachycardia.

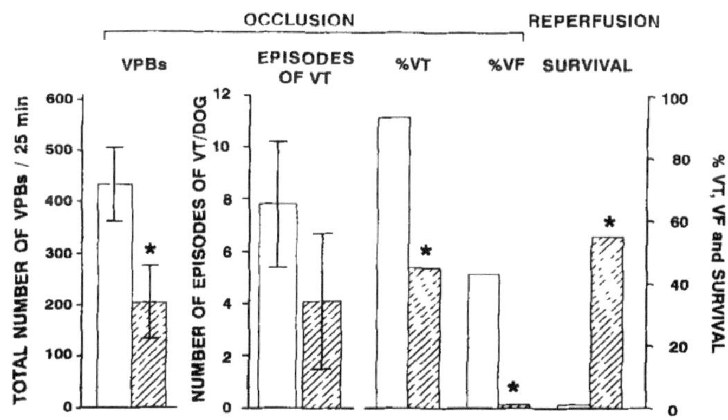

Figure 2. *The severity of ventricular arrhythmias during a 25-minute coronary artery* occlusion in control dogs (open columns) and in dogs subjected to right ventricular pacing 5 minutes previously (striped columns). Cardiac pacing (four times for 5 minutes at a rate of 220 beats-minute^{-1}) 5 minutes before occlusion of the left anterior descending coronary artery, markedly reduces the number of ventricular premature beats (VPBs), the number of episodes of ventricular tachycardia (VT), and the incidences of VT and ventricular fibrillation (VF) during occlusion, and also increases survival from the combined ischemia–reperfusion insult. **P* < .05 cp. control, nonpaced dogs.

In summary, the characteristics of the antiarrhythmic effects of ischemic preconditioning are as follows. They are *marked* (e.g., absence of ventricular fibrillation in the canine model), *real* (there is an absolute reduction in arrhythmia severity no matter how long the occlusion is maintained) but *transient* (the protective effects are lost if the interval between the preconditioning stimulus and the coronary artery occlusion is prolonged). However, as we will discuss later, the protection returns many hours later; that is, there are two phases of protection, both immediate and delayed.

III. DELAYED ANTIARRHYTHMIC PROTECTION INDUCED BY PERIODS OF CARDIAC PACING

As we have seen, the protection afforded by "classical preconditioning" is powerful but transient since most of the protection is lost if the interval between the preconditioning stimulus and the prolonged coronary artery occlusion is extended to 1 hour (Vegh et al., 1992a). It was first demonstrated in 1992 (Yamashita et al., 1992) at a poster demonstration at the World Congress of the International Society for Heart Research in Kobe, Japan, that protection by preconditioning against myocardial ischemic damage (limitation of infarct size) occurs in two phases. There is an early protection, as first demonstrated by Murry et al. (1986), which is then lost but returns 20 to 24 hours after the initial preconditioning stimulus (Kuzuya et al., 1993; Marber et al., 1993; and recently reviewed by (Yellon and Baxter, 1996). None of these studies examined a possible influence of preconditioning on arrhythmias during the so-called second window of protection.

We have investigated this influence using cardiac pacing as the preconditioning stimulus. In this study dogs were paced, at a rate of 220 beats-minute^{-1} for four 5-minute periods with 5-minute rest (reperfusion) periods between the pacing stimuli. The controls were dogs in which the bipolar pacing electrode was introduced into the right ventricle, left for the same period of time, and then withdrawn; these dogs were not paced. At various times after the pacing stimulus (from 5 minutes to 72 hours; Kaszala et al., 1996), the dogs were reanesthetized, thoracotomized, and prepared for a 25-minute occlusion of the left coronary artery. There was no evidence of protection by pacing against arrhythmias when the occlusion was carried out between 15 minutes and 6 hours after the termination of the pacing period. However, at 24, 48, and 72 hours, there was a marked suppression in ventricular ectopic activity and, of greater importance, a marked reduction in ventricular fibrillation at 24 hours (Figure 3). Although protection against ventricular fibrillation was not observed at 48 or 72 hours after the pacing stimulus, if the dogs were repaced 48 hours after the initial period of pacing and subjected to coronary occlusion 48 hours after that, protection against ventricular fibrillation was still observed (Kis et al., 1996). This suggests that repacing can extend the period of protection against occlusion-induced ventricular fibrillation. As yet un-

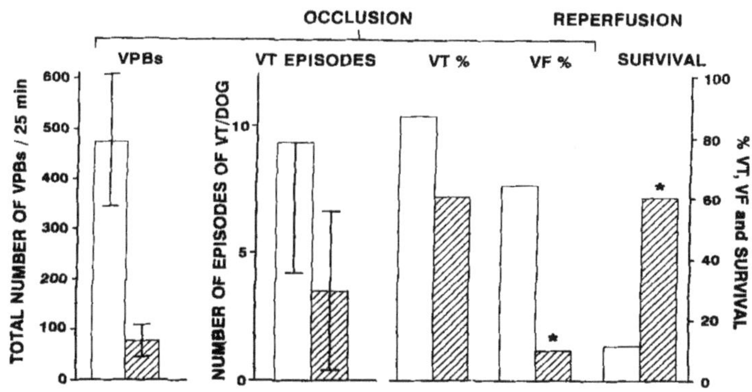

Figure 3. The incidence and severity of ventricular arrhythmias during a 25-minute coronary artery (left anterior descending occlusion in control dogs (open columns) and in dogs subjected to right ventricular pacing 20 to 24 hours previously (striped columns). Cardiac pacing markedly reduced the number of ventricular premature beats (VPBs). the number of episodes of ventricular tachycardia (VT), and the incidences of VT and ventricular fibrillation (VF) in these dogs when they were subjected to a 25-minute occlusion 20 to 24 hours later. Sixty percent of these dogs survived the combined ischemia–reperfusion insult; in contrast, there were few survivors in the control group. $^{*}P < .05$ cp. control, nonpaced dogs.

published studies show that protection can still be demonstrated 72 hours after the second period of pacing. This raises the possibility of being able to keep the heart protected for prolonged periods of time as a result of right ventricular pacing (or perhaps exercise).

IV. MECHANISMS OF ANTIARRHYTHMIC PROTECTION INDUCED BY PRECONDITIONING AND PACING

Although the precise mechanisms have still to be elucidated, they seem to involve the generation of endogenous myocardial protective substances, probably derived from the coronary vascular endothelium. Certainly, preconditioning results in less severe ischemia during the prolonged coronary artery occlusion, whether the preconditioning is induced by brief periods of coronary artery occlusion or by rapid ventricular pacing. This is clear from analysis of records of epicardial ST-segment elevation and of changes in the inhomogeneity of electrical activation within the ischemic area. Both these indices of ischemia severity are reduced as a result of preconditioning, and this is illustrated in Figures 4 (for short coronary artery occlusions) and 5. The precise relationship between ischemia and arrhythmia severity is unclear and it is unlikely that the reduction in these two particular indices of ische-

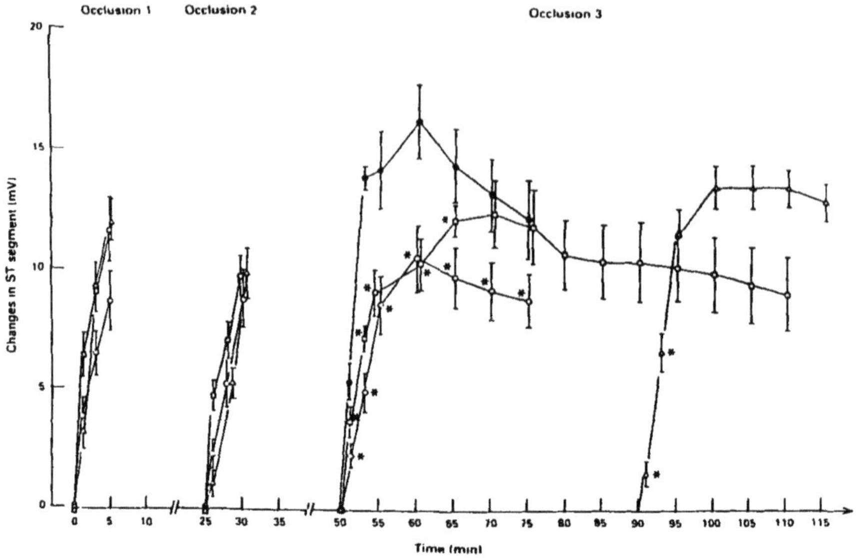

Figure 4. Changes in ST-segment elevation, recorded from epicardial electrodes, in anesthetized dogs subjected to a 25-minute occlusion of the left anterior descending coronary artery (●) and in dogs in which this occlusion (occlusion 3) was preceded by two short 5-minute preconditioning occlusions (occlusions 1 and 2) and in which animals were reperfused at the end of the 25-minute occlusion period (○) or at the end of a 60-minute occlusion period (□). The severity of the ST-segment changes was less pronounced in preconditioned dogs provided the reperfusion time was 20 minutes but not if the reperfusion time between the preconditioning occlusions and the prolonged occlusion was increased to 1 hour. (Δ) although, even at this time, there was some delay in the generation of the ST-segment change. Adapted from Vegh et al. (1992), with permission.

mia severity completely accounts for the much more marked reduction in arrhythmia severity. However, this must certainly be a contributing factor.

Two other factors thought to be critical as mediators of the reduction in infarct size achieved by ischemic preconditioning, protein kinase C (PKC) translocation to the sarcolemmal membrane and the opening of adenosine triphosphate–dependent potassium (K^+_{ATP}) channels are probably not important for protection by preconditioning against arrhythmias. PKC translocation certainly occurs, in our hands, 24 hours after a pacing stimulus (Wilson et al., 1996) but the significance of this is unclear. Certainly the translocation of most isoforms of PKC to the membrane would be proarrhythmic rather than antiarrhythmic (phorboylesters are arrhythmogenic in isolated perfused hearts), and we could find no clear evidence for a role of K^+_{ATP} channels in the acute antiarrhythmic effects of preconditioning (Vegh et al., 1993a). In contrast, evidence seems to be accumulating that the antiarrhythmic effects are

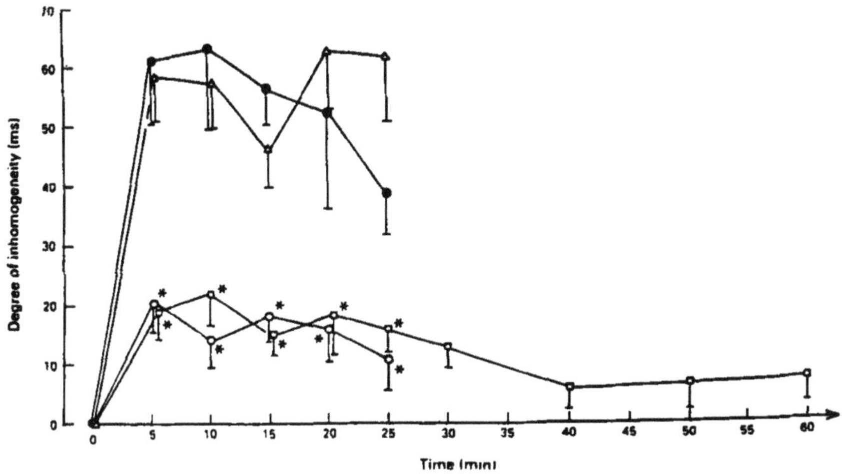

Figure 5. The degree of inhomogeneity of activation within the ischemic area in dogs subjected to a 25-minute occlusion of the left anterior descending coronary artery (●), and in dogs subjected to two 5-minute preconditioning occlusions (○ and □). Preconditioning markedly reduces the degree of inhomogeneity, but this is not seen if the artery is reoccluded 1 hour after the last preconditioning occlusion (Δ). *$P < .05$ versus changes in inhomogeneity during coronary occlusion and at the same time point in dogs that were not preconditioned. From Vegh et al. (1992), with permission.

mediated through the generation of nitric oxide by endothelial cells and this particular hypothesis, and the evidence for it, will now be discussed.

A. A Hypothesis to Explain the Mechanism of the Antiarrhythmic Effects of Ischemic preconditioning

We began with the supposition that the antiarrhythmic protection originates in the vascular wall. This supposition derives form the fact that it is not only the heart that can be preconditioned, but other organs such as the brain, kidney, and skeletal muscle as well. The one thing that these organs have in common is, of course, a vascular supply. We conclude, therefore, that the "target" for preconditioning in the heart is the coronary vasculature and, in particular, the endothelium. The evidence for this will now be outline.

1. Substances Derived from the Endothelium Appear to Modify Arrhythmia Severity

Removal of the endothelium in rat isolated perfused hearts results in a marked increase in arrhythmia severity when a coronary artery is occluded (Fatehi-

Hassanabad et al., 1996). This is illustrated in Figure 6. Not only is there a greatly increased number of ventricular premature beats in endothelium-denuded hearts but the arrhythmias are more severe; for example, ventricular fibrillation during reperfusion following a 30-minute period of ischemia in rat isolated hearts is rare but is quite common following endothelium denudation with the detergent triton X (no ventricular tachycardias followed reperfusion in control hearts but there was a 40% incidence in endothelium-denuded hearts) These experiments suggest that substances derived from the endothelium modify arrhythmia severity both during ischemia and during reperfusion. What are these substances likely to be? Endothelial cells generate a variety of potent vasoactive substances (prostanoids such as prostacyclin, bradykinin, nitric oxide, endothelin) and influence vascular activity by the release of such diffusable vasoactive substances as originally demonstrated in the elegant studies of Furchgott and Zawadzki (1980). The particular relevance of these findings to the coronary circulation has been reviewed on a number of occasions (Bassenge, 1995; Fleming et al., 1996).

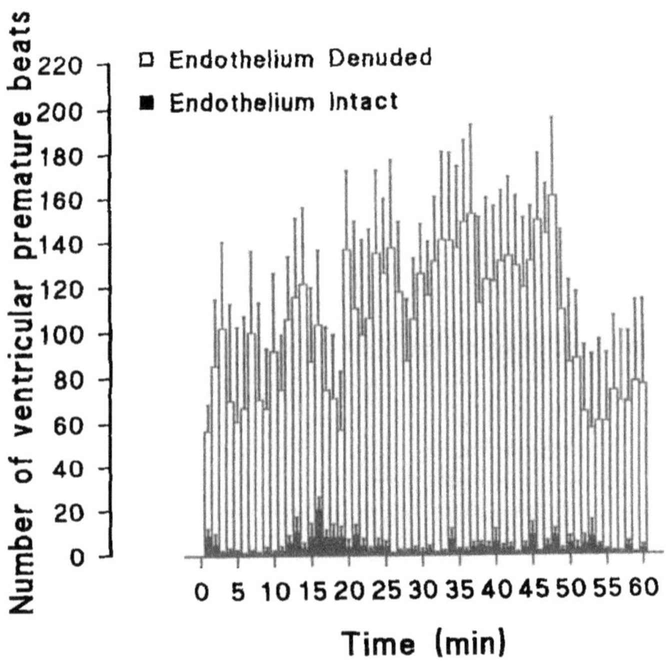

Figure 6. Ventricular arrhythmias (number of ventricular premature beats at 1-minute intervals) following coronary artery occlusion in rat isolated perfused hearts in which the endothelium was intact (closed histograms) or following treatment with triton X100 to denude the endothelium (open histograms). Arrhythmias are much more severe in endothelium-denuded hearts.

More recently it has been recognized that coronary vascular (and endocardial) endothelial cells communicate with cardiac myocytes as well as with vascular smooth muscle cells. This "cross-talk" between endothelial cells and cardiac myocytes serves to modulate myocardial contractility through the release of various diffusable endothelium-derived substances. This effect was first demonstrated by Brutsaert and colleagues (1988) in studies in which they selectively damaged the endocardial endothelium of cat isolated papillary muscle preparations using a controlled, transient exposure to a dilute detergent solution. This action modified twitch contraction in a characteristic manner; twitch contraction was abbreviated, with an earlier onset of mechanical relaxation and a small decrease in peak isometric force. In vivo, the main effect of impairing endocardial endothelium is abbreviation of the left ventricular pressure time curve as a result of earlier and more pronounced isometric relaxation (recently reviewed in Sys and Brutsaert [1995]). Some of our own in vivo studies in anesthetized dogs show that administration of an inhibitor of the L-arginine–nitric oxide pathway invariably results in an immediate increase in positive and negative left ventricular dP/dt max and marked increases in left ventricular diameter and in the end-systolic pressure diameter relationship (ESPDR; Kaszaki et al., 1996). This is illustrated in Figure 7. We suggest that this

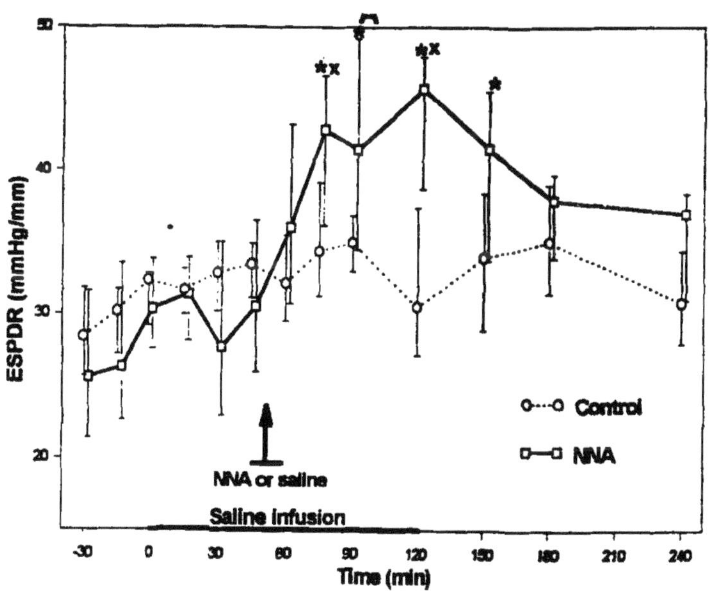

Figure 7. Changes in myocardial contractility (end-systolic pressure diameter relationship [ESPDR]) in saline and N^G-nitro-L-arginine (L-NNA) treated dogs. Inhibition of the L-arginine nitric–oxide pathway results in an immediate increase in this index of myocardial contractility. From Kaszaki et al. (1996), with permission.

relationship between coronary vascular endothelial cells and cardiac myocytes not only modulates cardiac function (contractility) but also modifies, in a protective function, those life-threatening arrhythmias that arise during a period of myocardial ischemia and reperfusion. There are thus two aspects of this cell-to-cell communication in which nitric oxide is particularly involved.

2. *Release of Endothelium-derived Mediators Appears to be Increased Under Conditions of Ischemic Preconditioning*

We have recently examined the release of bradykinin and nitric oxide under conditions of myocardial ischemia and ischemic preconditioning (Parratt et al., 1997). The protocol was to sample from the coronary sinus at various times during the preconditioning procedure (by two 5-minute periods of coronary artery occlusion) and throughout the subsequent prolonged occlusion. Unfortunately, in these dogs we were unable to sample directly from the ischemic region using the local coronary venous sampling technique that we have exploited earlier (e.g., Coker et al., 1981). However, even in mixed coronary venous blood draining from both the ischemic and essentially normal nonischemic regions of the left ventricular wall, elevated bradykinin levels could be demonstrated, particularly in preconditioned dogs immediately prior to the prolonged occlusion. Furthermore, the levels were considerably higher throughout the prolonged occlusion in preconditioned dogs as compared with controls (Parratt et al., 1997). These results suggest that bradykinin release is more pronounced under conditions of ischemic preconditioning. Similar results have been demonstrated in patients undergoing coronary angioplasty; during the balloon inflation (and subsequent deflation) bradykinin levels in coronary sinus blood are greatly elevated, indicating that this bradykinin release under conditions of ischemia is both marked and rapid (Parratt et al., 1997).

More recently, we have estimated nitric oxide generation under experimental conditions of ischemia and preconditioning by measuring nitrosyl-hem complexes in coronary sinus blood using electron paramagnetic resonance spectroscopy. Nitrosyl-hem complexes were often observed in coronary sinus blood of preconditioned dogs but never in dogs subjected to a single 25-minute coronary artery occlusion. Furthermore, methemoglobin levels were greatly elevated under conditions of preconditioning. One of the explanations for this could be that nitric oxide, generated by the ischemic myocardium, displaces oxygen from oxyhemoglobin with the resultant ultimate formation of nitrate. These results suggest that bradykinin release occurs as a trigger for ischemic preconditioning, resulting in the subsequent generation and release of nitric oxide.

3. *Antiarrhythmic Effects Appear to be Mediated by Generation of Nitric Oxide*

Several pharmacological studies have been designed that interfere with various aspects of the bradykinin–nitric oxide–cyclic guanosine monophosphate (cGMP) pathway. We have used three approaches to interfere with the formation of cGMP in cardiomyocytes, which we believe to be the final protective mediator (Figure 8). The first approach concerns the effects of blocking bradykinin B_2 receptors with icatibant (Vegh et al., 1994a; Parratt et al., 1995). The results of such experiments demonstrate that preconditioning is difficult to achieve in dogs in which the effects of bradykinin on B_2 receptors have been prevented by icatibant; the incidence of arrhythmias during the preconditioning occlusions themselves is much more pronounced than during the normal preconditioning procedure. This suggests that bradykinin is being released early as a trigger for the protection. In those dogs in which preconditioning was achieved in the presence of icatibant, the reduction in arrhythmia severity was markedly attenuated. This is illustrated in Figure 9. Even more dramatic effects of icatibant occur in dogs subjected to rapid ventricular pacing. In these experiments, and in con-

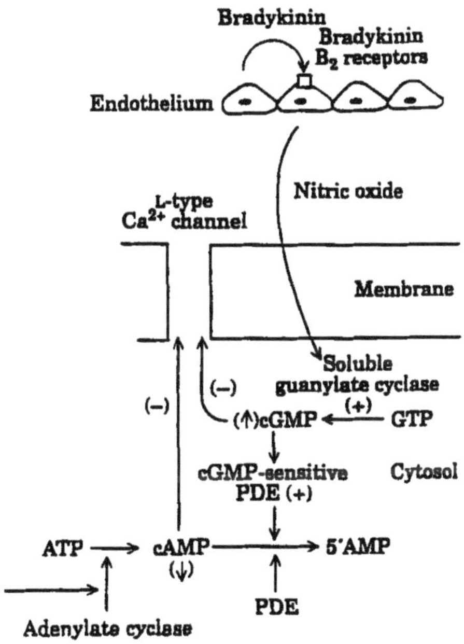

Figure 8. Role of endothelium-derived endogenous protective mediators in ischemic preconditioning—a hypothesis. Bradykinin is released, probably from endothelial cells (which have the mechanisms for generating and releasing kinins). It then acts on B_2 receptors on the endothelial surface to increase the calcium transient within these cells and to activate the L arginine–nitric oxide pathway. Nitric oxide then "talks" to the cardiac myocyte, stimulates soluble guanylyl cyclase, and elevates cyclic (cGMP). This stimulates a cGMP-sensitive phosphodiesterase (PDP), thus reducing cyclic AMP levels, inhibits calcium entry through L-type calcium, channels, and depresses myocardial contractility. GTP, guanosine triphosphate. From Parratt and Vegh (1996), with permission.

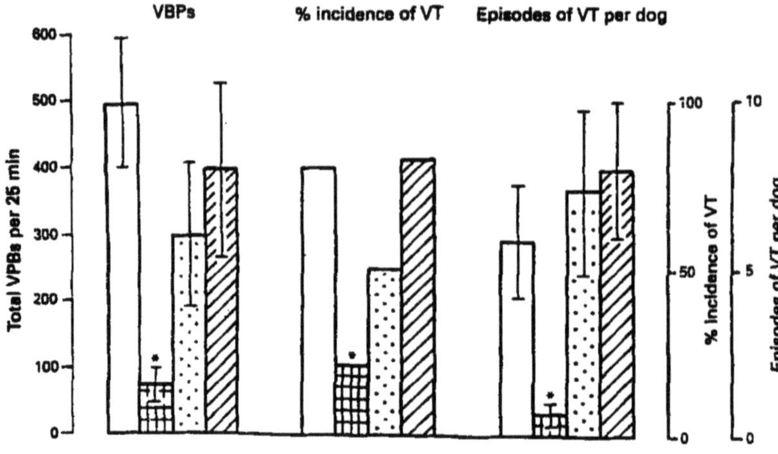

Figure 9. The number of ventricular premature beats (VPBs), the incidence of ventricular tachycardia (VT), and the number of episodes of VT that occur during a 25-minute occlusion of the left anterior descending coronary artery in control dogs (open columns) and in preconditioned dogs (crosshatched columns) in the absence, or presence, of the bradykinin B_2 antagonist icatibant. Icatibant was given either before the preconditioning procedures (stippled columns) or after preconditioning but prior to the prolonged occlusion (striped columns). From Vegh et al., (1994a), with permission.

trast to normal dog, arrhythmias were common immediately after the cessation of pacing in the presence of icatibant and no dogs survived the resultant coronary artery occlusion (Vegh et al., 1995). It should be noted in this respect that bradykinin infused locally into a side branch of the left anterior descending coronary artery is itself markedly antiarrhythmic (Vegh et al., 1991a) and that this effect is mediated through the generation of nitric oxide since it is prevented by inhibitors of the L-arginine–nitric oxide pathway (Vegh et al., 1993b).

The evidence for a role for nitric oxide (presumably generated as a result of bradykinin release) comes from studies in which dogs were preconditioned in the presence of an inhibitor of the L-arginine–nitric oxide pathway (Vegh et al. 1992b) Under these conditions, the protective effects of ischemic preconditioning are markedly attenuated and the reduction, by preconditioning, of various indices of ischemia severity (epicardial ST-segment elevation; increased inhomogeneity of electrical activation within the ischemic area) were also completely reversed in dogs in which the L-arginine–nitric oxide pathway had been inhibited.

A further piece of evidence that nitric oxide is involved comes from studies in which the inhibitor of guanylyl cyclase, methylene blue, was infused locally into the coronary circulation prior to, and during, the preconditioning stimulus, as well as during the prolonged period of ischemia (Vegh et al., 1993c). Methylene blue

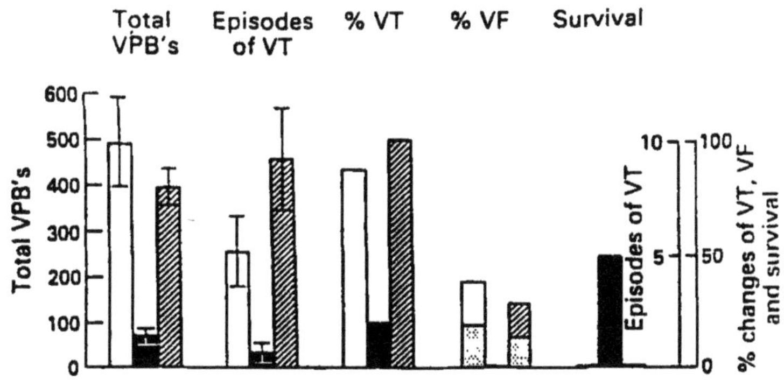

Figure 10. The effect of infusing methylene blue (by intracoronary infusion and given both during preconditioning and the prolonged occlusion in a total dose of 325 mg; shaded columns) on the protective effects of ischemic preconditioning (solid columns) in anesthetized mongrel dogs. The control data are seen in the initial open columns. Shown are the total number of ventricular premature beats during the 25-minute occlusion period (VPBs), the number of episodes and incidence of ventricular tachycardia (VT), the incidence of ventricular fibrillation (VF), and survival from the combined ischemia–reperfusion insult. The incidence of VF is given both as the total incidence throughout the 25-minute occlusion period and during the first 5 minutes (stippled column). From Vegh et al. 1993c, with permission.

completely reversed the protective effects of ischemic preconditioning (Figure 10), a result which again suggests that various endothelium-derived substances, particularly nitric oxide, contribute to the protection afforded against life-threatening ventricular arrhythmias by ischemic preconditioning.

V. EVIDENCE FOR A ROLE IN PRECONDITIONING OF CYCLO-OXYGENASE PRODUCTS

Some of our earlier studies (Coker and Parratt, 1983; and reviewed by Parratt, 1987) demonstrated that locally infused prostacyclin is markedly antiarrhythmic in the canine model and that there was some relationship between the amount of prostacyclin generated under conditions of ischemia and the resultant arrhythmia severity (Coker et al., 1981). Furthermore, under conditions in which prostacyclin generation was potentiated, by inhibition of breakdown using nafazatrom, protection also resulted against ischemia-induced ventricular arrhythmias (Coker and Parratt, 1984). These results suggest that prostacyclin, too, can be regarded as an "endogenous myocardial protective substance." Although, as yet, we have no evidence that prostacyclin is released in increased amounts under conditions of ischemic preconditioning we do know that blockade of the cyclooxygenase pathway, and hence prevention of the formation of prostacyclin, results in attenuation of the protective antiarrhythmic effects of ischemic preconditioning (Vegh et al., 1990). In addition, more recent studies in which both the cyclo-oxygenase and L-arginine–nitric oxide pathways have been simultaneously inhibited, results in a complete reversal of the protection afforded by preconditioning against ischemia-induced ventricular arrhythmias. Indeed, it is almost impossible to precondition dogs in the presence of such dual blockade (Kis et al., 1997).

In summary, these results suggest that the antiarrhythmic effects of ischemic preconditioning are due to the generation, by coronary vascular endothelial cells, of a variety of endogenous myocardial protective substances of which the most important appear to be bradykinin, nitric oxide, and prostacyclin. This result may have important clinical implications. One might expect that when there is coronary vascular endothelial dysfunction (as in hypertension, atherosclerosis, ventricular hypertrophy, and ischemic heart disease) arrhythmia severity might be increased, although the precise relationship between endothelial dysfunction and arrhythmia severity in patients is almost impossible to document. Certainly, in spontaneously hypertensive rats, where there is clear evidence of endothelial dysfunction, there is increased arrhythmia severity following coronary artery occlusion (Kolar and Parratt, 1997).

VI. MECHANISMS OF THE DELAYED PROTECTION AFFORDED BY CARDIAC PACING

We have much less information about possible mechanisms of delayed protection. Indeed, there are only two studies that throw any light on this. The first (Vegh et al., 1994b) demonstrated that the delayed protective effects of cardiac pacing were not seen in dogs pretreated with dexamethasone, suggesting the possibility of the induction of cyclo-oxygenase II or nitric oxide synthase II by cardiac pacing. There are, however, a number of other possible explanations for the detrimental effects of dexamethasone. The second study comes from the use of icatibant to block bradykinin B_2 receptors. This action also markedly attenuates the delayed protection afforded by cardiac pacing (Vegh et al., 1995). These results suggest that there may be similar mechanisms for the early and delayed antiarrhythmic effects of ischemic preconditioning and cardiac pacing.

VII. CONCLUSIONS

We have summarized the evidence for the marked antiarrhythmic effects of ischemic preconditioning and cardiac pacing which are as pronounced, or more so, than with any pharmacological intervention. Particularly important appear to be the possibility of prolonging protection for normal dogs by repeated cardiac pacing. One explanation for this might be the induction of protective proteins in enzymes. In classical preconditioning, the mechanism appears to involve endothelial–cardiac myocyte cross-talk by bradykinin, nitric oxide, and prostacyclin.

ACKNOWLEDGMENTS

The studies emanating from the authors' laboratories have been variously supported by the Royal Society, the Scottish Home and Health Department, the Wellcome Trust, the European Community (Network CIPA CT-92-4009, BIOMED I BMHI CT-92-1893, and BIOMED 2 BMH4-CT96-0979) the British Council, in association with the Hungarian Committee for Technical Development, and the Hungarian State Government (OTKA). We wish to acknowledge the former and present Chairmen of the Department of Pharmacology in Szeged (Professors Laszlo Szekeres and Julius Papp) for their support and encouragement and a number of our younger co-workers, especially Wu Song, Zahra Fatehi-Hassanabad, Adrian Kis, and Karoly Kaszala for their contagious enthusiasm. It is also a particular pleasure to acknowledge the help of Mrs. Margaret Laird in the preparation of the manuscript.

REFERENCES

Barber, M.J. (1983). Effect of time interval between repeated brief coronary artery occlusions on arrhythmias, electrical activity and myocardial blood flow. J. Am. Cell. Cardiol. 2, 699-705.

Bassenge, E. (1995). Control of coronary blood flow by autacoids. Basic Res. Cardiol. 90, 125-141.

Brutsaert, D.L., Meulemans, A.L., Sipido, K.R., and Sys, S.U. (1988). Effects of damaging the endocardial surface on the mechanical performance of isolated heart muscle. Circulation Res. 62, 358-366.

Coker, S.J., and Parratt, J.R. (1983). Prostacyclin—antiarrhythmic or arrhythmogenic? Comparison of the effect of intravenous and intracoronary prostacyclin and ZK 36374 during coronary artery occlusion and reperfusion in anaesthetised greyhounds. J. Cardiovasc. Pharmacol. 5,557-567.

Coker, S.J., Parratt, J.R. (1984). The effects of nafazatrom on arrhythmias and prostanoid release during coronary artery occlusion and reperfusion in anaesthetised greyhounds. J. Mol. Cell Cardiol. 16, 43-52.

Coker, S.J., Parratt, J.R., Ledingham, I. McA., and Zeitlin, I.J. (1981). Thromboxane and prostacyclin release from ischaemic myocardium in relation to arrhythmias. Nature 291, 323-324.

Connaughton, M., Lawson, C.S., and Hearse, D.J. (1996). Ischaemic preconditioning and arrhythmias induced by ischaemia and reperfusion. In: Ischaemia, Preconditioning and Adaptation. (Marber, M.S., and Yellon, D.M., Eds.), pp. 59-84. BIOS Scientific Publishers, Oxford.

Fatehi-Hassanabad, Z., Furman, B.L., Parratt, J.R. (1996). The effect of the endothelium on coronary artery occlusion-induced arrhythmias in rat isolated perfused hearts. J. Physiol. 494, 112-113P.

Fleming, I., Bauersachs, J., and Busse, R. (1996). Paracrine functions of the coronary vascular endothelium. Mol. Cell Biochem. 157, 137-145.

Furchgott, R.F., Zawadzki, J.V. (1980). The obligatory role of endothelial cells in the relaxation of arterial smooth muscle by acetylcholine. Nature (London) 288, 373-376.

Gulker, H., Kramer, B., Stephan, K., and Meesmann, W. (1977). Changes in ventricular fibrillation threshold during repeated short-term coronary occlusion and release. Basic Res. Cardiol. 72, 547-562.

Harris, A.G. (1950). Delayed development of ventricular ectopic rhythms following experimental coronary occlusion. Circulation 1, 1318-1326.

Kaszaki, J., Wolfard, A., Bari, F., Boros, M., Parratt, J.R., and Nagy, S. (1996). Effect of nitric oxide synthase inhibition on myocardial contractility in anesthetised normal and endotoxemic dogs. Shock 6, 279-285.

Kaszala, K., Vegh, A., Papp, J.G., and Parratt, J.R. (1996). Time-course of the protection against ischaemia and reperfusion-induced ventricular arrhythmias resulting from brief periods of cardiac pacing. J. Mol. Cell Cardiol. 28, 2085-2095.

Kis, A., Vegh, A., Papp, J.G., and Parratt, J.R. (1997). Simultaneous blockade of the cyclooxygenase and L-arginine nitric oxide pathways prevents the antiarrhythmic effect of preconditioning. Exper. Clin. Cardiol. 2, 112-119.

Kis, A., Vegh, A., Papp, J.G., and Parratt, J.R. (1996). Repeating pacing widens the time window of delayed protection against ventricular arrhyhtmias in dogs. J. Mol. Cell Cardiol. 28, 229 [Abstr.).

Kloner, R.A., Shook, T., and Przyklenk, K. (1995). Previous angina alters in-hospital outcome in TIMI 4: clinical correlate to preconditioning? Circulation 91, 37-45.

Kolar, F., and Parratt, J.R. (1997). Antiarrhythmic effect of ischaemic preconditioning in hearts of spontaneously hypertensive rats. Exper. Clin. Cardiol. 2, 124-128.

Komori, S., Fujimaki, S., Ijili, H., Asakawa, T., Watanabe, Y., Tamura, Y., and Parratt, J.R. (1990a). Inhibitory effect of ischemic preconditioning on ischemic arrhythmias using a rat coronary artery ligation model. Japan. J. Electrocardiol. 10, 774-782.

Komori, S., Vegh, A., Szekeres, L., and Parratt, J.R. (1990b). Preconditioning reduces the severity of ischaemia and reperfusion-induced arrhythmias in both anaesthetised rats and dogs. J. Physiol. 423, 16P.

Kuzuya, T., Hoshida, S., Yamashita, N., Fuji, H., Oe, H., Hori, M., Kamada, T., and Tada, M., (1993). Delayed effects of sublethal-ischemia on the acquisition of tolerance to ischemia. Circ. Res. 72, 1293-1299.

Lawson, C.S., Avkiran, M., Shattock, M.J., Coltart, D.J., and Hearse, D.J. (1993a). Preconditioning and reperfusion arrhythmias in the isolated rat heart: true protection to temporal shift in vulnerability? Cardiovasc Res. 27, 2274-2281.

Lawson, C.S., Coltart, D.J., and Hearse, D.J. (1993b). Dose-dependency and temporary characteristics of protection by ischaemic preconditioning against ischaemia-induced arrhythmias in rat hearts. J. Mol. Cell. Cardiol. 25, 1391-1402.

Marber, M.S., Latchman, D.S., Walker, J.M., and Yellon, D.M. (1993). Cardiac stress protein elevation 24 hours after brief ischemia or heat stress is associated with resistance to myocardial infarction. Circulation 88, 1264-1272.

Marshall, R.J., and Parratt, J.R. (1974). Drug-induced changes in blood flow in the acutely ischaemic canine myocardium: relationship to subendocardial driving pressure. Clin. Exper. Pharmacol. Physiol. 1, 99-112.

Marshall, R.J., and Parratt, J.R. (1980). The early consequences of myocardial ischaemia and their modification. J. Physiologie (Paris) 76, 699-715.

Murry, C.E., Jennings, R.B., and Reimer, K.A. (1986). Preconditioning with ischaemia: a delay of lethal cell injury in ischaemic myocardium. Circulation 74, 1124-1136.

Parratt, J.R. (1987). Modification of the thromboxane/prostacyclin balance as an approach to antiarrhythmic therapy during myocardial ischaemia and reperfusion; the concept of endogenous antiarrhythmic substances. In: Myocardial Ischaemia (Dhalla, N.S., Innes, I.K., and Beamish, R.E., Eds.), pp. 21-35. Martinus Nijhoff, Boston.

Parratt, J.R. (1993). Endogenous myocardial protective (antiarrhythmic) substances. Cardiovasc. Res. 27, 693-702.

Parratt, J.R., and Vegh, A. (1996). Endothelial cells, nitric oxide and ischaemic preconditioning. Basic Res. Cardiol. 91, 27-30.

Parratt, J.R., Vegh, A., and Papp, J.G. (1995). Bradykinin as an endogenous myocardial protective substance with particular reference to ischemic preconditioning—a brief review of the evidence. Canad. J. Physiol. Pharmacol. 73, 837-842.

Parratt, J.R., Vegh, A., Zeitlin, I.J., Ahmad, M., Oldroyd, K., Kaszala, K., and Papp, J.G. (1997). Bradykinin and endothelial-cardiac myocyte interactions in ischaemic preconditioning—a review. Am. J. Cardiol 80 (3A), 124A-132A.

Piacentini, L., Wainwright, C.L., and Parratt, J.R. (1993). The antiarrhythmic effect of ischaemic preconditioning in isolated rat hearts involves a pertussis toxin sensitive mechanism. Cardiovasc. Res. 27, 674-680.

Podzuweit, T., Binz, K-H., Nennstiel, P., and Flaig, W. (1989). The anti-arrhythmic effects of myocardial ischaemia. Relation to reperfusion arrhythmias? Cardiovasc. Res. 23, 81-90.

Przyklenk, K., and Kloner, R.A. (1995). Preconditioning: a balanced perspective. Br. Heart J. 74, 575-577.

Shiki, K., and Hearse, D.J. (1987). Preconditioning of ischemic myocardium; reperfusion-induced arrhythmias. Am. J. Physiol. 253, H1470-H1476.

Sys, S.U., and Brutsaert, D.L. (1995). Endocardial Endothelium: Control of Cardiac Performance. Heidelberg, Springer.

Vegh, A., Kaszala, K., Papp, J.G., and Parratt, J.R. (1995). Delayed myocardial protection by pacing-induced preconditioning: a possible role for bradykinin. Br. J. Pharmacol. 116, 228P.

Vegh, A., Komori, S., Szekereas, L., and Parratt, J.R. (1992a). Antiarrhythmic effects of preconditioning in anaesthetised dogs and rat. Cadiovasc. Res. 26, 486-495.

Vegh, A., Szekeres, L., and Parratt, J.R. (1992b). Preconditioning of the ischaemic myocardium; involvement of the L-arginine-nitric oxide pathway. Br. J. Pharmacol. 107, 648-652.

Vegh, A., Papp, J.G., and Parratt, J.R. (1994a). Attenuation of the antiarrhythmic effects of ischaemic preconditioning by blockage of bradykinin B_2 receptors. (Br. J. Pharmacol. 113, 1167-1172.

Vegh, A., Papp, J.G., and Parratt, J.R. (1994b). Prevention by dexamethasone of the marked antiarrhythmic effects of preconditioning induced 20 h after rapid cardiac pacing. Br. J. Pharmacol. 113, 1081-1082.

Vegh, A., Papp, J.G., Szekeres, L., and Parratt, J.R. (1993a). Are ATP sensitive potassium channels involved in the pronounced antiarrhythmic effects of preconditioning? Cardiovasc. Res. 27, 638-643.

Vegh, A., Papp, J.G., Szekeres, L., and Parratt, J.R. (1993b). Prevention by an inhibitor of the L-arginine nitric oxide pathway of the antiarrhythmic effects of bradykinin in anaesthetised dogs. Br. J. Pharmacol. 110, 18-19.

Vegh, A., Papp, J.G., Szekeres, L., and Parratt, J.R. (1993c). The local intracoronary administration of methylene blue prevents the pronounced antiarrhythmic effect of ischaemic preconditioning. Br. J. Pharmacol. 107, 910-911.

Vegh, A., Szekeres, L., and Parratt, J.R. (1990). Protective effect of preconditioning of the ischaemic myocardium involves cyclooxygenase products. Cardiovasc. Res. 12, 1020-1022.

Vegh, A., Szekeres, L., and Parratt, J.R. (1991a). Local coronary infusions of bradykinin profoundly reduce the severity of ischaemia-induced arrhythmias in anaesthetised dogs. Br. J. Pharmacol. 104, 294-295.

Vegh, A., Szekeres, L., and Parratt, J.R. (1991b). Transient ischaemia induced by rapid cardiac pacing results in myocardial preconditioning. Cardiovasc. Res. 25, 1051-1053.

Wilson, S., Song, W., Karoly, K., Ravingerova, T., Vegh, A., Papp, J., Tomisawa, S., Parratt, J.R., and Pyne, N.J. (1996). Delayed cardioprotection is associated with the subcellular relocalisation of ventricular protein kinase Cε, but not p42/44 MAPK. Molec. Cell. Biochem. 106/161, 225-230.

Yamashita, N., Kuzuya, T., Hoshida, S., Fuji, H., Oe, H., Kitabatake, A., Tada, M., and Kamada, T. (1992). Relationship between time interval from preconditioning to sustained ischemia and its effect on limiting infarct size. J. Mol. Cell. Cardiol. 24 (Suppl 1), P-01-41.

Yellon, D.M., and Baxter, G.F., (1996). "The second window of protection" associated with ischaemic preconditioning. In: Ischaemia, Preconditioning and Adaptation. (Marber, M.S., and Yellon, D.M., Eds.) pp. 113-130. BIOS Scientific Publishers. Oxford.

MYOCARDIAL PROTECTION BY BRIEF ISCHEMIC AND NONISCHEMIC STRESS

Pieter D. Verdouw, Ben C.G. Gho, and Mirella A. van den Doel

Advances in Organ Biology
Volume 6, pages 21-41.

ISBN: 0-7623-0391-3

I. INTRODUCTION

Since Murry et al. (1986) showed that in dogs myocardial infarct size produced by a 45-minute coronary artery occlusion was limited from 29% to 7% of the area at risk when that 45-minute occlusion was preceded by four sequences of 5-minute occlusions and 5 minutes of reperfusion, cardioprotection by ischemic preconditioning has been demonstrated in every species in which the phenomenon has been investigated (Lawson and Downey, 1993). Subsequent studies have also shown that protection by ischemic preconditioning is not restricted to the heart but also occurs in other organs, although in some of these studies other end points than infarct size limitation were used (Zager et al., 1984; Kitagawa et al., 1990; Mounsey et al., 1992; Lloris-Carsi et al., 1993; Neeley and Keith, 1995; Hotter et al., 1996).

In the following paragraphs, we will first review the evidence for protection by ischemic preconditioning in other organs than the heart. We will then summarize a number of studies from our laboratory, in which we investigated cardioprotection in models that mimic the clinical situation more closely than the abrupt occlusion–reperfusion models that are commonly used to investigate ischemic preconditioning (Koning et al., 1994a, 1994b, 1995) or in which stimuli were used that do not produce myocardial ischemia but also protect the myocardium (Koning et al., 1996; Verdouw et al., 1996a). In the last section, we address whether brief ischemia in other organs affects development of myocardial infarct size produced by a prolonged coronary artery occlusion (Gho et al., 1996; Verdouw et al., 1996b). These experiments appear to be rational because adaptation to ischemic stress occurs not only in the heart but also in other organs (Kitagawa et al., 1990; Mounsey et al., 1992; Lloris-Carsi et al., 1993; Neely and Keith, 1995; Hotter et al., 1996) and because myocardial tissue can also increase its tolerance to irreversible damage by other types of stress than brief ischemia (Koning et al., 1996; Verdouw et al., 1996a; Przryklenk et al., 1993; Ovize et al., 1994). We therefore hypothesize that brief ischemia in these organs might also be capable of triggering a mechanism that protects the myocardium against irreversible damage by a sustained coronary artery occlusion.

II. ISCHEMIC PRECONDITIONING IN ORGANS OTHER THAN THE HEART

Zager et al. (1984) investigated the susceptibility of kidneys to additional acute ischemic events by submitting rats to a 40-minute bilateral renal artery occlusion (RAO) and either 18 or 48 hours later, when morphological injury was maximal, challenging the kidneys again with either a 25- or 40-minute RAO. In these studies, changes in renal function (glomerular filtration rate and renal blood flow), adenine nucleotides, and morphology occurring in response to the second occlusion were compared to those occurring in response to a single occlusion. From these and additional experiments, the authors concluded that kidneys that had been subjected to an episode of ischemia did not have an increased susceptibility but rather showed protection when challenged again with an additional ischemic event at the maximum of morphological injury. The mechanism of protection was not further investigated, but the authors hypothesized that the renal failure–induced increments in solute loads per nephron conferred protection to previously injured nephrons.

Lloris-Carsi et al. (1993) showed that survival of livers in rats after 90 minutes of ischemia produced by cross-clamping the hepatic artery, portal vein, and bile duct was increased from 40% to 80% when the livers were preconditioned with a single period of 5-minute ischemia, 10 minutes before the 90-minute period of liver ischemia. Interestingly, protection was less when preconditioning was achieved by two sequences of 5-minute ischemia and 10 minutes of reperfusion, while it was absent after three such ischemia–reperfusion sequences. The mechanism underlying the protective action of preclamping was not explained, nor was an explanation proposed for the decrease in protection observed when multiple occlusion–reperfusion sequences were used. The data might suggest, however, that in the liver the window of the duration of protection is very narrow and that a new stimulus applied during this window cannot reinstate protection by ischemic preconditioning.

Muscle flap survival has also been shown to benefit from ischemic preconditioning (Mounsey et al., 1992). It is noteworthy that in order to obtain this augmented skeletal muscle survival after 4 hours of ischemia (using the latissimus dorsi of pigs), the authors needed a preconditioning stimulus of 30 minutes of ischemia that was separated by 30 minutes of reperfusion from the 4 hours of ischemia, while protection was absent when 10 minutes of ischemia and 10 minutes of reperfusion was used as a stimulus. The number of animals in each of the experimental groups was rather small ($n = 5$), however, and substantially larger groups of animals may be needed to confirm these data.

Hotter et al. (1996) showed that the increase in lactate dehydrogenase release after 90 minutes of intestinal ischemia was almost completely prevented when the intestines were preconditioned by 10 minutes of intestinal ischemia starting 15 minutes prior to the 90-minute period of ischemia. The authors also showed that inhibition of nitric oxide synthesis with L-nitro-methylesterarginine abolished—and administration of nitric oxide mimicked—the protection by ischemic preconditioning. The role of nitric oxide in preconditioning was not explained, but the authors suggested that it was independent of prostaglandin synthesis as nitric oxide administration and inhibition had little effect on 6-keto-Prostaglandin $F_{I\alpha}$ release. The authors suggested that brief ischemia caused an increase in intracellular calcium concentrations, which activated the calcium-dependent constitutive nitric oxide synthase, resulting in an increased nitric oxide production.

Ischemic preconditioning has also been shown to occur in the lung of spontaneously breathing cats (Neely and Keith, 1995). In these animals, lung pathology was assessed after 2 hours of reperfusion following 2 hours of ischemia. Ischemic preconditioning with 10 minutes of ischemia and 10 minutes of reperfusion limited edematous alveoli to 7% as compared to 22% in the control group. The data of the preconditioned lungs were very similar to those observed after 2 hours of ischemia and 1 hour of reperfusion, suggesting that preconditioning limited reperfusion injury rather than ischemic injury, although it cannot be excluded that a longer duration of reperfusion may be necessary to fully assess damage. The authors tested the hypothesis that adenosine A_1 antagonists are also capable of limiting ischemia–reperfusion injury of the lung. This hypothesis was developed on the basis of earlier observations that adenosine produces vasoconstriction in the feline pulmonary artery by acting on A_1 receptors, thereby inducing release of thromboxane. The authors did not investigate, however, whether protection by ischemic preconditioning and adenosine A_1 antagonism had a common pathway.

Evidence is also accumulating that transient ischemia induces tolerance and protects the brain from subsequent ischemia (Chen and Simon, 1997). Kitagawa and colleagues (1990) showed in gerbils that 2-minute occlusions of the carotid artery at 24-hour intervals before a 5-minute occlusion protected against neuronal death, while a single 2-minute occlusion applied 24 or 48 hours before the 5-minute occlusion was only partially protective. The importance of the duration of the cardioprotective stimulus was illustrated by the observation that two 1-minute occlusions at 24-hour intervals were not protective. The authors speculated that, similar to hyperthermia and other agents that induce stress, synthesis of highly conserved proteins may play a role in the neuronal protection. In view of the long time interval between the preconditioning stimulus and the sustained period of ischemia, this type of cerebral protection appears to be similar to the second window of protection by ischemic myocardial preconditioning. Whether or not in this model a "first window of protection" also exists was not investigated. In view of this, it is important to know that Chen and colleagues (1996) observed in a rat model of focal ischemia that three periods of 10-minute ischemia (mid cerebral artery occlusion) and 45 minutes of reperfusion limited infarct volume produced by a 100-minute occlusion applied after 2–5 days but not after 1 day.

From these observations it is clear that adaptive responses to brief ischemia are not limited to the myocardium but also exist in other organs. However, the mechanism underlying this protection does not appear to be the same for all organs.

III. ISCHEMIC PRECONDITIONING BY PARTIAL CORONARY ARTERY OCCLUSION WITHOUT INTERVENING REPERFUSION

If stringent conditions, such as abrupt brief total coronary artery occlusion and intervening complete reperfusion, are required to precondition the myocardium, it is very unlikely that ischemic preconditioning has a clinical analog. However, myocardium can also be protected by moderate partial coronary artery occlusions in the presence of adrenergic stimulation (demand ischemia) or endothelial injury (Ovize et al., 1992a; Iwamoto et al., 1993). Ovize and colleagues (1992b) showed that sufficiently severe partial coronary artery occlusions alone can also precondition the myocardium but a period of complete reperfusion between the partial occlusion and the subsequent sustained occlusion was a prerequisite to obtain cardioprotection. We have challenged the generalization of that conclusion because it was based on only one degree of flow reduction (approximately 50%) and hypothesized that partial occlusions might be able to protect the myocardium without intervening reperfusion, but that the severity and the duration of the partial coronary occlusions could play a pivotal role.

The partial occlusions were chosen such that in open-chest anesthetized pigs, flow reductions were either 30% or 70% of baseline and maintained for either 30 minutes or 90 minutes before the artery was occluded totally for 60 minutes without intervening reperfusion (Koning et al., 1994b, 1995). Table 1 shows that with 70% coronary blood flow reductions lasting 30 minutes before the artery was occluded for 60 minutes without intervening reperfusion, infarct areas were smaller than for control animals. When the duration of the 70% flow reduction was extended to 90 minutes, infarcted areas/area at risk (IA/AR) was still smaller than in the control groups, although some of the myocardium became already irreversibly damaged by the 90-minute partial occlusion. Flow reductions of 30% were not capable of reducing infarct size irrespective of whether durations of 30 minutes or 90 minutes were used.

Because a flow reduction will affect perfusion of the inner half of the myocardium more severely than the outer half, we also investigated whether the degree of protection was different for the inner (endocardial) and outer (epicardial) halves of the myocardium. Analysis of our results showed that protection with the 70% flow reductions was more pronounced in the epicardial than in the endocardial half. Based on earlier observations with radioactive microspheres, we may assume that with the 30% flow reduction, the endocardium had at least become as ischemic as the epicardial half with the 70% flow reductions. Nevertheless, we did not observe any protection in the endocardial half with either the 30 minutes or 90 minutes of 30% flow reduction. A possible reason could be that the distribution of adenosine triphosphate–sensitive potassium (K^{+}_{ATP}) channels, the activation of which is a potential mechanism for protection by ischemic preconditioning, is more dense in the subepicardium than in the subendocardium (Miyoshi et al., 1996). Our results may also provide some new insight into the observation by Harris (1950) that a two-stage coronary artery occlusion reduced the incidence of ventricular fibrillation, which is very high during the first 30 minutes of a total coronary artery occlusion, but almost absent when the artery was totally occluded in two stages (i.e., after a 30-minute partial occlusion, the coronary artery is occluded totally without intervening reperfusion). With the current knowledge about the effect of ischemic preconditioning on ventricular fibrillation (Hagar et al., 1991; Vegh et al., 1992), we can now hypothesize that the myocardium became preconditioned during the 30-minute partial occlusion, thereby almost completely eliminating the incidence of ventricular fibrillation during the subsequent total coronary occlusion. To support this hypothesis it would be informative to use partial occlusions of shorter duration

Table 1. Effect of Partial Occlusion on Infarct Size Produced by 60-Minute Coronary Artery Occlusio

			AR			*IA*	
Stimulus	*n*	*Infarct size after 60-min CAO*	*Transmural (%LV mass)*	*Endocardial* ($\%LV_{endo}$ *mass)*	*Epicardial* ($\%LV_{epi}$ *mass)*	*Transmural*	*E*
—	19	+	33±2	34±2	32±2	81±3	
min CAO + 15-min Rep	10	+	32±3	33±3	31±3	53±8*	
min 70% FR	9	+	30±2	31±2	29±2	39±8*	
min 70% FR	7	+	38±3	39±3	38±3	60±6*	
min 70% FR	7	–	34±1	36±2	33±1	21±7	
min 30% FR	7	+	37±2	38±2	36±2	73±6	
min 30% FR	7	+	36±3	37±3	35±2	79±6	
min 30% FR	5	–	30±2	30±1	24±6	0±0	

R, area at risk, CAO, total coronary artery occlusion; endo,endocardial half; epi, epicardial half, FR, flow reduction; IA, infarcted area; Rep
as (not) followed by a 60-minute CAO.

= number of animals per group.

$P<.05$ versus group 1 (only 60-minute CAO); data are presented as mean ± SEM.

(approximately 10 minutes) before completely occluding the coronary artery. This is of interest as, following a total coronary artery occlusion, severe arrhythmias occur in two phases: a first phase between 4 and 9 minutes, and a second phase between 15 and 20 minutes. The shorter partial occlusion now allows us to study the incidence of ventricular fibrillation usually observed during the second phase with the artery totally occluded, instead of partially occluded, which occurs when the partial occlusion lasts 30 minutes.

Kapadia and colleagues (1997) studied protection by ischemic preconditioning in sedated closed-chest pigs in which a permanent artificial stenosis was implanted in the mid left anterior descending coronary artery via the right carotid artery. When the coronary artery was completely occluded (by balloon inflation) for 45 minutes and reperfused for two hours (in the presence of the artificial stenosis), infarct size was 69.0 ± 5.4%, which was not different from the infarct size in the animals (66.8 ±6.4%), which underwent the 45-minute occlusion in the absence of the stenosis. These data indicate that the presence of the stenosis alone did not protect the myocardium. Because the stenosis reduced baseline blood flow by approximately 30%, this is not a surprising finding in view of the earlier observations by Koning and colleagues (1995). The presence of the stenosis did not abolish preconditioning, however, as infarct size was 29.7 ± 7.1% when a preconditioning stimulus consisting of two, 10-minute balloon occlusions followed by 15 minutes of reperfusion preceded the 45 minute occlusion. Because the same preconditioning stimulus limited infarct size after 45 minutes of total ischemia to 15.1 ± 5.9% in the absence of the stenosis, it must be concluded that the presence of the mild stenosis attenuated the protection by ischemic preconditioning.

Ito (1995) studied myocardial infarct size in anesthetized pigs in which the left anterior descending coronary artery blood flow was abruptly reduced to 10% of baseline during 60 minutes and found that, after two hours of complete reperfusion, 31.4 ± 6.9% of the area at risk was infarcted. However, when coronary blood flow reduction to10% was not abrupt but gradual in a linear manner over a period of 70 minutes, only 6.6 ± 1.9% of the area at risk became infarcted during the 60 minutes in which the 90% coronary blood flow reduction was maintained.

In view of the aforementioned results, it is also of interest that Schulz et al. (1995) showed an increased tolerance to sustained low-flow ischemia when this was preceded by a brief episode of no-flow ischemia without intervening reperfusion. In this study, infarct size produced by 90-minute 90% left anterior descending coronary artery flow reduction in anesthetized pigs was smaller when 80 minutes of this flow reduction was preceded by a 10-minute episode of no-flow ischemia without intervening reperfusion. The importance of the models by Kapadia and colleagues (1997) and Schula and colleagues (1995) is that they resemble the clinical condition of acute myocardial infarction followed by incomplete reperfusion (e.g., thrombolysis with a severe residual stenosis). The observation by Schulz and colleagues (1995) has been confirmed by a recent report of Ferrari and colleagues (1996), who showed in isolated rabbit hearts that recovery of function (developed pressure) was less in a group of animals that underwent 240 minutes of hypoperfusion (90% flow reduction) than in a group that had 10 minutes of total ischemia and 230 minutes of hypoperfusion. The relevance of the data in this last study (Ferrari et al., 1996) is limited because of the short duration (30 minutes) of reperfusion, as this is too short to gain an insight into the ultimate irreversible damage and recovery of function.

IV. CARDIOPROTECTION WITHOUT PRIOR BRIEF MYOCARDIAL ISCHEMIA

Ischemic preconditioning led to the discovery that pretreatment with pharmacological agents such as adenosine and K^+_{ATP} channel activators mimicked the protection by ischemic preconditioning. Przyklenk et al. (1993) were the first to show that myocardium subjected to a sustained coronary artery occlusion can be protected by brief ischemia in adjacent myocardium. Thus, when anesthetized dogs were subjected to a 60-minute occlusion of the left anterior descending coronary artery, the infarcted area was smaller when, prior to that occlusion, the adjacent myocardium (distribution territory of the left circumflex coronary artery) had been subjected to brief ischemia ("preconditioning of virgin myocardium"). Moreover, Ovize et al. (1994) showed that stretching the left ventricular wall by volume loading prior to a total coronary artery occlusion was also an effective method to limit infarct size, probably by stimulating stretch-activated channels, as protection was lost when animals were pretreated with gadolinium, an inhibitor of stretch-activated channels. In view of the preceding finding, it is of interest that Przyklenk et al. (1993) used a 5-minute left circumflex coronary artery occlusion to precondition the distribution territory of the left anterior descending coronary artery. Because in dogs the left circumflex coronary artery nourishes a major fraction of the myocardium, it cannot be excluded that during the 5-minute left circumflex coronary artery occlusion, the distribution territory of the left anterior descending coronary artery became preconditioned by stretching.

From these (Przyklenk et al., 1993; Ovize et al., 1994) and several other studies using pharmacological agents, it became clear that myocardium can be protected against development of irreversible damage produced by a sustained coronary artery occlusion by stimuli other than brief local ischemia. We further addressed this issue by applying rapid ventricular pacing prior to a 60-minute coronary artery occlusion in anesthetized pigs (Koning et al., 1996; Verdouw et al., 1996b). Ventricular pacing was chosen as a stimulus because in earlier studies it was shown that ventricular pacing prior to a coronary artery occlusion attenuated the incidence of ventricular arrhythmias and fibrillation (Vegh et al., 1991). Because in that study (Vegh et al., 1991) ventricular pacing (at 300 beats per minute [bpm] in dogs) produced ST-segment changes, we first excluded the possibility that ischemia developed when the left ventricle was paced at 200 bpm by studying functional (e.g., systolic segment shortening and transmural distribution of coronary blood flow) and metabolic changes (myocardial adenosine triphosphate [ATP] tissue levels) (Koning et al., 1996).

Table 2. Effect of Rapid Ventricular Pacing on Infarct Size produced by 60-minute Coronary Artery Occlusion in Pigs.

Group	*Stimulus*	*n*	*Infarct size after 60-min CAO*	*AR (% LV mass)*	*IA/AR (%)*
1	—	12	+	34±2	84±2
2	10-min RVP + 15-min NSR	6	+	33±2	79±3
3	30-min RVP + 15-min NSR	7	+	36±2	71±2*
4	30-min RVP + Glib + 15-min NSR	7	+	33±2	73±3
5	30-min RVP	12	+	32±2	63±4*
6	Glib + 30-min RVP	8	+	33±3	78±4

Notes: Glib, glibenclamide 1 mg/kg IV administered immediately after (group 4) or 10 minutes before the onset of 30-minute RVP (group 6) ; NSR, normal sinus rhythm; RVP, rapid ventricular pacing at 200 bpm.
n = number of animals per group.
* $P<.05$ versus group 1 (60-minute CAO only); data are presented as mean ± SEM.
Adapted from Koning et al. (1996), with permission.

In Table 2 the effects of rapid ventricular pacing (RVP) on infarct size during a subsequent 60-minute coronary artery occlusion (CAO) are summarized. When 10-minute RVP and 60-minute CAO were separated by 15 minutes of normal sinus rhythm (NSR), IA/AR was not different from the control group. However, when the duration of RVP was extended to 30 minutes a small protection was observed. Without an intervening period of normal sinus rhythm between the 30-minute RVP and the 60-minute CAO, infarct size was about 25% smaller than in control animals. The protection was still less than obtained with ischemic preconditioning. The finding that 15 minutes of intervening NSR already attenuated protection by 30-minute RVP implies that the time course of protection by ventricular pacing is different from that by ischemic preconditioning (Koning et al., 1994a). K^+_{ATP} channel activation proved to play a role in the protection by rapid ventricular pacing as pretreatment with glibenclamide abolished the protection. However, administration of glibenclamide after ventricular pacing was terminated did not attenuate the protection, suggesting that in this model K^+_{ATP} channels do not have to remain activated after the protective stimulus has been applied, which is at variance with observations in ischemic precondition experiments.

Domenech and colleagues (1998) reported on a series of similar experiments performed in anesthetized dogs in which blood pressure was controlled at 80-90 mmHg to avoid effects of changes in hemodynamics during the five sequences of ventricular tachycardia at 213±12 cycles/min and the subsequent infarct-producing coronary artery occlusion. Based on the coronary flow reserve during the pacing period, it was excluded that the tachycardia caused myocardial ischemia. Ventricular pacing again protected the myocardium, but this protection was abolished by treatment with the nonselective adenosine blocker 8-phenyltheophylline (8-PT). Using microdialysis, the authors also showed that interstitial adenosine concentrations increased twofold during the tachycardia. Because dialysis data were not obtained in the other series of experiments, it is unknown how treatment with 8-PT affected the interstitial adenosine concentrations during and after the coronary artery occlusion. It is also of interest that Domenech and colleagues (1998) did not find any changes in cytosolic or particulate protein kinase C activitity or translocation of the α, β, ε, and ζ - isoenzymes, suggesting the (isoenzyme) protein kinase C activity or translocation does not play a role int he tachycardia-induced protection.

The results of the above described lend further support to the hypothesis that myocardium does not first have to become ischemic for a brief period in order to increase its tolerance to the development of irreversible damage during a sustained coronary artery occlusion.

V. CARDIOPROTECTION BY BRIEF ISCHEMIA IN REMOTE ORGANS

The possibility of interorgan protection was first addressed by McClanahan et al. (1993), who reported preliminary data showing that in rabbits infarct size produced by a 45-minute CAO was reduced when the occlusion was preceded by a brief period of renal ischemia. We observed a similar pattern in anesthetized rats who were subjected to a 60-minute CAO, but there was a considerable scatter in the data, which we ascribed to variations in body temperature during the course of the experiments (Gho et al., 1994, 1995). Temperature is not only an important factor determining infarct size (Chien et al., 1994; Ducker et al., 1996), but may also be a factor that determines the effectiveness of a pharmacological agent in limiting infarct size (McClanahan et al., 1994). We therefore used two temperatures (normothermia between 36.5°C and 37.5°C and hypothermia between 30°C and 31°C) to study the effect of brief left renal or anterior mesenteric artery occlusions (RAO and MAO, respectively) on infarct size produced by a 60-minute CAO. (Protocol I, Table 3) Details of the experimental procedures have been described extensively (Koning et al., 1996; Verdouw et al., 1996b).

Figure 1 shows the results of eight groups of animals that were all subjected to a 60-minute CAO and in which infarcted areas were determined after 3 hours of reperfusion. Figure 1 and Table 3 clearly show that ischemic preconditioning with a 15-minute CAO significantly limited infarct size at both normothermia and hypothermia as compared to a control group, which underwent the 60-minute CAO but received no preconditioning stimulus. A striking observation was that the protection by ischemic preconditioning was larger when the experiments were performed at hypothermia. At variance with earlier reports, we did not find an effect of temperature on infarct size produced by 60-minute CAO; a finding we ascribed to the difference in the duration of CAOs used in our and the other studies (Chien et al., 1994; Duncker et al., 1996; see below). Figure 1 and Table 3 also show that both a 15-minute MAO and a 15-minute RAO were capable of limiting infarct size produced by the 60-minute CAO, but the 15-minute RAO was only protective at hypothermia.

le 3. Effect of Remote Organ Ischemia on Infarct Size Produced by 60-Minute Coronary Artery Occlusion

	Normothermia (36.5–37.5° C)			Hypothermia (30	
	Group	AR(%LV_{mass})	IA/AR(%)	Group	AR(%LV_m
t)					
	1	31 ± 4 (n = 11)	68 ± 2	6	36 ± 4 (n =
+ 10-min Rep	2	47 ± 4 (n = 9)	50 ± 3*	7	40 ± 5 (n =
+ 10-min Rep	3	42 ± 4 (n = 10)	50 ± 3*	8	41 ± 3 (n =
- 10-min Rep	4	35 ± 8 (n = 8)	72 ± 5	9	37 ± 2 (n =
blockade with hexamethonium)					
	10	37 ± 5 (n = 7)	68 ± 3	13	35 ± 3 (n =
10-min Rep	11	45 ± 3 (n = 7)	54 ± 3*	14	35 ± 2 (n =
- 10-min Rep	12	40 ± 3 (n = 7)	74 ± 2	15	37 ± 3 (n =
)	16	36 ± 4 (n = 6)	70 ± 3	17	34 ± 2 (n =
	18	37 ± 6 (n = 6)	68 ± 5		

onary artery occlusion; Hex, hexamethonium (20 mg/kg IV); MAO, mesenteric artery occlusion; RAO,renal artery occlusion; Rep, reperf
5, infarct size was determined in four normothermic animals which underwent only the 15-minute CAO (3 ± 1%) (not presented in tab
ent MAO started 25 minutes before the onset of 60-minute CAO and was maintained until the end of the 3 hour reperfusion period.
te MAO started 14 minutes prior to the 60-minute CAO.
versus control; (60-minute CAO only).
ersus corresponding 36.5–37.5°C group; data are mean ± SEM.
rom Gho et al. (1996), with permission.

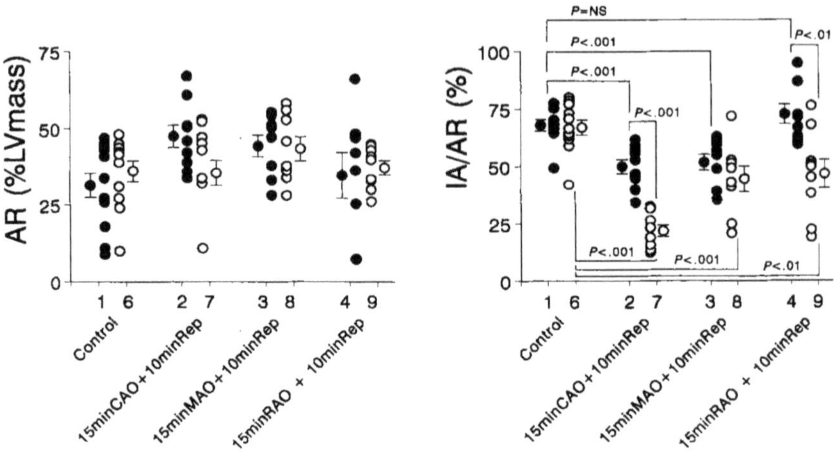

Figure 1. Scatterplots show the effects of 15-minute occlusions of a coronary artery (CAO), the anterior mesenteric artery (MAO) and left renal artery (RAO) on infarct size (IA/AR) produced by 60-minute CAO during both normothermia (•, 36.5 C–37.5°C) and hypothermia (O, 30 C–31°C). AR, area at risk; IA, infarcted area; Rep, reperfusion. See Table 3 for quantitative data.

We can only speculate on the reason why the 15-minute MAO was effective and the 15- minute RAO ineffective during normothermia. An argument could be made that the 15-minute RAO produced less severe ischemia than the 15-minute RAO. As only approximately 10% of renal artery flow (compared to 90% of mesenteric flow) is needed for nutritional flow, the less severe ischemia produced by the 15-minute RAO may therefore not have been severe enough to trigger myocardial protection.

In the following series of experiments, we investigated whether the mechanism of cardioprotection by remote brief organ ischemia differed from that by brief myocardial ischemia. Because protection by 15-minute MAO occurred during normothermia as well as hypothermia, we selected this stimulus for further investigation. We first evaluated whether a neurogenic mechanism could be involved in the protection by 15-minute MAO. To this end infarct size was determined after neurogenic blockade with hexamethonium (Protocol II). We found that hexamethonium affected IA/AR neither in the control animals nor in the animals preconditioned with 15-minute CAO, independent of whether the experiments were performed under normothermic or hypothermic endpoints. In contrast, protection by 15-minute MAO was completely abolished after pretreatment with hexamethonium (Figure 2 and Table 3). These data demonstrate that activation of the neurogenic pathway was involved in the protection by 15-minute MAO but not by 15-minute CAO.

In the next series of experiments, we investigated whether activation of the neurogenic pathway had occurred during early ischemia or during the 10-minute intervening reperfusion period, which separated the 15- minute MAO from the

60-minute CAO. We therefore studied the effect on myocardial infarct size of a permanent mesenteric artery occlusion starting 25 minutes prior to the onset of the 60-minute CAO (Protocol III). In these experiments, MAO failed to limit infarct size produced by 60-minute CAO; the neurogenic pathway therefore appears to be triggered during early reperfusion of the mesenteric artery (Figure 3 and Table 3).

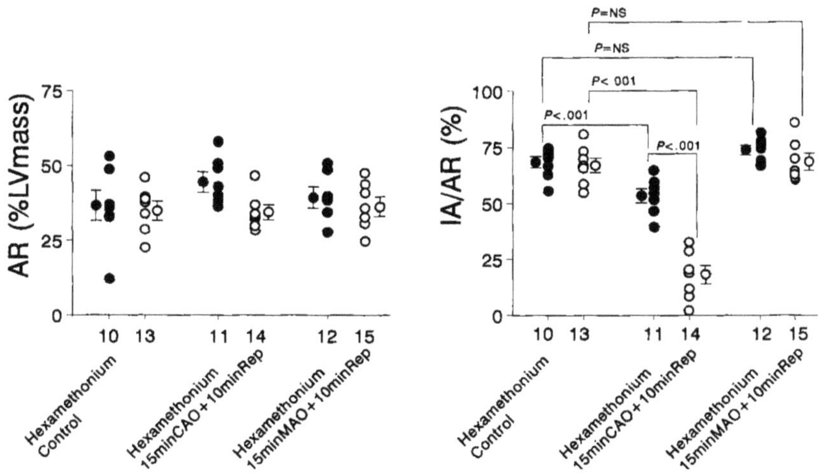

Figure 2. Scatterplots show that ganglion blockade with hexamethonium inhibits protection by 15-minute MAO but has no effect on protection by 15-minute coronary artery occlusion (CAO) during normothermia (●) and hypothermia (○). The plots also show that hexamethonium per se had no effect on myocardial infarct size. For further details see Figure 1.

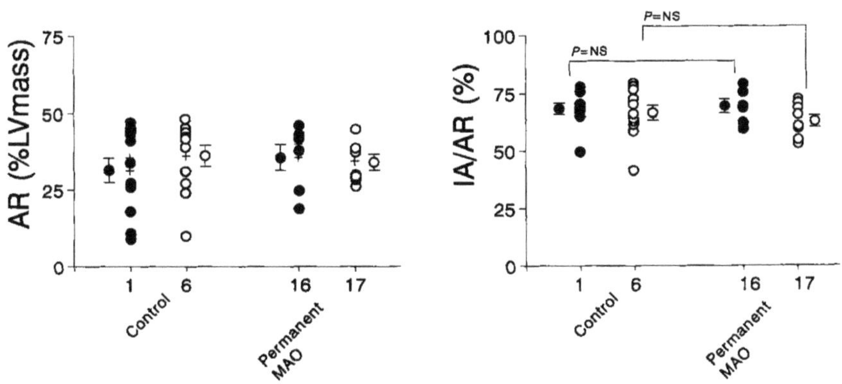

Figure 3. Scatterplots show that a permanent mesenteric artery occlusion (MAO) did not affect infarct size (IA/AR) produced by 60-minute CAO during both normothermia (●) and hypothermia (○). For further details see Figure 1.

In the last series of experiments (Protocol IV), we investigated whether reperfusion between the 15-minute MAO and 60-minute CAO was mandatory. To this end, the timing of the 15-minute MAO was chosen such that reperfusion of the intestines occurred 1 minute after the onset of 60-minute CAO. The data in Table 3 reveal that cardioprotection by the 15-minute MAO was not observed when there was no intervening reperfusion.

From our experiments we may not conclude that protection by 15-minute RAO also involves a neurogenic pathway. Furthermore, although protection by 15-minute MAO appears to differ from that by 15-minute CAO, we cannot exclude the possibility that these two stimuli exert their cardioprotection through a common end point.

Other groups of investigators have also found a cardioprotective effect of brief renal ischemia and suggest the involvement of adenosine in the protection. (Takaoka et al., 1997; Pell et al., 1998). The preliminary data by Pell and colleagues (1998), which were obtained in rabbits preconditioned with a 10-minute renal artery occlusion and 10 minutes of reperfusion prior to a 30-minute coronary artery occlusion, suggest that brief renal ischemia and preconditioning by local myocardial ischemia protect the myocardium by similar mechanisms as pretreatment with both the nonselective adenosine blocker 8-(*p*-sulphophenyl) theophylline and that the ATP-sensitive K^+ channel blocker 5-hydroxydecanoate abolished the cardioprotection.

Since not only total coronary occlusions can precondition the myocardium but also ischemia produced by an imbalance between oxygen supply and oxygen demand, Birnbaum and colleagues (1997) investigated whether a 55–65% reduction in femoral artery blood flow for 30-minutes and a 30-minute stimulation of the gastrocnemius muscle could also limit infarct size produced by a 30-minute coronary artery occlusion. The results revealed that if performed alone, neither of these two interventions limited myocardial infarct size when compared to the infarct size of a group of control animals (26 ± 3%). However, when both maneuvers were combined, infarct size was limited to 9 ± 2%. At variance with the abrupt brief renal and mesenteric artery occlusions and reperfusion, the combination of flow restriction in the femoral artery and stimulation of the gastrocnemius muscle may have a clinical analog in patients with peripheral vascular disease.

VI. THE MYOCARDIAL INFARCT SIZE–LIMITING EFFECT OF LOW BODY TEMPERATURE IN RATS DEPENDS ON THE DURATION OF THE CORONARY ARTERY OCCLUSION

As stated previously, our observation that infarct size after the 60-minute CAO was not affected by temperature appears to be in conflict with earlier reports. For instance, Chien et al. (1994) reported a steep relationship between body core temperature in the "normothermic" range (35–42°C) and myocardial infarct size in rabbits subjected to a 30-minute CAO and 3 hours of reperfusion, so that an increase of 1°C resulted in 12% infarction of the area at risk with no infarction occurring at a body core temperature of 34.5°C. Duncker et al. (1996) showed an even steeper relationship between body core temperature and infarct size produced by a 45-minute CAO and 4 hours of reperfusion in swine, as 20% of the area at risk became infarcted with a 1°C increase in temperature in the range of 35°C to 39°C. All three species (rat, rabbit, and pig) lack a significant coronary collateral circulation and this determinant of infarct size can thus be excluded as a possible explanation. However, between 30 and 45 minutes after CAO, infarction progresses rapidly; we therefore hypothesized that a reason for the apparent discrepancy could be the longer duration of coronary artery occlusion that we employed in rats. To test this hypothesis, we determined infarct sizes in rats produced by CAOs of different duration at both normothermia and hypothermia. Since hypothermia is associated with bradycardia, which has been suggested to limit infarct size per se, we also determined infarcted areas in a group of hypothermic rats in which heart rates (HR) were increased to baseline levels in normothermic animals.

Figure 4 shows that the duration of occlusion is indeed an important factor. Thus, although in CAOs of 60 minutes or longer a decrease in body temperature had no effect on IA/AR, there was a marked protection of hypothermia with CAOs of 30 minutes. Hypothermia was associated with bradycardia (HR = 355 ± 5 bpm and HR = 275 ± 8 bpm for 28 normothermic and 24 hypothermic animals, respectively) without an effect on mean arterial pressure (MAP = 89 ± 4 mmHg and MAP = 92 ± 3 mmHg, respectively) for 28 normothermic and 24 hypothermic animals, respec-

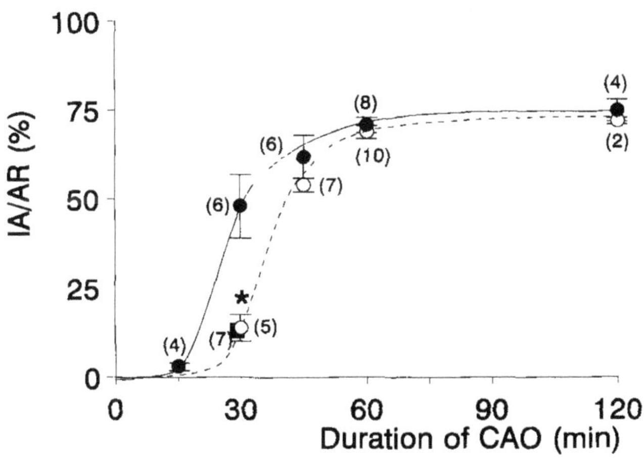

Figure 4. Effect of temperature on infarct size after coronary artery occlusions (CAOs) of different durations. Experiments were performed at normothermia (●, 36.5 C–37.5°C) and hypothermia (○, 30 C–31°C). ■ denotes the hypothermic animals in which heart rate was raised by atrial pacing such that the double product (heart rate * systolic arterial blood pressure) was not different from that of the relevant normothermic group. *$P<.05$ versus normothermic group. AR, area at risk; IA, infarcted area. Data are mean ± SEM. n = number of animals per group.

tively). Figure 4 also shows that for CAOs of 30 minutes, the cardioprotection by hypothermia was maintained when the product of heart rate and systolic arterial pressure was raised to match that of the normothermic animals (37,300 ± 1700 and 34,400 ± 2400 mmHg·bpm in normothermic [n = 6] and paced hypothermic animals [n = 7] [p = ns], respectively).

The infarct size–limiting effect of hypothermia thus depends on the duration of the CAO, so that when the coronary artery in rats was occluded for 30 minutes, hypothermia was protective (5.2% of the area at risk per 1°C), but when the duration of the CAO was extended to 60 minutes, the protective effect of hypothermia was lost. These data therefore reconcile the different observations regarding the effect of hypothermia on infarct size.

Myocardial oxygen demand at the onset of the coronary artery occlusion has been postulated as a determinant of infarct size, but the data are controversial. A positive correlation (Reimer et al., 1985; Miura et al., 1987) as well as no relation (Reimer et al., 1985; Miura et al., 1987, 1992) between the rate-pressure product and infarct size have been reported. Nienaber et al. (1983) produced bradycardia in dogs with a synthetic opiate to lower the metabolic demand at the onset of a 24-hour coronary artery occlusion, thereby producing a smaller infarction compared to a group of animals with a high metabolic demand at the onset of coronary artery occlusion. It cannot be excluded that the protection obtained by bradycardia was actually a direct result of μ-opoid receptor stimulation (Schultz et al., 1996). In collateral deficient species such as rabbit and swine, infarct size does not appear to be correlated with the rate-pressure product. In the study by Duncker et al. (1996), univariate or stepwise multivariate regression analysis did not reveal a significant correlation between temperature and systemic hemodynamic variables or myocardial blood flow under baseline conditions, suggesting that temperature did not exert its effect by altering myocardial oxygen demand at the onset of occlusion. Similarly, in rabbit hearts (Chien et al., 1994) and rat hearts (Figure 4), the infarct size–limiting effect of hypothermia was unmitigated when hypothermia-induced bradycardia was prevented.

Hale and colleagues (1997a, 1997b) studied infarct sizes in rabbits in which the entire anterior portion of the heart was cooled using a bag containing ice and water. This procedure permitted a local temperature reduction of approximately 6°C within 5 minutes. These studies showed that regional cooling can be protective even when started after the onset of ischemia and that changes in hemodynamic parameters are not responsible for the protection. As in man infarct size development is slower than in rabbits, it could very well be that the delay in the onset of cooling with respect to the onset of the development of infarction may be longer then 10 minutes. This approach of regional cooling therefore has therapeutic value if the procedure can be done safely and without surgical intervention in patients developing an infarction.

To investigate whether the protection by ischemic preconditioning also depends on the body temperature, we determined infarct sizes in anesthetized pigs with a body temperature of either 30-31°C or 36-37°C after 3 hours of reperfusion following coronary artery occlusions varying from 15 min to 120 min. Animals were preconditioned with a 15-min coronary artery occlusion and 10 min of reperfusion (Van der Doel et al., 1998). We observed that hypothermic animals still could benefit from ischemic preconditioning and that the duration of the coronary artery occlusion at which infarct size was limited could be expanded to 120 min when hypothermia and ischemic preconditioning were combined.

VII. CONCLUSION

If we try to put the preceding results in perspective, it is obvious that ischemic preconditioning is not restricted to the myocardium, but occurs in a large number of organs. It should be kept in mind, however, that in several of the studies involving other organs, infarct size was not always the end point. Moreover, in these organs the time course of protection has not been studied extensively enough to decide whether there is a similar pattern of protection (first and second window) as in the heart. For instance, in several studies organs were challenged with a sustained period of ischemia several hours after the ischemic stimulus was applied, and it is not clear if the organ would also be protected if the sustained period of ischemia followed the stimulus after a much shorter interval. The mechanism by which brief ischemia protects these other organs is also not completely understood, which is not surprising as precise knowledge about protection by ischemic preconditioning is also lacking for the much more extensively studied myocardium.

As ischemic preconditioning occurs in so many organs, the question naturally arose as to whether there could be an interaction between these organs. This question became even more relevant as several studies have now shown that the heart can also become protected by several types of stress (e.g., stretch, ventricular pacing) which do not lead to ischemia. We have now provided evidence that such an interaction between different organs is indeed possible under well-controlled experimental conditions. The clinical relevance of these observations is yet unclear, as the conditions by which brief renal or intestinal ischemia provided cardioprotection in these studies do not exist frequently in the clinical setting. The present observations, if they have a clinical analog, definitely complicate the search for proof of ischemic preconditioning in humans (Kloner and Yellon, 1994; Lawson, 1994; Verdouw et al., 1995) and the interpretation of clinical studies aiming to limit infarct size in situations such as thrombolysis, as protection by brief ischemia in remote organs or the heart itself may be confounding and uncontrolled factors.

VIII. SUMMARY

Brief periods of abrupt myocardial ischemia and reperfusion preceding a sustained coronary artery occlusion have been shown to limit myocardial infarct size, a phenomenon that has been termed ischemic preconditioning. Reperfusion between the preconditioning stimulus and the sustained coronary artery occlusion have proven not to be mandatory provided that partial occlusions of sufficient severity were used as a preconditioning stimulus. Evidence is also accumulating that ischemic preconditioning is not specific to the heart, but also occurs in other organs such as the kidney, liver, skeletal muscle, lung, and brain. Time course and mechanism of protection in these organs, although not studied in detail, may differ from that in the heart. Because myocardium can be protected by stimuli that do not produce local ischemia (e.g., pharmacological agents, stretch, and ventricular pacing), we have addressed the possibility that brief ischemia in other organs may trigger a cardioprotective action. Experiments in anesthetized rats revealed that brief occlusion of the left renal artery or anterior mesenteric artery is capable of protecting the myocardium, but that the degree of protection is temperature dependent. The mechanism of protection by brief mesenteric artery occlusion may be different from that of ischemic preconditioning, as ganglion blockade abolishes protection by mesenteric artery occlusion but has no effect on protection by ischemic preconditioning.

ACKNOWLEDGMENT

These studies have been supported by grants NHS 92.144, NHS 95.103, and D96.024 from the Netherlands Heart Foundation.

REFERENCES

Birnbaum, Y., Hale, S.L., and Kloner, R.A. (1997). Ischemic preconditioning at a distance: Reduction of myocardial infarct size by partial reduction of blood supply combined with rapid stimulation of the gastrocnemius muscle in the rabbit. Circulation. 96, 1641-1646.

Chen, J., Graham, S.H., Zhu, R.L., and Simon, R.P. (1996). Stress proteins and tolerance to focal cerebral ischemia. J. Cereb. Blood Flow Metab. 16, 566-577.

Chen, J. and Simon, R. (1997). Ischemic tolerance in the brain. Neurology 48, 306-311.

Chien, G.L., Wolff, R.A., Davis, R.F., and Van Winkle, D.M. (1994). "Normothermic range" temperature affects myocardial infarct size. Cardiovasc. Res. 28, 1014-1017.

Domenech, R.J., Macho, P., Velez, D., Sanchez, G., Liu, X., and Dhalla, N. (1998). Tachycardia preconditions infarct size in dogs: Role of adenosine and protein kinase C. Circulation. 97, 786-794.

Duncker, D.J., Klassen, C.L., Herrlinger, S.H., Pavek, T.J., Ishibashi, Y., and Bache, R.J. (1996). Effect of temperature on myocardial infarction in swine. Am. J. Physiol. (Heart Circ. Physiol. 39) 270, H1189-H1199.

Ferrari, R., Cargnoni, A., Bernocchi, P., Pasini, E., Curello, S., Ceconi, C., and Ruigrok, T.J.C. (1996). Metabolic adaptation during a sequence of no-flow and low-flow ischemia. Circulation 94, 2587-2596.

Garcia-Dorado, D., Theroux, P., Elizaga, J., Aviles, F.F., Alonso, J., and Solares, J. (1988). Influence of tachycardia and arterial hypertension on infarct size in the pig. Cardiovasc. Res. 22, 620-626.

Gho, B.C.G., Schoemaker, R.G., van den Doel, M.A., Duncker, D.J., and Verdouw, P.D. (1996). Myocardial protection by brief ischemia in noncardiac tissue. Circulation 94, 2193-2200.

Gho, B.C.G., Schoemaker, R.G., Van der Lee, C., Sharma, H.S., and Verdouw, P.D. (1994). Cardioprotection by transient renal ischemia, an in vivo study in rats. Circulation 90, 1476 [Abstr.].

Gho, B.C.G., Schoemaker, R.G., and Verdouw, P.D. (1995). Myocardial protection by transient renal ischemia in rats is temperature dependent. J. Mol. Cell. Cardiol. 27, A148 [Abstr.].

Hagar, J.M., Hale, S.L., and Koner, R.A. (1991). Effect of preconditioning ischemia on reperfusion arrhythmias after coronary artery occlusion and reperfusion in the rat. Circ. Res. 68, 61-68.

Hale, S.L., Dave, R.H., and Kloner, R.A. (1997a). Regional hypothermia reduces myocardial necrosis even when instituted after the onset of ischemia. Basic Res. Cardiol. 92, 351-357.

Hale, S.L. and Kloner, R.A. (1997b). Myocardial temperature in acute myocardial infarction: Protection with mild regional hypothermia. Am. J. Physiol. 273, H220-H227.

Harris, A.S. (1950). Delayed development of ventricular ectopic rhythms following experimental coronary occlusion. Circulation 1, 1318-1328.

Hotter G., Closa, D., Prados, M., Fernández-Cruz, L., Prats, N., Gelpí, E., and Roselló-Catafau, J. (1996). Intestinal preconditioning is mediated by a transient increase in nitric oxide. Biochem. Biophys. Res. Commun. 222, 27-32.

Ito, B.R. (1995). Gradual onset of myocardial ischemia results in reduced myocardial infarction. Association with reduced contractile function and metabolic downregulation. Circulation 91, 2058-2070.

Iwamoto, T., Bai, X.J., and Downey, H.F. (1993). Preconditioning with supply-demand imbalance limits infarct size in dog heart. Cardiovasc. Res. 27, 2071-2076.

Kapadia, S.J., Terlato, J.S., and Most, A.S. (1997). Presence of a critical coronary artery stenosis does not abolish the protective effect of ischemic preconditioning. Circulation 95, 1286-1292.

Kitagawa, K., Matsumoto, M., Tagaya, M., Hata, R., Ueda, H., Niinobe, M., Handa, N., Fukunaga, R., Kimura, K., Mikoshiba, K., and Kamada, T. (1990). "Ischemic tolerance" phenomenon found in the brain. Brain Res. 528, 21-24.

Kloner, R.A., and Yellon, D. (1994). Does ischemic preconditioning occur in patients? J. Am. Coll. Cardiol. 24, 1133-1142.

Koning, M.M.G., de Zeeuw, S., Nieukoop, S., De Jong, J.W., and Verdouw, P.D. (1994a). Is myocardial infarct size limitation by ischemic preconditioning an "all or nothing" phenomenon? In: Cellular, Biochemical and Molecular Aspects of Reperfusion Injury (Das, D.K., Ed.). Ann. N.Y. Acad. Sci. 723, 333-336.

Koning, M.M.G., Gho, B.C.G., Van Klaarwater, E., Duncker, D.J., and Verdouw, P.D. (1995). Endocardial and epicardial infarct size after preconditioning by a partial coronary artery occlusion without intervening reperfusion. Importance of the degree and duration of flow reduction. Cardiovasc. Res. 30, 1017-1027.

Koning, M.M.G., Gho, B.C.G., Van Klaarwater, E., Opstal, R.L.J., Duncker, D.J., and Verdouw, P.D. (1996). Rapid ventricular pacing produces myocardial protection by non-ischemic activation of K^+_{ATP} channels. Circulation 93, 178-186.

Koning, M.M.G., Simonis, L.A.J., de Zeeuw, S., Nieukoop, S., Post, S., and Verdouw, P.D. (1994b). Ischaemic preconditioning by partial occlusion without intermittent reperfusion. Cardiovasc. Res. 28, 1146-1151.

Lawson, C.S. (1994). Does ischemic preconditioning occur in the human heart? Cardiovasc. Res. 28, 1461-1466.

Lawson, C.S., and Downey J.M. (1993). Preconditioning: state of the art myocardial protection. Cardiovasc. Res. 27, 542-550.

Lloris-Carsi, J.M., Cejalvo, D., Toledo-Pereyra, L.H., Calvo, M.A., Suzuki, S. (1993). Preconditioning: effect upon lesion modulation in warm liver ischemia. Transplant. Proc. 25, 3303-3304.

McClanahan, T.B., Mertz, T.E., Martin, B.J., and Gallagher, K.P. (1994). Pentostatin reduces infarct size in pigs only when combined with mild hypothermia. Circulation 90, I478 [Abstr.].

McClanahan, T.B., Nao, B.S., Wolke, L.J., Martin, B.J., Metz, T.E., and Gallagher, K.P. (1993). Brief renal occlusion and reperfusion reduces myocardial infarct size in rabbits. FASEB J. 7, A118,682 [Abstr.].

Miura, T., Ogawa, T., Iwamoto, T., Shimamoto, K., and Iimura, O. (1992). Dipyridamole potentiates the myocardial infarct size-limiting effect of ischemic preconditioning. Circulation 86, 979-985.

Miura, T., Yellon, D.M., Hearse, D.J., and Downey, M. (1987). Determinants of infarct size during permanent occlusion of a coronary artery in the closed chest dog. J. Am. Coll. Cardiol. 9, 647-654.

Miyoshi, S., Myazaki, T., Moritani, K., and Ogawa, S. (1996). Different responses of epicardium and endocardium to K_{ATP} channel modulators during regional ischemia. Am. J. Physiol. (Heart Circ. Physiol. 40) 271, H140-H147.

Mounsey, R.A., Pang, C.Y., and Forrest, C. (1992). Preconditioning: a new technique for improved muscle flap survival. Otolaryngol. Head Neck Surg. 107, 549-552.

Murry, C.E., Jennings, R.B., and Reimer, K.A. (1986). Preconditioning with ischemia: a delay of lethal cell injury in ischemic myocardium. Circulation 75, 1124-1136.

Neely, C.F., and Keith, I.M. (1995). A1 adenosine receptor antagonists block ischemia-reperfusion injury of the lung. Am. J. Physiol. (Lung Cell. Mol. Physiol. 12) 268, L1036-L1046.

Nienaber, C.H., Gottwik, M., Winkler, B., and Schaper, W. (1983). The relationship between the perfusion deficit, infarct size and time after experimental coronary artery occlusion. Basic Res. Cardiol. 78, 210-225.

Ovize, M., Kloner, R.A., Hale, S.L., and Przyklenk, K. (1992a). Coronary cyclic flow variations "precondition" ischemic myocardium. Circulation 85, 779-789.

Ovize, M., Kloner, R.A., and Przyklenk, K. (1994). Stretch preconditions canine myocardium. Am. J. Physiol. 266, H137-H146.

Ovize, M., Przyklenk, K., and Kloner, R.A. (1992b). Partial coronary stenosis is sufficient and complete reperfusion is mandatory for preconditioning the canine heart. Circ. Res. 71, 1165-1173.

Pell, T.J., Baxter, G.F., Yellon, D.M., and Drew, G.M. (1998). Adenosine receptors and K_{ATP} channels are involved in renal preconditioning of the myocardium J. Mol. Cell. Cardiol. 30, A19 (Abstr.)

Przyklenk, K., Bauer, B., Ovize, M., Kloner, R.A., and Whittaker, P. (1993). Regional ischemic "preconditioning" protects remote virgin myocardium from subsequent sustained coronary occlusion. Circulation 87, 893-899.

Reimer, K.A., Jennings, R.B., Cobb, F.R., Murdock, R.H., Greenfield, J.C. Jr., Becker, L.C., Bulkley, B.H., Hutchins, G.M., Schwartz, R.P., Bailey, K.R., and Passamani, E.R. (1985). Animal models for protecting ischemic myocardium: results of the NHLBI Cooperative Study. Comparison of unconscious and conscious dog models. Circ. Res. 56, 651-665.

Schultz, J.E., Hsu, A.K., and Gross, G.J. (1996). Morphine mimics the cardioprotective effect of ischemic preconditioning via a glibenclamide-sensitive mechanism in the rat heart. Circ. Res. 78, 1100-1004.

Schulz, R., Post, H., Sakka, S., Wallbridge, D.R., and Heusch, G. (1995). Intraischemic preconditioning: increased tolerance to sustained low-flow ischemia by a brief episode of no-flow ischemia without intermittent reperfusion. Circ. Res. 76, 942-950.

Takaoka, A., Nakae, I., Liu, Q., Yabe, T., Mitsunami, K., Morikawa, S., Inubushi, T., and Kinoshita, M. (1997). Renal preconditioning remotely augments myocardial ischemic tolerance via adenosine receptors. Circulation. 96, 1-253 (Abstr.)

Van den Doel, M.A., Gho, B.C.G., Duval, S.Y., Schoemaker, R.G., Duncker, D.J., and Verdouw, P.D. (1998). Hypothermia extemds the cardioportection by ischaemic preconditioning to coronary artery occlusions of longer duration. Cardiovasc. Res. 37, 76-81.

Vegh, A., Komoro, S., Szekers, L., and Parratt, J.R. (1992). Antiarrythmic effects of preconditioning in anaesthetised dogs and rats. Cardiovasc. Res. 26, 487-495.

Vegh, A., Szekeres, L., and Parratt, J.R. (1991). Transient ischaemia induced by rapid cardiac pacing results in myocardial preconditioning. Cardiovasc. Res. 25, 1051-1053.

Verdouw, P.D., Gho, B.C.G., and Duncker, D.J. (1995). Ischaemic preconditioning: is it clinically relevant? Eur. Heart J. 16, 1169-1176.

Verdouw, P.D., Gho, B.C.G., and Duncker, D.J. (1996a). Cardioprotection by organs in stress or distress. Basic Res. Cardiol. 91, 44-46.

Verdouw, P.D., Gho, B.C.G., Schoemaker, R., and Duncker, D.J. (1996b). Cardioprotection by ischemic and non-ischemic myocardial stress and ischemia in remote organs: implications for the concept of ischemic preconditioning. Ann. N.Y. Acad. Sci. 793, 27-42.

Zager, R.A., Baltes, L.A., Sharma, H.M., and Jurkowitz, M.S. (1984). Responses of the ischemic acute renal failure kidney to additional ischemic events. Kidney Int. 26, 689-700.

CARDIAC ADAPTATION TO CHRONIC HYPOXIA

Bohuslav Ostadal, Ivana Ostadalova, Frantisek Kolar, Vaclav Pelouch, and Naranjan S. Dhalla

I. INTRODUCTION

Cardiovascular disturbances are among the most frequent diseases of modern times and hence the most widely studied. Among the most serious of these disturbances are hypoxic

Advances in Organ Biology
Volume 6, pages 43-60.

ISBN: 0-7623-0391-3

states of the cardiopulmonary system resulting from disproportion between the amount of oxygen supplied to the cell and the amount actually required by the cell. The degree of hypoxic injury depends not only on the intensity and duration of the hypoxic stimulus, but also on the level of cardiac tolerance to oxygen deprivation. It is, therefore, not surprising that the interest of many experimental and clinical cardiologists during the past 35 years has been centered around the question of increasing cardiac tolerance to oxygen deprivation. In this regard, there are two possibilities: (1) pharmacological interventions that either increase myocardial oxygen supply (vasodilators) or reduce oxygen demand (negative inotrops), and (2) physiological interventions that may be short-lasting (i.e., preconditioning) or long-lasting (i.e., adaptation) to counteract the adverse effects of reduced oxygen supply.

As ischemic preconditioning, described by Murry et al. (1986), represents the most efficient form of temporal protection, it has attracted a great deal of attention and, in fact, considerable progress has been made in understanding this phenomenon over the past decade. At the same time, however, other forms of adaptation have been largely ignored (Kolar, 1996). Among them, the most common and clinically important is the cardiac adaptation to chronic hypoxia. This phenomenon is characterized by a variety of functional changes to maintain homeostasis with a minimal expenditure of energy (Durand, 1982). Such adjustment may protect the heart under conditions that require enhanced work and consequently increased metabolism. In addition to the protective effects, adaptation to chronic hypoxia may also exert adverse influences on the cardiopulmonary system, including pulmonary hypertension and right ventricular hypertrophy, which may result in congestive heart failure. Therefore, in this review, we will deal with the pathogenic mechanisms participating in the development of both beneficial and adverse effects of cardiopulmonary adaptation to chronic hypoxia. Particular attention will be paid to chronic hypoxia–induced pulmonary hypertension and right ventricular hypertrophy as well as to the possible regression of chronic hypoxia–induced cardiopulmonary changes.

Chronic myocardial hypoxia, the result of disproportion between oxygen supply and demand at the tissue level, may be induced by several mechanisms. The most common causes are undoubtedly (1) *ischemic hypoxia,* induced by the reduction or interruption of the coronary blood flow, and (2) *systemic hypoxia,* characterized by a drop in PO_2 in the arterial blood. For the sake of completeness we could add (3) *anemic hypoxia,* in which the arterial PO_2 is normal, but the oxygen transport capacity of the blood is decreased. On the other hand, the most frequent causes of raised oxygen consumption are increased physical activity, mental stress, or administration of substances with positive inotropic and chronotropic effects. In terms of the relevant chronic clinical syndromes, ischemic hypoxia is manifested primarily in chronic ischemic heart disease whereas systemic hypoxia is associated with chronic cor pulmonale of varying origin, cyanosis due to a hypoxemic congenital heart disease, and changes in the cardiopulmonary system induced by a decrease in barometric pressure at a high altitude. In two cases, however, systemic hypoxia can be considered as physiological: (1) the fetal myocardium adapted to hypoxia corresponding to an altitude of 8000 m, and (2) the myocardium of subjects living permanently at high altitudes. In both situations the myocardium is significantly more resistant to acute oxygen deficiency, but in populations living in lowlands, this property is lost soon after birth (Moret, 1980; Heath and Williams, 1995). Although the heart obviously has the capability to adapt to various forms of hypoxia, this chapter relates only to effects of chronic systemic hypoxia.

It should be pointed out that the term "adaptation" has been described in different ways which occasionally leads to semantic problems in biology. According to the glossary edited by the International Union of Physiological Sciences (Bligh and Johnson, 1973), adaptation is "a change which reduces the physiological strain produced by a stressful component of the total environment." In contrast, the definition proposed by Adolph (1956) discards the notion of benefit: "adaptations are modifications of organisms that occur in the presence of particular environments and circumstances . . . not limited, as is often done, to modifications that seem favorable to the individual." In fact, adaptation to chronic systemic hypoxia is an adjustment that does not imply, in an obligatory sense, that it is beneficial. Functional adaptive changes require time to materialize. They can occur through (1) *genotypic adaptations*, which result from genetically fixed attributes in those species that have lived for generations in their environment, and (2) *phenotypic adaptations* (including accommodation, acclimation, and acclimatization), which are labile processes occurring within the lifetime of an organism and decay when these circumstances no longer exist (Bouverot, 1985). The term "adaptation," as used in this chapter, refers to changes in cardiac structure and function that result from acute, prolonged, or chronic exposure to oxygen deprivation.

The most frequently used experimental model in research on chronic hypoxia is that of high altitude, either as seen in the mountain environment or as simulated under laboratory conditions in a normobaric- or hypobaric chamber. This model permits the study of the time-course of development of beneficial and adverse adaptive changes, the possibility of their spontaneous reversibility when the animals are removed from the hypoxic atmosphere, and/or the pharmacological protection of unwanted manifestations. Sensitivity to hypoxia is characterized by marked interspecies differences; this raises the question of suitable experimental animals. Cattle and pigs are among the most sensitive animals; sheep and dogs seem less liable to develop hypoxic pulmonary hypertension and right ventricular hypertrophy, while rats and rabbits fall between these two groups (Tucker et al., 1975). The significance of the experimental results for clinical practice depends on the extent to which observed changes are comparable to findings in patients: pulmonary hypertension and right ventricular hypertrophy, muscularization of the pulmonary arterioles, and the enlargement of the carotid body occur in both rats and humans; the development of their ventilatory adaptation to chronic hypoxia is comparable. On the other hand, polycythemia, which is always found in rats, occurs in only a few patients with chronic obstructive lung disease.

II. PROTECTIVE EFFECTS OF ADAPTATION

A. Increased Cardiac Tolerance to Oxygen Deprivation

In chronic high-altitude hypoxia, the myocardium must maintain adequate contractile function in spite of lowered oxygen tension in the coronary circulation. Such an environment requires genotypical adaptation or acclimatization (in lowlanders after prolonged residence at high altitude) which may have cardioprotective effects. It was reported in the late 1950s (Hurtado, 1960) that the incidence of myocardial infarction is lower in people who live at high altitude (Peru, 4000 m). An epidemiological survey from New Mexico (Mortimer et al., 1977) gave some evidence that even living at moderate elevations (2100 m) could result in protection against death from ischemic heart disease. In addition to chronic hypoxia, however, relatively increased physical activity and reduced obesity have to be taken into consideration while explaining the protective effects of living at high altitudes.

Epidemiological observations on the protective effect of high altitude on the heart are consistent with experimental studies using *anoxia in vitro* and *necrogenic doses of isoproterenol* for testing the myocardial resistance in acclimatized animals. In this connection, it is pointed out that the first experimental studies on the protective effect of high altitude on the cardiac muscle were carried out in Prague in 1958 by Kopecky and Daum (acute anoxia in vitro) and in 1966 by Poupa et al. (acute anoxia in vitro, as well as isoproterenol). These findings were later confirmed by McGrath and Bullard (1968, anoxia in vitro). Furthermore, it has been reported (Widimsky et al., 1973; McGrath et al., 1973) that a similar protective effect can be induced by a relatively short intermittent exposure of rats to simulated high attitude (4 hours per day, a total of 24 exposures up to 7000 m) (Figure 1). However, a significant sex difference was demonstrated in the resistance of the cardiac muscle to acute anoxia in vitro; the myocardium of female control rats proved to be more resistant to oxygen deficiency. Chronic hypoxia resulted in markedly enhanced resistance in both sexes, yet the sex difference was maintained (Ostadal et al., 1984b) (Figure 2). Similarly, Ou and Smith (1984) have demonstrated that the cardiopulmonary system in female rats tolerated simulated altitude of 18,000 ft better than that in males. Whether such differences in male and female hearts are due to differences in the expression of heat shock or antioxidant proteins upon exposure to hypoxia needs to be investigated.

Adaptation to chronic hypoxia also protects the heart against *experimental ischemia*. Meerson et al. (1973) described how rats acclimatized to intermittent high altitude developed smaller myocardial necrosis following coronary occlusion than control animals kept at sea level. In another study, however, this effect was ob-

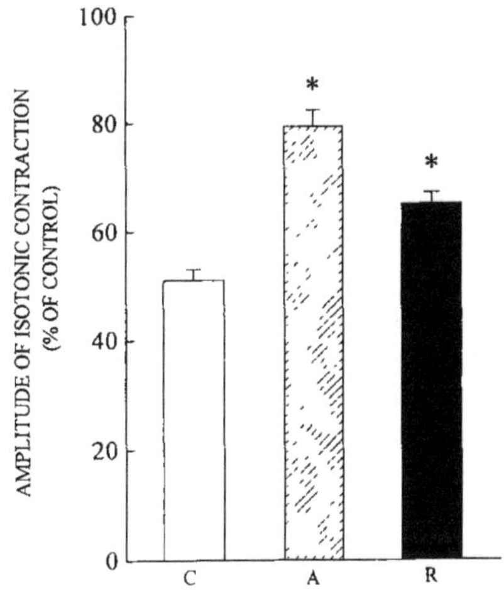

Figure 1. Cardiac resistance to acute anoxia in vitro (expressed as percentage of recovery of the amplitude of isotonic contraction of isolated right ventricle) in control rats (C), in rats adapted to intermittent high altitude hypoxia (A), and in rats 4 months after the last hypoxic exposure (R). *Significantly different from controls ($p < .01$). Data taken from Ostadal et al. (1985).

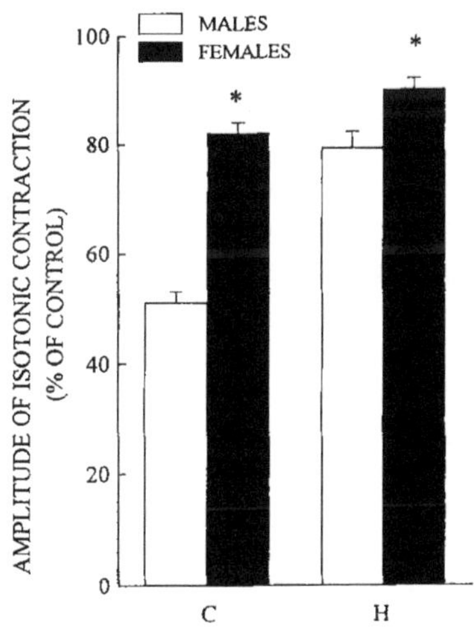

Figure 2. Cardiac resistance to acute anoxia in vitro (expressed as percentage of recovery of the amplitude of isotonic contraction of isolated right ventricle) in male and female rats exposed to sea level (C) and to intermittent high-altitude hypoxia (H). *Statistical significance of sexual difference ($p < .01$). Data taken from Ostadal et al. (1984b).

served only in rats permanently exposed to the same altitude, but not in those acclimatized intermittently (Turek et al., 1980). Opie et al. (1978) have observed that the isolated perfused hearts of rats acclimatized to natural hypoxic conditions at 3500 m maintained higher myocardial levels of adenosine triphosphate (ATP), phosphocreatine, and glycogen content after coronary occlusion, but their function did not differ from controls. Recently, an increased tolerance to global ischemia, manifested as improved recovery of the contractile function following reperfusion, was shown in isolated perfused hearts of rats acclimatized to permanent normobaric hypoxia (Tajima et al., 1994). Meerson et al. (1989), however, did not find any improvement in the recovery of heart function by intermittent hypoxia in a similar ischemia–reperfusion model. Adaptation to high altitude hypoxia had a pronounced antiarrhythmic effect under conditions of acute myocardial ischemia (Meerson et al., 1989) and attenuated the development of systemic hypertension and left ventricular hypertrophy in spontaneously hypertensive rats (Henley et al., 1992).

An important feature of adaptation to chronic high-altitude hypoxia is that the protective effect persists even 4 months after removal of rats from the hypoxic environment (Ostadal and Widimsky, 1985) (see Figure 1). This finding needs further detailed experimental investigation; nevertheless, it suggests some possibilities in the search for effective protection of the myocardium against various types of hypoxic injury. It may be concluded that the majority of studies have clearly shown the cardioprotective effects of adaptation to chronic hypoxia against different types of injury. Since most of these effects were observed in acclimatized rats, the behavior of larger species under conditions of chronic hypoxia needs be examined in future investigations. The variability in results obtained by different investigators may be due to technical differences as well as to the selection of degree, duration, and type (e.g., permanent, intermittent, natural) of hypoxic exposure, in addition to the type of damage determined at the end point (for a review, see Kolar [1996]).

B. Possible Protective Mechanisms

1. *Blood Oxygen Transport*

An increase in hemoglobin will increase the blood O_2 concentration. Indeed, mammals and birds introduced to high altitudes develop variable degrees of polycythemia (Reynafarje, 1958), associated with a shift of the oxygen dissociation curve to the right due to an increased concentration of 2,3-diphosphoglycerate (Monge and Leon-Velarde, 1991). This response mainly results from the stimulation of erythroid precursors cells in the bone marrow by erythropoietin, produced in hypoxic kidneys. On the other hand, mammals and birds genotypically adapted to high altitude show a modest or no increase in hematocrit (Banchero et al., 1971). It can therefore be concluded that an increased blood O_2 content may not necessarily be an adaptive parameter. The role of *myoglobin* consists of storing oxygen and facilitating its transport to tissue. Several studies demonstrated an increased concentration of myoglobin in chronically hypoxic myocardium (for a review, see Moret [1980]), but significant differences exist depending upon species and age.

2. *Tissue Oxygen Transport*

There are disagreements with respect to the development of *myocardial capillaries* in animals exposed to hypobaric hypoxia. Whereas Rotta (1943), Clark and Smith (1978), and Smith and Clark (1979) found a decrease in the ventricular capillary density in chronically hypoxic guinea pigs and rats, Miller and Hale (1970) found an increased capillary density in both ventricles. Turek et al. (1972) described an increase in the hypertrophied right ventricle but observed no change in the left ventricle; Moravec et al. (1983) as well as Banchero et al. (1985) found an increase in the left ventricle, only. Finally, Rakusan et al. (1981) as well as Pietschmann and Bartels (1985) found no evidence of increased myocardial capillary density in animals exposed to high altitude. Mathieu-Costello (1988, 1989) and Mathieu-Costello et al. (1989) showed that correction for sarcomere length is critical to the interpretation of results; capillary length per volume of muscle fiber was not significantly different between lowland and highland deer mouse. Turek et al. (1975) and Scheel et al. (1990) have found a greater *coronary flow* in rats and dogs, whereas functional measurements in dogs did not reveal any effect of chronic hypoxia on coronary collateral flow (Scheel et al., 1990). Such a variability in results may be due to the selection of experimental animals, their age, and the degree as well as duration of hypoxia. Nevertheless, it seems that coronary angiogenesis and increased coronary flow are effective compensatory mechanisms at the beginning of the process of acclimatization; later their importance may decrease. This view is supported by normal or even decreased coronary blood flow in residents living at high altitude (Moret, 1980).

3. *Energetic Metabolism*

Paul Bert was the first to postulate that tissues may gradually alter their cellular metabolism during acclimatization of the body to high altitude (for further information, see Bouverot [1985]). According to Moret (1980), this protective effect may involve increased capacity of cardiac anaerobic metabolism, increased energy utilization capacity, and possibly, selection of metabolic pathways or substrates with a higher energy efficiency, which would decrease the oxygen requirements. This view is supported by the findings in chronically hypoxic rats (Bass et al., 1989) in which both ventricles had significantly increased the capacity for glucose utilization as well as for the synthesis and degradation of lactate.

On the other hand, the ability of the heart to break down fatty acids was found to decrease significantly. Chronic hypobaric hypoxia increased the number of cardiac mitochondria and slightly decreased their mean volume (Costa et al., 1988). Recently, Novel-Chate et al. (1995) have shown that chronic hypoxia increased the cardiac function/oxygen consumption ratio and induced phosphocreatine overshoot upon β-adrenoceptor stimulation. Furthermore, acclimatization to high altitude lowers the specific activities of several sarcolemmal adenosine triphosphatases (ATPases) and at the same time increases their affinity for ATP (Ziegelhoffer et al., 1987). These effects can be considered as an adaptive mechanism at the enzyme level that enables more efficient utilization of ATP and helps to prevent changes in membrane transport function under conditions of low energy production. It should be mentioned, however, that it is often impossible to distinguish the direct effects of hypoxia from those of hypertrophy and other factors, including the nutritional state associated with exposure to high altitude (Barrie and Harris, 1976). The experiments of McGrath and Bullard (1968) showing that the protective effect of acclimatization was removed by the treatment of animals with a glyceraldehyde-3-phosphate-dehydrogenase inhibitor, iodoacetate, seem to support the view that adaptive changes in energy metabolism may be responsible for the cardioprotective effect of chronic hypoxia.

4. *Neural Regulations*

High-altitude hypoxia produces sustained stimulation of the *sympathetic nervous system*. Initially this increases heart rate, but with time, the responsiveness of the heart decreases, so the initial tachycardia may not be sustained (Grover et al., 1986). Despite increased sympathetic activity, chronic hypoxia is associated with impaired response to adrenergic stimulation due to the down-regulation of myocardial β-adrenergic receptors (Bernstein et al., 1992; Kacimi et al., 1992). One potential mechanism of desensitization to catecholamines in chronic hypoxia appears to involve deceased functional activity of the G-protein α subunit (Kacimi et al., 1995). Chronic hypoxia may also lead to increased degradation of catecholamines, as indicated by the elevated activity of catechol-o-methyl transferase (Maher et al., 1978). Furthermore, a chronic hypoxia–induced decrease in sensitivity to catecholamines can be seen to be antiarrhythmic and may be responsible for the protective effect of adaptation on isoproterenol-induced necrotic lesions (Poupa et al., 1966). The *parasympathetic nervous system* may also be stimulated at high altitude (Grover et al., 1986), but its role is poorly understood. Kacimi et al. (1993) have found that chronic hypoxia increases muscarinic receptor affinity and density, which may contribute to the blunted responsiveness of the heart to catecholamines.

5. *Humoral Factors*

Chronic hypoxia–induced *hypothyroidism* (Harris, 1981) may be one of the cardioprotective mechanisms since hypothyroid myocardium is less susceptible to anoxia (Martin et al., 1971). On the other hand, hyperthyroidism increases cardiac sensitivity to hypoxia (Palacios et al., 1979). Adaptation to high altitude hypoxia has been reported to increase the myocardial level of *prostaglandins* significantly (Pshennikova et al., 1992). A single dose of 7-oxo-prostacyclin was observed to increase cardiac tolerance to hypoxia in control and hypoxic rats (Ziegelhoffer et al., 1993). The protective effect of acclimatization and prostacyclin were additive, suggesting that different mechanisms may be involved; however, it is not clear whether prostacyclin contributes to the cardioprotective effect of adaptation to chronic hypoxia. Adaptation to chronic hypoxia has also been shown to induce a decrease in *adenosine* receptor density without any change in the affinity for the agonist (Kacimi et al., 1993), but the possible involvement of adenosine in acclimatization to chronic hypoxia is poorly understood.

From the foregoing discussion it may be concluded that despite the fact that the first experimental study on the cardioprotective effect of adaptation to high altitude was published more than 35 years ago, no satisfactory explanation of this important phenomenon has yet been found. Moreover, suggested possible mechanisms are not concise and in fact are often controversial. The role of heat shock proteins (Meerson et al., 1992), remodeling of the ventricles (Pelouch et al., 1994), sex hormones, and antioxidant reserve in the cardioprotective effect of adaptation to chronic hypoxia is not fully established. Furthermore, the optimal degree of hypoxia and length of acclimatization for cardioprotection are still unclear.

III. ADVERSE EFFECTS OF ADAPTATION

A. Pulmonary Hypertension

Sustained hypoxia exerts opposite effects on the systemic and pulmonary vascular smooth muscles, bringing about vasodilatation in the systemic, but vasoconstriction and consequently structural remodeling in the pulmonary circulation. The low PO_2 in alveoli and blood is known to affect the vascular smooth muscle cells, but the actual mechanism of hypoxic pulmonary vasoconstriction is still not fully understood (for a review, see Fishman [1990]). Similarly, the extent to which constriction of pulmonary vessels is capable of redistributing blood flow and thus improving pulmonary gas exchange needs further investigation. Pulmonary hypertension develops as a result of chronic hypoxia, whereas the systemic blood pressure is normal or even below normal in well-adapted subjects (Heath and Williams, 1995). The pulmonary hypertension can be seen to induce right heart failure, which may have serious consequences in individuals exposed to chronic hypoxia.

Reliable investigations of the effect of high altitude on the cardiopulmonary system started some 40 years ago. Rotta et al. (1956) first reported that healthy men and women living at high altitude have some degree of pulmonary hypertension and right ventricular hypertrophy. These observations in people living in the Peruvian Andes were later confirmed by Penaloza et al. (1962) and Sime et al. (1963) for the same geographical region, as well as by Vogel et al. (1962) for residents living at high altitudes in the United States, and by Singh et al. (1965) for temporary residents in the Himalayas. The critical altitude for the development of pulmonary hypertension and right ventricular hypertrophy in men, was specified to be 3000 m.

Since hypoxic pulmonary vasoconstriction exists in all adult mammals, it is not surprising that exposure to high altitude, either natural or simulated in a barochamber, may induce pulmonary hypertension and right ventricular hypertrophy in different animal species including guinea pig, rat, cow, rabbit, pig, dog, mouse, and sheep (for a review, see Herget and Palecek [1978]; Reeves et al., [1979]). There are, however, significant interspecies variations in the pulmonary hemodynamic response to chronic hypoxia. On the other hand, the native high-altitude mammals including llamas, alpacas, and vicunas have largely lost their hypoxic pulmonary vasoconstrictor response (Heath, 1988). They have thin-walled elastic and muscular pulmonary arteries and do not develop muscularization of the pulmonary arterioles. Such a thin-walled vasculature in the lung is associated with low pulmonary artery resistance and pressure; thus, these animals do not develop right ventricular hypertrophy. Durmowicz et al. (1993) have demonstrated that endothelial cells from the pulmonary arteries of native yaks at high altitude were much longer, wider, and rounder than those of domestic cows. Accordingly, it was suggested that adaptation to high altitude may include changes not only in the pulmonary vascular smooth muscle but also in the endothelial cell structure and function.

As mentioned above, both chronic hypoxia and subsequent right ventricular hypertrophy may be of an intermittent nature (e.g., in exacerbation of chronic obstructive lung disease during acute respiratory infection, or in repeated ascents of mountains). Hypoxia is likewise not continual in myocardial ischemia, particularly when it depends on the actual coronary blood flow (Ostadal et al., 1994; Weitzenblum, 1994). Therefore, a model of chronic but intermittent hypoxia was developed in which animals were exposed to the hypoxic environment for only a given period of the day (4 to 8 hours per day, 5 days a week, stepwise up to 7000 m; for a review, see Ostadal et al. [1994]). Twenty-four such exposures for 4 hours per day did not lead to a significant change in the right ventricular systolic pressure; however, prolongation to 75 exposures induced a highly significant increase of this parameter. When the daily hypoxic exposure was prolonged from 4 to 8 hours, marked pulmonary hypertension was observed after 24 exposures; an extension to 60 exposures did not lead to further significant changes in the pressure (Widimsky et al., 1973) (Figure 3). The development of pulmonary hypertension is connected with structural changes in the pulmonary vessels (Urbanova et al., 1975); the most significant alterations were observed in distal pulmonary arteries which are situated in closest neighborhood of the alveoli. Further commonly discussed factors that may participate in the development of chronic pulmonary hypertension (i.e., increased viscosity of blood and higher cardiac output) were also observed in our experiments (Ostadal et al., 1975; Kasalicky et al., 1977).

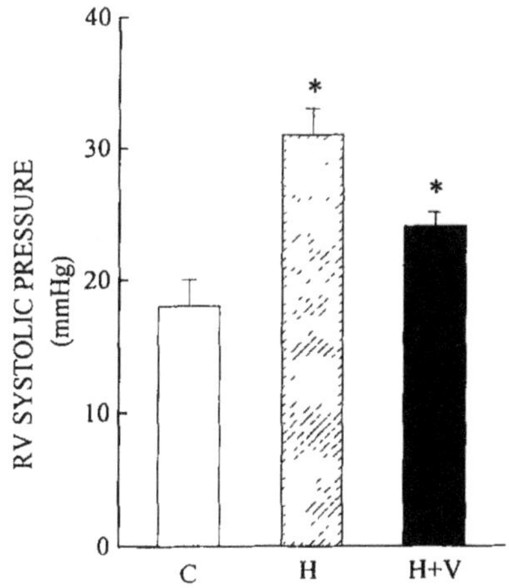

Figure 3. Effect of the preventive administration of verapamil on right ventricular (RV) systolic pressure. C, control; H, intermittent high altitude hypoxia; H + V, hypoxia + verapamil (8 mg/kg before each exposure). *Significantly different from control ($p < .01$). Data taken from Ostadal et al. (1981).

B. Right Ventricular Hypertrophy

The hypoxic pulmonary hypertension leads to right ventricular enlargement, including ventricular hypertrophy and dilatation. This is a beneficial adaptation, allowing the right ventricle to cope with an increased afterload and to maintain a normal cardiac output. On the other hand, right ventricular failure is observed with prolongation of exposure or degree of hypoxia in which worsening of hypoxemia induces a marked increase in afterload (Weitzenblum, 1994). We have observed that 24 exposures to high altitude did not significantly change right ventricular systolic pressure, but right ventricular hypertrophy could already be seen (Widimsky et al., 1973). This finding suggests that right ventricular hypertrophy can be induced by intermittent pulmonary hypertension that is present only during the stay of experimental animals in the barochamber (Figure 4). Adaptation to chronic hypoxia induced significant changes in the protein profile of the myocardium: myosin isoforms were shifted from the high activity V1 ATPase to the low activity V3 form (Pelouch et al., 1985). Chronic hypoxia also modulates qualitative and quantitative changes of collagenous proteins (Pelouch et al., 1985; Ostadal et al., 1978) as the proportion of this protein fraction is increased, whereas the collagen I/collagen III ratio is decreased, suggesting an increased synthesis of collagen III. The development of hypoxic right ventricular hypertrophy is accompanied by an increase of ventricular atrial natriuretic peptide, but its role has not been clarified at present (McKenzie et al., 1994).

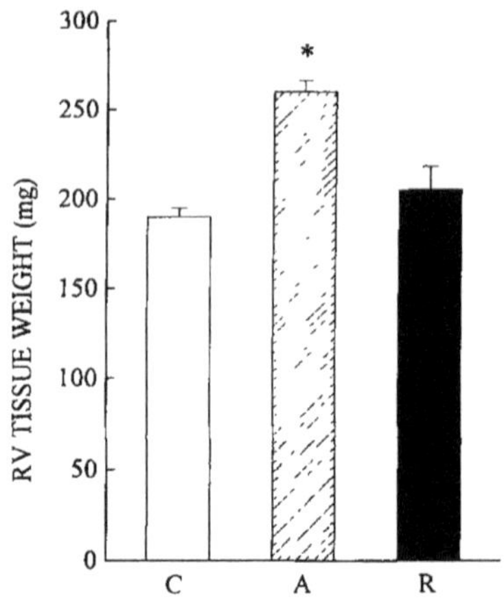

Figure 4. Right ventricular (RV) hypertrophy and its reversibility in rats adapted to intermittent high altitude hypoxia. A, intermittent high altitude hypoxia; C, control; R, 4 months after the last hypoxic exposure. *Significantly different from controls ($p < .01$). Data taken from Ostadal et al. (1985).

The increased right ventricular afterload would imply possible negative consequences for the intrinsic ventricular function. Therefore, the function of the hypertrophic right ventricle was studied in adult rats with chronic hypoxia–induced pulmonary hypertension (Kolar and Ostadal, 1991). For this purpose an isolated preparation of the right ventricular working heart was employed. The peak indices of mechanical performance were almost doubled in chronic hypoxia–exposed animals when compared with the normoxic group, while the index of contractility remained unchanged. Maximum ventricular performance was found to be a linear function of the relative right ventricular weight. No evidence of right ventricular pump dysfunction was detected; moreover, the ability of the ventricle to maintain cardiac output against increased pulmonary resistance was markedly improved. It seems that the increase in right ventricular mass of chronic hypoxia–exposed rats serves to improve maximum ventricular performance, which aids in overcoming an elevated pulmonary resistance without disturbing the pump function.

Interestingly, the right ventricular contractile function is also generally preserved in patients with chronic obstructive lung disease suffering from pulmonary hypertension and right ventricular hypertrophy (Weitzenblum, 1994). In addition, chronic hypoxia–induced focal myocardial necrotic lesions localized predominantly in the right ventricular myocardium may lead to the development of disseminated fibrosis or chronic aneurysm of the right ventricular wall. With the prolongation of exposures, no further new acute necroses was observed (Ostadal and Widimsky, 1985). The mechanisms of chronic hypoxia–induced necrotic myocardial changes are complex; besides the effect of hypoxia, the influence of stress due to the stay of animals in the barochamber cannot be excluded (Ostadal et al., 1984a).

IV. REGRESSION OF ADAPTIVE CHANGES

A. Spontaneous Reversibility

One of the most characteristic features of phenotypical adaptations is their reversible nature. It has been found that even severe chronic hypoxia–induced changes (body weight loss, polycythemia, pulmonary hypertension, right ventricular hypertrophy, and alterations in cardiac function and metabolism) were completely reversible after removal of rats from the hypoxic atmosphere for a sufficiently long period of time (Bass et al., 1989; Kolar and Ostadal, 1991; Ressl et al., 1974; Ostadal et al., 1985). The regression of muscularization of distal pulmonary arterioles was, however, incomplete. Similar relationships were described by Leach et al. (1977), Herget et al. (1978), and Varosian et al. (1990). The first fully normalized parameter was the body weight of experimental animals; it returned to control value 2 weeks after the end of hypoxia. Hemoglobin and right ventricular weight were normalized 2 weeks later (Ostadal et al., 1985). The protein composition of the ventricular myocardium was, however, significantly different from controls; an increased proportion of the collagenous fraction persisted even when the right ventricular weight was already normal (Ostadal et al., 1978). As mentioned earlier, the cardioprotective effect of adaptation (i.e., increased tolerance to oxygen deprivation), persists even 4 months after removal of the animals from the hypoxic environment (Leach et al., 1977). On the other hand, the attenuation of systemic hypertension in adapted spontaneously hypertensive rats dissipated when the rats returned to normoxic conditions (Henley et al., 1992).

B. Pharmacological Treatment

In view of the potential value of chronic hypoxia for cardioprotection, it would be ideal to reduce the development of the adverse changes and simultaneously preserve the beneficial signs of the process of adaptation. From the relatively narrow spectrum of promising drugs, calcium antagonists were selected for their combined vasodilatatory and cardioprotective effects (Leach et al., 1977; Ostadal et al., 1981). Administration of verapamil (before each of the hypoxic exposures) significantly reduced the degree of pulmonary hypertension and right ventricular hypertrophy (both by approximately 20%) and partially prevented hypertensive changes in the pulmonary vasculature. Although verapamil also reduced the incidence of necrotic lesions in the myocardium significantly, the beneficial sign of acclimatization, (i.e., cardiac resistance to acute anoxia) is diminished. Therapeutic administration of verapamil or nifedipine (when the cardiopulmonary changes were already developed) was, however, without any effect. For explanation of these findings, possible differences in calcium metabolism of pulmonary vascular smooth muscle under different hypoxic conditions should be taken into consideration.

Since angiotensin-converting enzyme (ACE) inhibitors are known to reduce left ventricular hypertrophy and fibrosis in experimental systemic hypertension (for a review, see Weber et al. [1991]), an attempt was made to determine whether treatment with an ACE inhibitor, enalapril, could also reduce the ventricular content of collagen in animals recovering from hypoxia-induced pulmonary hypertension (Kolar et al., 1993; Pelouch et al., 1996). It was shown that enalapril significantly decreased heart rate, systemic arterial pressure, and left ventricular weight in both hypoxic and control animals; however, the pulmonary blood pressure and right ventricular weight remained unchanged. On the other hand, the content of collagen was reduced in both ventricles from the enalapril-treated animals. These data suggest that the regression of cardiac fibrosis due to enalapril may be independent of hemodynamic load. Further studies using other ACE inhibitors and angiotensin II receptor antagonists are required if we are to fully appreciate the role of the renin-angiotensin system in the process of adaptation due to chronic hypoxia.

It may be concluded that adaptation to chronic hypoxia, like other physiological adaptations, may induce both positive and negative responses of the cardiopulmonary system. However, reduction of the negative consequences also leads to the diminution of the positive effects by both calcium antagonists and ACE inhibitors. An extensive investigation involving different types of pharmacological agents as a means of reducing the adverse effects of chronic hypoxia may yield valuable information in the future. In this regard, the use of β-adrenoceptor blocking drugs and antioxidants may provide some valuable information. Nevertheless, the fact that the protective sign of adaptation to chronic hypoxia (i.e., increased cardiac tolerance to oxygen deprivation) persists long after the other chronic hypoxia–induced changes are already normalized, offers a more optimistic view of the future of effective protection of the hypoxic myocardium.

REFERENCES

Adolph, E.F. (1956). General and specific characteristics of physiological adaptations. Am. J. Physiol. 184, 18-28.

Banchero, N., Grover, R.F., and Will, J.A. (1971). Oxygen transport in the llama *(Lama glama)*. Respir. Physiol. 13, 102-115.

Banchero, N., Kayar, S.R., and Lechner, A.J. (1985). Increased capillarity in skeletal muscle of growing guinea pigs acclimated to cold and hypoxia. Respir. Physiol. 52, 245-255.

Barrie, S.E., and Harris, P. (1976). Effects of chronic hypoxia and dietary restriction on myocardial enzyme activities. Am. J. Physiol. 231, 1308-1313.

Bass, A., Ostadal, B., Prochazka, J., Pelouch, V., Samanek, M., and Stejskalova, M. (1989). Intermittent high altitude-induced changes in energy metabolism in the rat myocardium and their reversibility. Physiol. Bohemoslov. 38, 155-161.

Bernstein, D., Doshi, R., Huang, S., Strandness, E., and Jasper, J.R. (1992). Transcriptional regulation of left ventricular beta-adrenergic receptors during chronic hypoxia. Circ. Res. 71, 1465-1471.

Bligh, J., Johnson, K.G. (1973). Glossary of terms for thermal physiology. J. Appl. Physiol. 35, 941-961.

Bouverot, P. (1985). Adaptation to Altitude—Hypoxia in Vertebrates. Springer-Verlag, Berlin.

Clark, D.R., and Smith, P. (1978). Capillary density and muscle fibre size in the hearts of rats subjected to simulated high altitude. Cardiovasc. Res. 12, 578-584.

Costa, L.E., Boveris, A., Koch, O.R., and Taquini, A.C. (1988). Liver and heart mitochondria in rats submitted to chronic hypobaric hypoxia. Am. J. Physiol. 255, C123-C129.

Durand, J. (1982). Physiologic adaptation to altitude and hyperexis. In: High Altitude Physiology and Medicine (Brendel, W., and Zink, R.A., Eds.), pp. 209-211. Springer-Verlag, New York.

Durmowicz, A.G., Hofmeister, S., Kadyraliev, T.K., Aldashev, A.A., Stenmark, K.R. (1993). Functional and structural adaptation of the yak pulmonary circulation to residence at high altitude. J. Appl. Physiol. 74, 2276-2285.

Fishman, A.P. (1990). The Pulmonary Circulation: Normal and Abnormal. Mechanisms, Management, and the National Registry. University of Pennsylvania Press, Philadelphia.

Grover, R.F., Weil, J.V., and Reeves, J.T. (1986). Cardiovascular adaptation to exercise at high altitude. Exerc. Sport. Sci. Rev. 14, 269-302.

Harris, P. (1981). Man at high altitude. In: Myocardial Metabolism (Heath, D., and Williams, D.R., Eds.), pp. 196-208. Churchill Livingstone, Edinburgh.

Heath, D. (1988). The pathology of high altitude. Ann. Sports. Med. 4, 203-212.

Heath, D., and Williams, D.R. (1995). High Altitude Medicine and Pathology. Oxford University Press, Oxford.

Henley, W.N., Belush, L.L., and Notestine, M.A. (1992). Reemergence of spontaneous hypertension in hypoxia-protected rats returned to normoxia as adults. Brain Res. 579, 211- 218.

Herget, J., and Palecek, F. (1978). Experimental chronic pulmonary hypertension. Int. Rev. Exp. Pathol. 18, 347-406.

Herget, J., Suggett, A.J., Leach, E., and Barer, G.R. (1978). Resolution of pulmonary hypertension and other features induced by chronic hypoxia in rats during complete and intermittent normoxia. Thorax 33, 468-473.

Hurtado, A. (1960). Some clinical aspects of life at high altitudes. Ann. Intern. Med. 53, 247- 258.

Kacimi, R., Molic, J.M., Aldashev, A., Vatner, D.E., Richalet, J.P., and Crozatier, B. (1995). Differential regulation of G protein expression in rat hearts exposed to chronic hypoxia. Am. J. Physiol. 269, H1865-H1873.

Kacimi, R., Richalet, J.-P., and Corsin, A. (1992). Hypoxia-induced down regulation of beta- adrenergic receptors in rat heart. J. Appl. Physiol. 73, 1377-1382.

Kacimi, R., Richalet, J.-P., and Crozatier, B. (1993). Hypoxia-induced differential modulation of adenosinergic and muscarinic receptors in rat heart. J. Appl. Physiol. 75, 1123-1128.

Kasalicky, J., Ressl, J., Urbanova, D., Ostadal, B., Prochazka, J., Pelouch, V., and Widimsky, J. (1977). Relative organ blood flow in rats exposed to intermittent high altitude hypoxia. Pfugers Arch. 368, 111-115.

Kolar, F. (1996). Cardioprotective effects of chronic hypoxia: relation to preconditioning. In: Myocardial Preconditioning (Wainwright, C.L., and Parratt, J.R., Eds.), pp. 261-275, Springer, Berlin.

Kolar, F., and Ostadal, B. (1991). Right ventricular function in rats with hypoxic pulmonary hypertension. Pflugers Arch. 419, 121-126.

Kolar, F., Pelouch, V., Papousek, F., Ostadal, B., Cihak, R., and Widimsky, J. (1993). Regression of myocardial collagen due to enalapril in chronically hypoxic rats. J. Mol. Cell. Cardiol. 25, S41.

Kopecky, M., and Daum, S. (1958). Tissue adaptation to anoxia in rat myocardium (in Czech). Cs. Fysiol. 7, 518-521.

Leach, E., Howard, P., and Barer, G.R. (1977). Resolution of hypoxic changes in the heart and pulmonary arterioles of rats during intermittent correction of hypoxia. Clin. Sci. Mol. Med. 52, 153-162.

Maher, J.T., Denniston, J.C., and Wolfe, D.L. (1978). Mechanism of the attenuated cardiac response to beta-adrenergic stimulation in chronic hypoxia. J. Appl. Physiol. 44, 647-651.

Martin, L.G., Westenberger, G.E., and Bullard, R.W. (1971). Thyroidal changes in the rat during acclimatization to simulated high altitude. Am. J. Physiol. 221, 1057-1063.

Mathieu-Costello, O. (1988). Capillary configuration in contracted muscles: comparative aspects. Adv. Exp. Med. Biol. 227, 229-236.

Mathieu-Costello, O. (1989). Muscle capillary tortuosity in high altitude mice depends on sarcomere length. Respir. Physiol. 76, 289-302.

Mathieu-Costello, O., Poole, D.C., and Logemann, R.B. (1989). Muscle fibre size and chronic exposure to hypoxia. Adv. Exp. Med. Biol. 248, 305-311.

McGrath, J.J., and Bullard, R.W. (1968). Altered myocardial performance in response to anoxia after high-altitude exposure. J. Appl. Physiol. 25, 761-764.

McGrath, J.J., Prochazka, J., Pelouch, V., and Ostadal, B. (1973). Physiological response of rats to intermittent high altitude stress: effect of age. J. Appl. Physiol. 34, 289-293.

McKenzie, J.C., Kelley, K.B., Merisko-Liversidge, E.M., Kennedy, J., and Klein, R.M. (1994). Developmental pattern of ventricular atrial natriuretic peptide (ANP) expression in chronically hypoxic rats as an indicator of the hypertrophic process. J. Mol. Cell. Cardiol. 26, 753-767.

Meerson, F.Z., Gomazkov, G.A., and Shimkovich, M.V. (1973). Adaptation to high altitude hypoxia as a factor preventing development of myocardial ischemic necrosis. Am. J. Cardiol. 31, 30-34.

Meerson, F.Z., Malyshev, I.Y., and Zamotrinsky, A.V. (1992). Differences in adaptive stabilization of structures in response to stress and hypoxia relate with the accumulation of hsp70 isoforms. Mol. Cell. Biochem. 111, 87-95.

Meerson, F.Z., Ustinova, E.E., and Manukhina, E.B. (1989). Prevention of cardiac arrhythmias by adaptation: regulatory mechanisms and cardiotropic effect. Biomed. Biochim. Acta 48, 583-588.

Miller, A.T., and Hale, D.M. (1970). Increased vascularity of brain, heart, and skeletal muscle of polycythemic rats. Am. J. Physiol. 219, 702-704.

Monge, C., and Leon-Velarde, F. (1991). Physiological adaptation to high altitude: oxygen transport in mammals and birds. Physiol. Rev. 71, 1135-1172.

Moravec, J., Cluzeaud, F., Rakusan, K., and Turek, Z. (1983). Capillary supply and utilization of intracellular oxygen in the left ventricular myocardium from rats adapted to high altitude. Adv. Exp. Med. Biol. 159, 243-252.

Moret, P.R. (1980). Hypoxia and the heart. In: Heart and Heart-like Organs (Bourne, G.H., Ed.), pp. 239-387. Academic Press, New York.

Mortimer, E.A. Jr., Monson, R.R., and McMahon, B. (1977). Reduction in mortality from coronary heart disease in men residing at high altitude. N. Engl. J. Med. 296, 581-585.

Murry, C.E., Jennings, R.B., and Reimer, K.A. (1986). Preconditioning with ischemia: a delay of lethal cell injury in ischemic myocardium. Circulation 74, 1124-1136.

Novel-Chate, V., Aussedat, J., Saks, V.A., and Rossi, A. (1995). Adaptation to chronic hypoxia alters cardiac metabolic response to beta stimulation: novel face of phosphocreatine overshoot phenomenon. J. Mol. Cell. Cardiol. 27, 1679-1687.

Opie, L.H., Duchosal, F., and Moret, P.R. (1978). Effect of increased left ventricular work, hypoxia, or coronary artery ligation on hearts from rats at high altitude. Eur. J. Clin. Invest. 8, 309-315.

Ostadal, B., Kolar, F., Pelouch, V., Prochazka, J., and Widimsky, J. (1994). Intermittent high altitude and the cardiopulmonary system. In: The Adapted Heart (Nagano, M., Takeda, N., and Dhalla, N.S., Eds.), pp. 173-182. Raven Press, New York.

Ostadal, B., Kvetnansky, R., Prochazka, J., and Pelouch, V. (1984a). Effect of intermittent high altitude stress on epinephrine levels in the right and left ventricular myocardium of rats. In: The Role of Catecholamines and Other Neurotransmitters Under Stress (Kvetnansky, R., Ed.), pp. 669-674. Gordon and Breach, New York.

Ostadal, B., Mirejovska, E., Hurych, J., Pelouch, V., and Prochazka, J. (1978). Effect of intermittent high altitude hypoxia on the synthesis of collagenous and non-collagenous proteins of the right and left ventricular myocardium. Cardiovasc. Res. 12, 303-308.

Ostadal, B., Prochazka, J., Pelouch, V., Urbanova, D., and Widimsky, J. (1984b). Comparison of cardiopulmonary responses of male and female rats to intermittent high altitude hypoxia. Physiol. Bohemoslov. 33, 129-138.

Ostadal, B., Prochazka, J., Pelouch, V., Urbanova, D., Widimsky, J., and Stanek, V. (1985). Pharmacological treatment and spontaneous reversibility of cardiopulmonary changes induced by intermittent high altitude hypoxia. Prog. Respir. Res. 29, 17-25.

Ostadal, B., Ressl, J., Urbanova, D., Pelouch, V., Prochazka, J., and Widimsky, J. (1975). The role of polycythemia in the development of experimental high altitude hypertension and right ventricular hypertrophy. Prog. Respir. Res. 9, 130-137.

Ostadal, B., Ressl, J., Urbanova, D., Prochazka, J., Pelouch, V., and Widimsky, J. (1981). Effect of verapamil on pulmonary hypertension and right ventricular hypertrophy induced in rats by intermittent high altitude hypoxia. Respiration 42, 221-227.

Ostadal, B., and Widimsky, J. (1985). Intermittent Hypoxia and Cardiopulmonary System. Academia, Prague.

Ou, L.C., and Smith, R.P. (1984). Strain and sex differences in the cardiopulmonary adaptation of rats to high altitude. Proc. Soc. Exp. Biol. Med. 177, 308-311.

Palacios, I., Sagar, K., and Powell, W.J. (1979). Effect of hypoxia on mechanical properties of hyperthyroid cat papillary muscle. Am. J. Physiol. 237, H293-H298.

Pelouch, V., Dixon, I., Golfman, L., Beamish, R.E., and Dhalla, N.S. (1994). Role of extracellular matrix proteins in heart function. Mol. Cell. Biochem. 129, 101-120.

Pelouch, V., Kolar, F., Ostadal, B., Milerova, M., Cihak, R., and Widimsky, J. (1997). Regression of chronic hypoxia–induced pulmonary hypertension, right ventricular hypertrophy and fibrosis: effect of enalapril. Cardiovasc. Drug Therap. 11, 177-185.

Pelouch, V., Ostadal, B., Prochazka, J., Urbanova, D., and Widimsky, J. (1985). Effect of high altitude hypoxia on the protein composition of right ventricular myocardium. Prog. Respir. Res. 20, 41-48.

Penaloza, D., Sime, F., Banchero, N., and Gamboa, R. (1962). Pulmonary hypertension in healthy men born and living at high altitude. Med. Thorac. 19, 449-460.

Pietschmann, M., and Bartels, H. (1985). Cellular hyperplasia and hypertrophy, capillary proliferation and myoglobin concentration in the heart of newborn and adult rats at high altitude. Respir. Physiol. 59, 347-360.

Poupa, O., Krofta, K., Prochazka, J., and Turek, Z. (1966). Acclimatization to simulated high altitude and acute cardiac necrosis. Fed. Proc. 25, 1243-1246.

Pshennikova, M.G., Kuznetsova, V.A., Kopylov, I.N., Shimkovich, M.V., Vovk, V.I., Saprygin, D.V., and Meerson, F.Z. (1992). The role of prostaglandine system in the cardioprotective effect of adaptation to hypoxia in stress (in Russian). Kardiologiya 32, 61- 64.

Rakusan, K., Turek, Z., and Kreuzer, F. (1981). Myocardial capillaries in guinea pigs native to high altitude (Junin, Peru, 4,105 m). Pflugers Arch. 391, 22-24.

Reeves, J.T., Wagner, W.W. Jr., McMurtry, I.F., and Grover, R.F. (1979). Physiological effects of high altitude on the pulmonary circulation. In: Environmental Physiology III (Robertshow, D., Ed.), pp. 289-310. University Park Press, Baltimore.

Ressl, J., Urbanova, D., Widimsky, J., Ostadal, B., Pelouch, V., and Prochazka, J. (1974). Reversibility of pulmonary hypertension and right ventricular hypertrophy induced by intermittent high altitude hypoxia in rats. Respiration 31, 38-46.

Reynafarje, C. (1958). The influence of high altitude on erythropoietic activity. In: Homeostatic Mechanisms, pp. 132-146. Brookhaven National Lab., Upton, NY.

Rotta, A. (1943). Peso del corazon y numero de capilares en cobayos de diferentes alturas. Rev. Argent. Cardiol. 10, 186-199.

Rotta, A., Canepa, A., Hurtado, T., and Chavez, R. (1956). Pulmonary circulation at sea level and at high altitude. J. Appl. Physiol. 9, 328-336.

Scheel, K.W., Seavey, E., Gaugl, J.F., and Williams, S.E. (1990). Coronary and myocardial adaptations to high altitude in dogs. Am. J. Physiol. 259, H1667-H1673.

Sime, F., Banchero, N., Penaloza, D., Gamboa, R., Cruz, J., and Marticorena, E. (1963). Pulmonary hypertension in children born and living at high altitude. Am. J. Cardiol. 11, 143- 149.

Singh, I., Capila, C.C., Khanna, P.K., Nanda, R.B., and Rao, B.D.P. (1965). High-altitude pulmonary oedema. Lancet 1, 229-234.

Smith, P., and Clark, D.R. (1979). Myocardial capillary density and muscle fibre size in rats born and raised at simulated high altitude. Br. J. Exp. Path. 60, 225-230.

Tajima, M., Katayose, D., Bessho, M., and Isoyama, S. (1994). Acute ischaemic preconditioning and chronic hypoxia independently increase myocardial tolerance to ischaemia. Cardiovasc. Res. 28, 312-319.

Tucker, A., McMurtry, I.F., Reeves, J.T., Alexander, A.F., Will, D.H., and Grover, R.F. (1975). Lung vascular smooth muscle as a determinant of pulmonary hypertension at high altitude. Am. J. Physiol. 228, 762-767.

Turek, Z., Grandtner, M., and Kreuzer, F. (1972). Cardiac hypertrophy, capillary and muscle fiber density, muscle fiber diameter, capillary radius and diffusion distance in the myocardium of growing rats adapted to a simulated altitude of 3,500 m. Pflugers Arch. 335, 19-28.

Turek, Z., Kubat, K., and Ringnalda, B.E.M. (1980). Experimental myocardial infarction in rats acclimated to simulated high altitude. Basic Res. Cardiol. 75, 544-553.

Turek, Z., Turek-Maischeider, M., Claessens, R.A., Ringnalda, B.E.M., and Kreuzer, F. (1975). Coronary blood flow in rats native to simulated high altitude and in rats exposed to it later in life. Pflugers Arch. 355, 49-62.

Urbanova, D., Pelouch, V., Ostadal, B., Widimsky, J., Ressl, J., and Prochazka, J. (1975). The development of myocardial changes during intermittent high altitude hypoxia in rats. Cor. Vasa. 19, 246-250.

Varosian, M.A., Martirosjan, M.F., Tatinian, N.G., and Petrosian, V.P. (1990). Various characteristics of the contractile function of the heart in cardiac hypertrophy during readaptation after adaptation of the body to high-altitude hypoxia (in Russian). Kosm. Biol. Aviakosm. Med. 24, 42-45.

Vogel, J.A., Weaver, W.F., Rose, R.L., Blount, S.G., and Grover, R.F. (1962). Pulmonary hypertension in exertion in normal man living 10,150 feet. Med. Thorac, 19, 461-477.

Weber, K.T., Brilla, C.G., and Janicki, J.S. (1991). Signals for the remodeling of the cardiac interstitium in systemic hypertension. J. Cardiovasc. Pharmacol. 17, 514-519.

Weitzenblum, E. (1994). The pulmonary circulation and the heart in chronic lung disease. Monaldi. Arch. Chest Dis. 49, 231-234.

Widimsky, J., Urbanova, D., Ressl, J., Ostadal, B., Pelouch, V., and Prochazka, J. (1973). Effect of intermittent altitude hypoxia on the myocardium and lesser circulation in the rat. Cardiovasc. Res. 7, 798-808.

Ziegelhoffer, A., Grunermel, J., Dzurba, A., Prochazka, J., Kolar, F., Vrbjar, N., Pelouch, V., Ostadal, B., and Szekeres, L. (1993). Sarcolemmal cation transport systems in rat hearts acclimatized to high altitude hypoxia: influence of 7-oxo-prostacycline. In: Heart Function in Health and Disease (Ostadal, B., and Dhalla, N.S., Eds.), pp. 219-228. Kluwer Academic Publishers, Boston.

Ziegelhoffer, A., Prochazka, J., Pelouch, V., Ostadal, B., Dzurba, A., and Vrbjar, N. (1987). Increased affinity to substrate in sarcolemmal ATPases from hearts acclimatized to high altitude hypoxia. Physiol. Bohemoslov. 36, 403-415.

ANALYSIS OF ALTERED GENE EXPRESSION DURING ISCHEMIC PRECONDITIONING

Nilanjana Maulik

Advances in Organ Biology
Volume 6, pages 61-80.

ISBN: 0-7623-0391-3

I. INTRODUCTION

Cellular and molecular mechanisms of ischemic preconditioning (PC) leading to the myocardial adaptation to ischemia are based on the fact that enhancement of the endogenous cellular defense system provides each cell with new protein synthesis and thereby the means to protect itself when it is more susceptible to injury. It has long been known that cells exhibit specific responses when confronted with sudden changes in their environmental conditions. The ability of the cells to acclimate to their new environment is the integral driving force for the adaptive modification of the cells. Such adaptation involves a number of cellular and biochemical alterations, including metabolic homeostasis and reprogramming of gene expression. The changes in the metabolic pathways are generally short lived and reversible, whereas the consequences of gene expression are a long-term process and may lead to the permanent alteration in the pattern of gene expression.

Using this concept, it has been shown that preconditioning of the heart by repeated stunning can delay the onset of further irreversible injury (Murry et al., 1990; Li et al., 1990), or even reduce the subsequent postischemic ventricular dysfunction (Flack et al., 1991; Kimura et al., 1992; Schott et al., 1990, Hendrikx et al., 1993) and incidence of ventricular arrhythmias (Tosaki et al., 1994; Lawson et al., 1992). Our laboratory has demonstrated that repeated ischemia, distinguished from a single ischemic insult, can reduce subsequent ischemia reperfusion injury (Flack et al., 1991; Kimura et al., 1992), infarct size (Kimura et al., 1992) and postischemic ventricular fibrillation (Tosaki et al., 1994). Such myocardial preservation by repeated short-term reversible ischemia led to the development of the concept of stress adaptation. Consequently, new ideas of preconditioning have been developed that include adenosine (Liu et al., 1991), potassium channel opening (Gross and Auchampach, 1992), α_1-receptor (Banerjee et al., 1993; Tosaki et al., 1995), hypoxia (Schizukuda et al., 1993; Engelman et al., 1995), oxidative stress (Maulik et al., 1995b; Maulik et al., 1995c; Maulik et al., 1993), drug (Maulik et al., 1994), and heat shock (Liu et al., 1992).

The precise mechanism of ischemic preconditioning is far from clear. It is generally believed that ischemic preconditioning occurs in two different steps: (1) early effect (short-term adaptation) triggered between seconds to minutes, which is likely to be receptor-mediated and potentiated by the release of some endogenous compound(s) and may last up to several hours (*ischemic preconditioning*); and (2) late effect (long-term adaptation), which may occur after several hours and may last days to months. It is now known that multiple kinases involving tyrosine kinase–mitogen-activated protein (MAP) kinases–MAPKAP kinase 2 play a major role in ischemic preconditioning (Maulik et al., 1996c; Das et al., 1996). Inhibition of tyrosine kinase results in the inhibition of phospholipase D and abolishes preconditioning-mediated activation of MAP kinases and MAPKAP kinase 2. The long-term adaptation is believed to be mediated by the transcription of genes and their subsequent translation into proteins and has been termed *myocardial adaptation to ischemia* (Das, 1993; Das et al., 1994). The subject of this review is the transcription regulation of the genes that are expressed during the myocardial adaptation to ischemia.

II. ALTERED GENE EXPRESSION—THE THIRD LINE OF DEFENSE

Much evidence exists to support the idea that ischemic preconditioning/myocardial adaptation rapidly induces the expression of a wide variety of the mRNAs of the stress-related proteins in mammalian hearts. These include mRNAs of heat shock proteins (Maulik et al., 1996; Andres et al., 1993; Donnelly et al., 1992; Knowlton et al., 1991; Hutter et al., 1994; Mehta et al., 1988), antioxidants (Maulik et al., 1996; Das et al., 1993), Ca^{2+}-regulated proteins (Frass et al., 1993), and growth hormones (Sharma et al., 1989). Most of these stress genes are also induced when hearts are subjected to heat shock or oxidative stress (Engelman et al., 1995; Maulik et al., 1995a; Maulik et al., 1995b; Maulik et al., 1993; Maulik et al., 1994; Liu et al., 1992; Maulik et al., 1996, Andres et al., 1993; Donnelly et al., 1992; Knowlton et al., 1991; Hutter et al., 1994; Mehta et al., 1988). It seems, therefore, reasonable to postulate that there may be a common inducible pathway for the stress-mediated induction of gene expression.

It is believed that constitutive cellular protection against acute stress such as ischemia is provided by a variety of intracellular components, including antioxidants and perhaps heat shock proteins. These components are the integral part of the defense system of the heart. When myocardial cells sense stress, they readily react by increasing the elements of the defense system. Antioxidants and antioxidant enzymes are believed to comprise the *first line of defense* because they undergo rapid changes as a consequence of the development of oxidative stress associated with a large number of cardiovascular diseases, including ischemia and reperfusion. Hearts are also protected by a *second line of defense,* consisting of lipolytic and proteolytic enzymes, proteases, and phospholipases, which are involved in the systematic recognition and removal of the injured cell components. Recent evidence suggests that myocardial cells possess an inducible pathway for cellular defense, presumably translated by the signal transduction pathway. The intracellular signaling now believed to be mediated by the action of multiple kinases is likely to be instrumental for induction of the expression of the mRNAs of a variety of stress proteins, including heat shock and oxidative stress–inducible proteins. This may lead to the systhesis of the related proteins involved in cellular protection or repair of injury. This phenomenon related to specific gene expression has been viewed as the *third line of defense* and may reflect the ultimate adaptive stress responses (Das et al., 1995) (Figure 1).

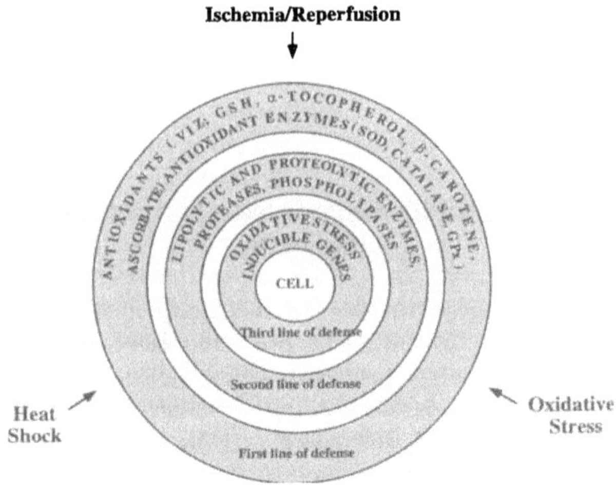

Figure 1. Cellular defense against acute stress.

III. TECHNIQUES FOR STRESS ADAPTATION

A. Ischemic Preconditioning

Various methods are available to adapt the heart to ischemia. The one that was originally used consisted of subjecting the heart to repeated episodes of short durations of ischemia and reperfusion (Murry et al., 1990). This technique is very reproducible and has been found to be valid for many animal models, including the dog, pig, rabbit, and rat (Flack et al., 1991; Kimura et al., 1992; Tosaki et al., 1994; Das et al., 1993). I will describe here the method used for the rat heart. Hearts can be excised from properly anesthetized rats, which are perfused for 10 minutes with nonrecirculating Krebs–Henseleit bicarbonate (KHB) buffer containing 3% bovine serum albumin. The KHB buffer consists of the following ion concentrations (in mM): 119.0 NaCL, 25.0 $NaHCO_3$, 4.6 KCl, 1.2 KH_2PO_4, 1.2 $MgSO_4$, 2.5 $CaCl_2$, and 11.0 glucose. Coronary perfusion pressure and perfusate temperature are generally maintained at 100 cm H_2O and 37°C, respectively. Hearts can be made globally ischemic by terminating the aortic flow for 5 minutes followed by 10 minutes of reperfusion (1 × PC). To induce repeated episodes of ischemia and reperfusion, this procedure may be repeated four times (4 × PC). Each preconditioning (1 × PC or 4 × PC) is followed by 20 to 30 minutes of ischemia and 60 minutes of reperfusion. Experiments are terminated at various points, namely, prior to preconditioning (baseline), after preconditioning, and after reperfusion. Heart biopsies are then examined for the expression of oncogenes and stress-related genes, as well as for antioxidative enzymes.

B. Heat Stress

Hearts can be adapted by subjecting them to direct heat shock. Among many methods, the one most commonly used consists of allowing the animals to swim in warm water to raise their body temperature (Donnelly et al., 1992; Hutter et al., 1994). Studies are also known to exist where animals were heated using an electric blanket or heating pad. In the case of isolated heart, the heart can be directly heated by perfusing it with warm buffer or blood (Liu et al., 1992).

C. Oxidative Stress

Similar to myocardial adaptation to ischemic or heat shock stress, the heart can be adapted to oxidative stress by a variety of methods. For example, animals can be subjected to oxidative stress by treating them to a low dose of endotoxin or its active component lipid A (Maulik et al., 1995b; Maulik et al., 1995c). Oxidative stress can also be induced by cytokine treatment. For instance, oxidative stress was induced by injecting rats with 30 μg/kg body weight of human recombinant interleukin-1α, 24 hours before the isolation of the hearts (Maulik et al., 1993). The development of oxidative stress was confirmed by the increased amount of malonaldehyde formation in the heart. Control and experimental hearts were subjected to 30 minutes of ischemia followed by 60 minutes of reperfusion.

D. Drug

Recently, a novel approach was used in the author's laboratory to induce heat shock (Liu et al., 1992). A sympathomimetic drug, amphetamine, was used to raise the body temperature. Rats were injected with amphetamine sulfate (3 mg/kg, IM), and the rectal temperature was monitored. The control animals (n = 6) received saline injection only. The rectal temperature of the rats before amphetamine injection ranged from 37°C to 38°C. After 1 hour of amphetamine injection, the rectal temperature increased markedly, in the range of 41.5°C to 42.5°C. This hyperthermic state lasted for up to 3 hours, then it gradually returned to the basal temperature in another 3 to 4 hours. Forty hours after the injection, the animals were anesthetized with an intervaneous injection of sodium pentobarbital (Nembutal, 25 mg/ml), and the isolated perfused heart was prepared as described above. Both control and experimental hearts were subjected to 30 minutes of ischemia followed by 60 minutes of reperfusion.

IV. METHODS FOR STUDYING GENE EXPRESSION

For most of the studies related to myocardial stress mentioned above, the investigations depended entirely on the expression of certain genes for which indirect, circumstantial, or sometimes purely speculative evidence existed suggesting their possible induction as a consequence of the particular experimental model of stress studied. In other words, the experimental design was to screen the expression of various genes that had already been found to be induced in response to stress—such as proto-oncogenes, heat shock proteins, and genes for antioxidants—during specific protocols of acute stress such as ischemia–reperfusion, hyperthermia, and oxidative stress.

A more direct but far more difficult approach, which was initiated in our laboratory, is to actually identify newly expressed genes by subtraction analysis. This technique was initially pioneered in cultured cell lines in an attempt to compare the phenotypes of closely related cell types. Although there are many variants of this method, the basic principle is to generate a cDNA population of specific messages present in one experimental sample (i.e., cell, tissue, organ) but not in another (control) by eliminating all common sequences that hybridize to control mRNA. Usually ss cDNA is generated from the test sample ("tester") and hybridized extensively to excess mRNA from the control sample ("driver") and the unhybridized, specific, ss cDNA separated from cDNA/mRNA hybrids by hydroxiapatite column chromatography. After two or more rounds of such "subtraction," a labeled probe is generated to screen cDNA libraries from the test sample, identifying clones of the specifically induced (or highly overexpressed) genes, which are then amplified, sequenced, and possibly identified by searching DNA sequence databases.

Many of the technical problems related to subtraction analysis may be overcome by using some recent improvements such as generation of biotinylated DNA driver populations and use of streptavidin-coated tubes for separation, as well as by using subtracted cDNA not as a probe but for cloning (thus generating subtracted cDNA libraries). There is no general agreement at this time about the clear superiority of any one of these methods. The significant effort and cost notwithstanding, this approach can lead to very interesting new insights about phenotypical changes during acute myocardial stress, not to mention the almost certain byproduct of identifying and cloning some unknown genes.

Recently, we have also initiated another sophisticated approach to study gene expression; that is, the differential display method. Polymerase chain reaction (PCR)–DDR is a powerful technique for anlyzing differences in gene expression that can be used to identify novel genes and their functions during preconditioning. Differential gene expression can result from a number of factors, including mutations, viral infections, cellular differentiation, as well as in response to stress and environmental agents, such as hormones, drugs, and metals. These influencing factors make differential display especially beneficial in analyzing disease states, characterizing pharmacologically active compounds, discovering novel drug targets, and determining the functions of oligonucleotides or gene therapy agents. Differential display functions by selectively amplifying large number of expressed sequences (cDNAs) in individual analysis and then displaying these sequences (cDNA fragments) by gel electrophoresis. The PCR differential display technology used in our study was developed by Drs. Arthur B. Pardee and Peng Liang (1992).

This method involves reverse transcription of the mRNAs with oligo-dT primers anchored to the beginning of the poly (A) tail followed by the PCR in the presence of a second 13 mer arbitrary in sequence. The amplified cDNA subpopulations of 3' termini of mRNAs as defined by this pair of primers are distributed on a sequencing gel. By changing the primer combination, it is possible to sequence approximately 95% of all mRNA species of total RNA preparation from a given tissue or cell type.

A. Use of cDNA Probes

For obvious reasons, earlier studies related to stress-induced gene expression focused primarily on the stress-related genes such as genes of heat shock proteins. Subsequent studies demonstrated that many other related genes are also expressed, of which the genes of antioxidant enzymes have proven to be extremely important. All of these studies, however, depended on speculation about the nature of the genes. Accordingly, a specific cDNA probe was used to identify the corresponding induction of the gene expression.

When myocardial cells are subjected to acute stress such as ischemia–reperfusion, oxidative stress, and hyperthermia, they readily react by inducing genes that encode antioxidant enzymes and related proteins in addition to other stress proteins. In most cases, mRNA for catalase is transcribed in the mammalian heart after it has been subjected to a stress insult (Mehta et al., 1988). In addition, increased expression of manganese superoxide dismutase (Mn-SOD) mRNA has also been documented in the heart. Mn-SOD constitutes one of the major cellular defense mechanisms against the toxic effects of the superoxide radical. Repeated ischemia and reperfusion was associated with the enhancement of the antioxidant defense system. It was shown that repeated ischemia (four times 5 minutes of ischemia each separated by 10 minutes of reperfusion; 4 × PC) enhanced the expression of catalase and Mn-SOD genes (Das et al., 1993). Interestingly, the 4 × PC hearts showed relatively higher amount of the expression of these genes as compared to those present in the 1 × PC (5 minutes of ischemia followed by 10 minutes of reperfusion) hearts. We also observed enhanced peroxisomal catalase activity after 60 minutes of reperfusion following repeated ischemia and reperfusion. In addition, glutathione peroxidase, glutathione reductase, and Mn-SOD activities were also higher in the 4 × PC hearts after 60 minutes of reperfusion compared to those in 1 × PC and control hearts, suggesting that these enzymes could have been modulated by the stress induced by repeated ischemia and reperfusion.

The development of oxygen adaptation has been related to an increase in enzyme activities such as glucose-6-phosphate dehydrogenase, glutathione peroxidase, as well as catalase and SOD (Lu et al., 1993). Oxidative stress induced by cytokines such as interleukin-1, interleukin-6, or tumor necrosis factor (TNF) can induce the expression of mRNA levels of Mn-SOD in human hepatoma cells (Liang et al., 1994). Oxidative stress induced by endotoxin, interleukin-1 and interleukin-6 also increased the Mn-SOD mRNA levels in rat livers. SOD plays a key role in protection against oxygen radicals, and SOD gene expression is highly induced during stress. Recently our laboratory has shown that oxidative stress induced by interleukin-1α caused the enhancement of several antioxidant enzymes in rat hearts, including catalase, copper/zinc superoxide dismutase (Cu/Zn-SOD), Mn-SOD, glutathione peroxidase, and glucose-6-phosphate dehydrogenase. Oxidative stress is likely to shift the cellular redox equilibrium toward the oxidized status. Surprisingly, the expression of the Cu/Zn-SOD gene is induced by sulfhydryl antioxidants such as reduced glutathione, cysteine, and dithiothreitol. It seems likely, therefore, that reduced glutathione acts directly as an antioxidant and simultaneously activates the Cu/Zn-SOD gene during oxidative stress.

Another antioxidative enzyme, glutathione S-transferase (GST) has recently been found to be transcriptionally regulated by oxidative stress induced by reactive oxygen intermediates. The authors speculated that the gene product of primary GR target genes directly or indirectly affects the redox state of the cell. A recent study from our laboratory demonstrated activation of SOD when an isolated in situ pig heart was subjected to heat shock by warm blood cardioplegia (Liu et al., 1992). In contrast, Currie and Tanguay found no detectable change in mRNA for catalase after heat shock or during recovery following the heat shock. However, they demonstrated an increase in catalase enzyme activity at 24 and 48 hours after the induction of a heat shock response. The authors concluded that catalase activity is either translationally or post-translationally regulated.

These studies suggested that heat shock enhances the cellular antioxidative defense system. Thus, the stimulation of SOD and catalase could reflect a mechanism of myocardial adaptation to the stress. Although it is tempting to speculate that stress-mediated stimulation of antioxidant enzymes is instrumental for the enhanced postischemic functional recovery of heart, further study is required to confirm such a possibility.

B. Subtractive Hybridization

1. *Construction of Ischemic Heart cDNA Library*

The method described here was used in the author's laboratory (Das et al., 1994; Currie and Tanguay, 1991). Total RNA was isolated from ventricular muscle fragments (500 mg) of hearts subjected to 30 minutes of ischemia as described above. Poly(A)$^+$ RNA (corresponding to mRNA) was further purified from the total RNA by hybridization to biotinylated oligo-dT, capture with streptavidin-coated paramagnetic particles, and elution (PolyAtract kit; Promega Corp., Madison, WI); 0.6% to 0.8% of total RNA was recovered as undegraded mRNA without significant rRNA contamination (as by denaturing agarose electrophoresis). A plasmid cDNA library was then constructed using the SuperScript System for cDNA Synthesis and Plasmid Cloning and MAX Efficiency DH5α competent *Escherichia coli* cells (Gibco BRL, Gaithersburg, MD). Briefly, first-strand cDNA was synthetized from 5 μg mRNA using a RNase H⁻ M-MLV reverse transcriptase and 1 μg of a *Not* I primer-adapter (44 bases long) having a protruding $(T)_{15}$ 3' end, followed by RNase H-primed, second-strand synthesis by *E. coli* DNA polymerase I in combination with *E. coli* DNA ligase. After addition of *Sal* I adapters, digestion with *Not* I was performed to obtain an asymmetrical double-stranded cDNA for directional ligation into *Not* I/*Sal* I precut pSPORT 1 plasmid vector (a pUC 19 derivative). The cDNA was size-fractionated on a Sephacryl S-500 HR column before ligation into the vector; the first 20 ng were ligated into a 50-ng precut vector, and an aliquot of the ligation reaction was used to determine transformation efficiency in a standard transformation protocol using MAX Efficiency DH5α competent cells. After transformation of the remainder of the ligation reaction, aliquots were plated on X-Gal-coated amphicillin liquid broth agar plates to the desired density; white/blue colony screening for *Lac*⁺ showed greater than 95% recombinants. A schematic diagram showing the important steps of subtractive hybridization is shown in Figure 2.

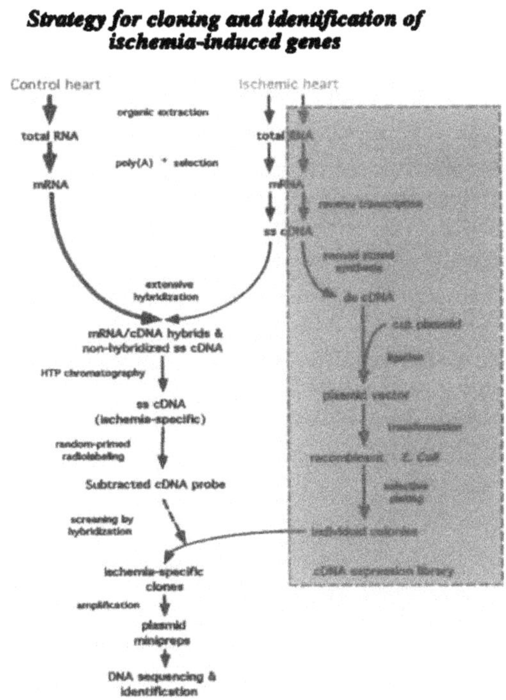

Figure 2. Strategy for cloning and identification of stress-induced genes.

2. *Subtractive Screening of Ischemic cDNA Library*

A low-density plating of the transformed cells was performed (10 plates at approximately 100 colonies per plate) for screening by hybridization with a subtracted probe to identify dominant ischemia-induced mRNA species. Briefly, 10 μg poly(A)$^+$ RNA from ischemia-subjected hearts was reverse transcribed as described above; after alkaline hydrolysis of the RNA strand with 0.1 volume 3 M NaOH (30 minutes at 65°C), neutralization, and phenol/chloroform extraction; a 10-fold excess (by weight) of driver mRNA (poly(A)$^+$ RNA from control hearts) was added to the single-stranded cDNA (ss cDNA). After ethanol precipitation, the ischemic cDNA/control mRNA mixture was resuspended in 30 μl 2 M phosphate buffer pH 6.8, 2 mM EDTA, 0.1% SDS, denatured for 10 minutes on boiling water, and hybridyzed at 68°C to a C_0t value of 500. Nonhybridized cDNA was separated from mRNA/cDNA hybrids by chromatography on BioGel HTP hydroxyapatite columns (Bio-Rad Laboratories, Melville, NY) at 60°C (elution with 0.16 M phosphate buffer, pH 6.8) and desalted on EconoPack 10DG columns (Bio-Rad); after addition of the same amount of driver mRNA and ethanol precipitation, a second round of subtractive hybridization was performed. The desalted, precipitated, remaining nonhybridized ss cDNA (highly enriched in ischemia-specific sequences) was resuspended in 10 μl H_2O, and random-primed synthesis of high specific activity ($> 0.8 \times 10^9$ cpm/μg) radiolabeled second strand was performed using standard kits (Boehringer-Mannheim, Indiana & Polis, IN). This probe was used to screen replicas of the LB plates as follows: colony lifts were performed using CP/L positively charged nylon membranes (Bio-Rad); colonies were lysed (0.1% SDS, 0.2 N NaOH), neutralized, and washed (2 × SSC) by capillary action on soaked extra-thick blotting paper (Bio-Rad); the plasmid DNA released was bound to the membrane by baking for 2 hours at 80°C in a vacuum oven; and the filters were blocked, hybridized to the probe, and washed in a Hybaid hybridization oven (National Labnet Co., Woodbridge, NJ). After drying and autoradiography, mylar masks were used to match the positive signals to the respective colonies on the master plates (regenerated after colony lifts by 5 hour incubation at 37°C).

3. *Validation and DNA Sequencing of Positive Colonies*

Positive colonies were amplified by an overnight culture in terrific broth, and 1-ml samples were used to isolate the plasmid (Magic Miniprep kit; Promega). Two μg aliquots were subjected to restriction cutting with *Bam*H I and *Pst* I; the fragment(s) corresponding to the cDNA insert were separated by electrophoresis on 0.6% submarine agarose gels, radiolabeled by random-primed DNA synthesis, and used as probes in standard Northern blotting experiments (see latter) against ischemic and control heart DNA to validate them as being specifically highly expressed in ischemic hearts. Aliquots containing 3 μg plasmid from confirmed positives (> 90% of positive colonies identified by subtractive hybridization) were then subjected to bidirectional partial DNA sequencing (using M13 and T7/T3α primers) by the chain-extension/chain-termination method (Maulik et al., 1996) using the Sequenase version 2.0 kit (U.S. Biochemical Corp., Cleveland, OH). Approximatively 250 to 300 bases were routinely read from both ends of the insert. These sequences were subjected to a computerized search for homologies against all sequences in the updated worldwide GenBank and EMBL sequence databases (release 76 and 33, respectively) using the FASTA program (Wilbur-Lipman algorithm; part of the GCG software package version 7.2 run on a LAN molecular biology computer group VAX server). In the case of previously unknown sequences, restriction mapping and reading frame identification were also performed. Using the subtractive hybridization technique, our laboratory demonstrated the induction of the expression of several mitochondrial genes in the ischemic myocardium (Moraru et al., 1994; Das et al., 1993) (Table 1).

C. Differential Display

The general strategy for differential display (Liang and Pardee, 1992; Liang et al., 1994; Zhao et al., 1995; Haag and Raman, 1994) seems to be straightforward, and the strategy of mRNA differential display is to resolve the 3' terminal portions of mRNAs on a DNA-sequencing gel using a primer designed to bind to the 5' boundary of a poly A tail for reverse transcription, which is followed by PCR amplification. Then the mRNA population can be resolved and compared on nondenaturing gels. The resulting cDNA patterns reflect differences in the mRNA levels and composition. Differentially displayed cDNAs are isolated, sequenced, and used as probes to confirm differences in the particular mRNA levels. The isolation of intact RNA that is free of chromosomal DNA is crucial for the success of differential display. DNAse I treatment to remove the chromosomal DNA contamination is essential to ensure that mRNAs and not genomic DNA products are displayed; this important point is often ignored. In fact, this contamination may be a major source of false positives or negatives.

To simplify the screening procedure of an entire mRNA pool, three one-base–anchored oligo-dT primers can be used for excellent selectivity in subdividing mRNA into three populations. A restriction site for Hind III at the 5' ends of both PCR primers should facilitate manipulation of the amplified cDNAs after cloning

Table 1. Expression of Myocardial ATP Syunthase Genes

	Baseline	*20-min Ischemia*	*20-min Ischemia + 60-min Reperfusion*
ATPase 6	3.1 (1.0)	74 (23.9)*	79 (25.5)*
ATPase	8 2.7 (1.0)	51 (18.9)*	18 (6.7)*
$F_{1\alpha}$ATPase	0.9 (1.0)	1.3 (1.4)	8.3 (2.2)*

Notes: Quantification of mRNA coding for subunits 6, 8, and $F_{1\alpha}$ of the mitochondrial adenosine triphosphate (ATP) synthase complex by northern hybridization, using β-actin as housekeeping internal control gene. Results are expressed as ratio to β-actin (×10^2); ( ) = ratio to baseline expression; *$P<.05$ as compared to baseline.

into the PCR cloning vector and more efficient amplification of the cDNA due to the larger primers used. The differential display technique consists of several major steps. These are descibed in the paragraphs that follow.

1. Reverse Transcription of RNA

PCR-DDR analysis carried out in our laboratory is based on the method described by Liang et al. (1994). Total RNA was used in our study to obtain a cleaner background signal, easy purification, and integrity verification as compared to poly (A^{+}) RNA. The flow diagram of the procedure is depicted in Figure 3. We used three one-base–anchored oligo-dT primers to subdivide the mRNA popula-

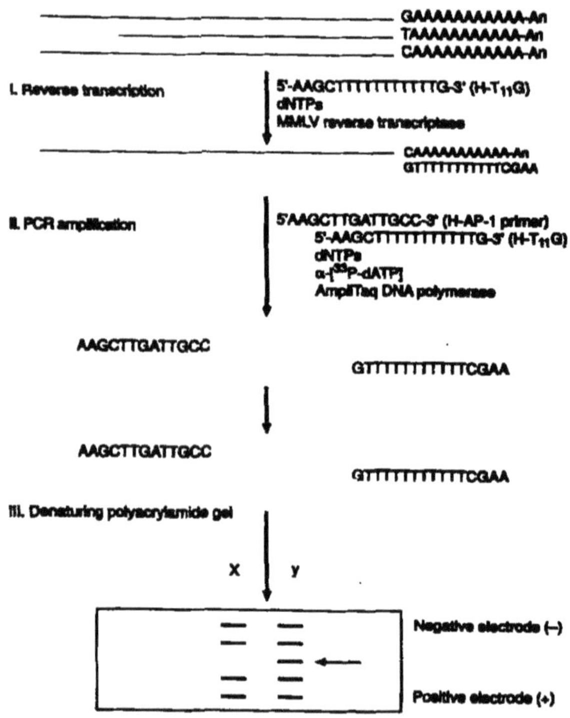

Figure 3. Schematic representation of differential display analysis. mRNAs from hearts are reverse transcribed using an anchored oligo-dT primer ($H\text{-}T_{11}G$) where G can be replaced by C or A to produce single stranded DNA complements of a subset of the mRNAs. The second primer, which is relatively short and arbitrary in sequence, is then used in combination with the anchored primer to amplify a subset of mRNA 3′ termini from the cDNAs generated by the reverse transcription. These cDNA fragments, each corresponding to a 3′ end of an mRNA species are displayed on a polyacrylamide gel. "An" stands for the end of the mRNA strand.

tion. The reverse transcription of mRNA and PCR reactions were done as follows: Total RNA (0.2 μg) was reverse transcribed in a 20-μl reaction mixture containing Superscript reverse transcriptase (Gibco/BRL, Grand Island, NY), dNTP mix (250 μM), 5 × RT buffer and oligo-dT primer, H-T_{11}M (2 μM) (where M may be G, A, or C). Three reverse transcription reactions for each RNA sample were prepared in PCR tubes (0.5-mL size), each containing one of the three different one-base–anchored H-T_{11}M primers in duplicates. Control reactions were performed in the absence of reverse transcriptase. The thermocycler was then programmed to operate as follows: 65°C for 5 minutes, 37°C for 60 minutes, and 75°C for 5 minutes. The tubes were then cooled down to 4°C. The cDNAs were amplified by PCR in the presence of $\alpha(^{33}P)$dATP (2000 Ci/mmol) using a Perkin-Elmer 9600 thermal cycler. Control experiments were performed by substituting water for cDNA.

2. *RNA Arbitrarily Primed–Polymerase Chain Reaction*

The reaction mixtures (20 μl) included arbitrary primers, H-AP_1 to H-AP_8. We used eight different arbitrary primers in combination with three-anchored primers in different reactions. The reaction mixture also contained 10 × PCR buffer (2 μM), dNTP mixture (25 μM), H-T_{11}M (2 μM), RT mixture from the reverse transcription reaction which contained the same H-T_{11}M used for PCR, as well as Ampli Taq DNA polymerase (Perkin Elmer, Foster City, CA). The PCR was programmed for 40 cycles as follows: denaturation at 94°C for 15 seconds (for Perkin Elmer's 9600 thermocycler), annealing at 40°C for 2 minutes, and extension at 72°C for 30 seconds for 40 cycles, at 72°C for 5 minutes one cycle. The tubes were then cooled down to 4°C. Radiolabeled PCR amplification products were analyzed by electrophoresis in denaturing 4.5% polyacrylamide gels. In our laboratory, a new programmable DNA sequencer was used for differential display. This Genomix LR sequencer (Genomyx Corp., Foster City, CA) provided better resolution of cDNA bands and better separation of large cDNA fragments. This sequencer can independently control voltage and temperature so that gels may be run at higher temperatures, which helps to enhance band separation. In this system, gels are dried directly on the glass plates, which provides accurate band excision by allowing the autoradiogram that serves as a template to be placed directly underneath the dried gel. The use of single base–anchored primers and longer arbitrary primers, in combination with the use of a 4.5% gel, results in higher resolution of bands which in turn decreases the possibility of isolating multiple cDNAs from a single band. An M_{13} cycle DNA sequencer (Genomyx Corp., Foster City, CA) was used to obtain DNA standard ladders on differential display gels. A variability of 5% to 15% was observed in the number and intensity of bands among given samples on repeated PCR analysis. To confirm the reproducibilty of amplification for selected bands, the reactions were repeated at least three times using different cDNA preparations.

3. *PCR Amplification*

PCR bands of interest were recovered from the sequencing gels and reamplified in a 40-cycle PCR (40-μl mixture) in the absence of isotope. For differential display gel band reamplification, expand PCR (Taq + Pwo) High Fidelity Ampli Taq (Boehringer-Mannheim, Indianapolis, IN) was used, consisting of a unique enzyme mixture containing thermostable Taq DNA and Pwo DNA polymerase.

Thirty μl of the reamplified cDNA was run on a 1.5% agarose gel using xylene cyanole as the loading dye. The gel was stained with ethidium bromide. The remaining PCR samples were saved at -20°C for cloning. About 90% of the probes were found after the first round of PCR. The size of the reamplified PCR products were the same as those on the DNA sequencing gel (not shown). The reamplified cDNA probes were cut out from the agarose gel and extracted by means of Qiaex kit (Qiagen, Chataworth, CA). The extracted cDNA probes were eluted in 20 μl H_2O and saved for Northern blot analysis.

4. *Validation by Northern Blot Analysis*

Northern blot analysis is used to verify the differential expression of the genes. Both cloned and reamplified cDNA probes can be used directly for Northern blot analysis to verify the differentially expressed gene. For Northern blot analysis, 10 μg total RNA is electrophoresed in 1% agarose formaldehyde-formamide gel and transferred to Gene Screen Plus hybridization transfer membrane (Biotech Systems, NEN Research Products, Boston, MA) by 18-hour standard capillary transfer. The membrane is then baked under vacuum at 80°C for 1 hour. The cDNA probe is labeled with (a-^{32}P)dCTP by the random-prime method.

Unincorporated isotope is removed by using a Sephadex G-50 spin column. Prehybridization and hybridization are performed in a hybridization oven (Hybaid, Labnet, NJ) using one membrane and 10 mL Quick Hyb aqueous exclusion rate enhancing solution (Stratagene, La Jolla, CA) per roller bottle, according to the manufacturer's instruction. QuickHyb hybridization solution can be used for Northern blot electrophoresis employing randomly labeled radioactive as well as nonradioactive nucleic acid probes. After hybridization, the membrane is dried and exposed to an X-ray film with an intensifying screen at –80°C.

5. *Cloning and Sequencing*

The cDNA fragments that showed a specific hybridization pattern or substantial differences by Northern blot analysis were cloned into the *Insert-Read PCR-TRAP Cloning Vector* by using the PCR-TRAP Cloning System (Gen Hunter). After cloning, the colonies were lysed, and the lysates were used as a template for PCR to identify the correct clones. The plasmids having corect inserts were isolated, and both strands of cDNA were sequenced with the Lgh Primer and Rgh Primer (Gen Hunter) by using a Sequenase version 2.0 kit (US Biochemical Corp. Cleveland, OH). The nucleotide sequences obtained were compared with known sequences by searching the GenBank and EMBL databases using the FASTA program software (Genetics Computer Group, Madison, WI). Figures 4 and 5 depict the differential display of the genes expressed during ischemic preconditioning.

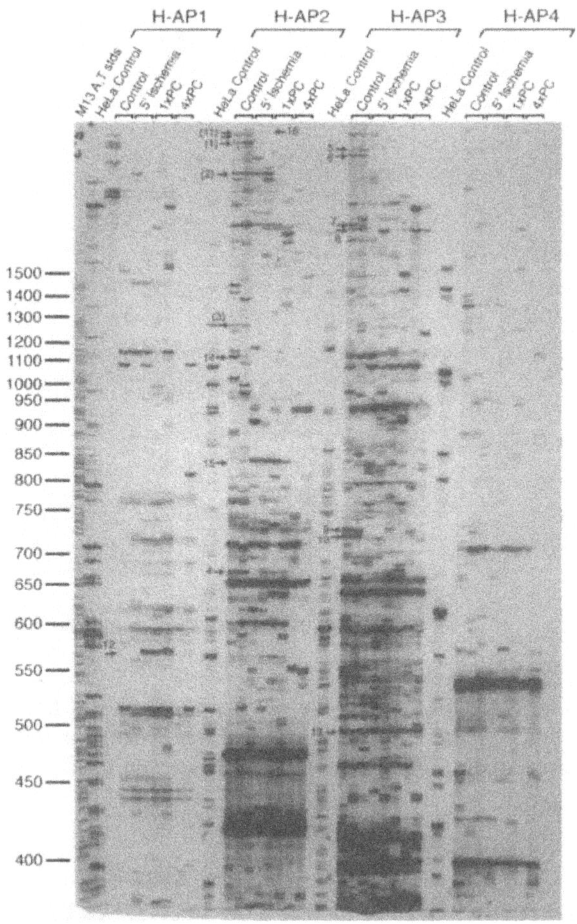

Figure 4. Differential display of expressed genes using h-t_{11}g-anchored primer in combination with four different arbitrary primers (H-AP_1-HAP_4). Note the up-regulation and down-regulation of the differentially expressed bands denoted by the numbers 1, 2, 3, and so on.

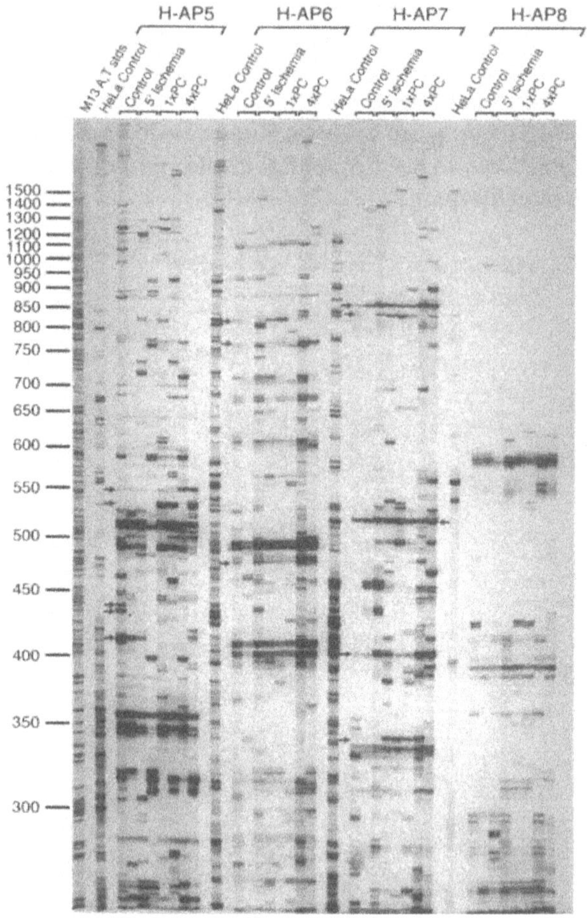

Figure 5. Differential display of expressed genes using h-t_{11}g anchored primer in combination with four different arbitrary primers (H-AP_5-HAP_8). Note the up-regulation and down-regulation of the differentially expressed bands denoted by the numbers 1, 2, 3, and so on.

V. RELATIVE ADVANTAGES AND DISADVANTAGES OF DIFFERENTIAL DISPLAY COMPARED TO SUBTRACTIVE HYBRIDIZATION

- Differential display detects changes in the expression patterns in response to either transfected genes, growth factors, or drug situations.

- Simultaneous display of all up- and down-regulated genes makes differential display a powerful technique compared with subtractive hybridization.
- Differential display allows side-by-side comparisons of mRNA from different sources, which is not possible by subtractive hybridization.
- Differential display requires only a few μg of RNA compared to 50 times more for subtractive hybridization.
- Differences can be used directly for identification and isolation of the corresponding genes, which makes differential display a powerful technique compared with 2D protein separation, which often results in frustration because of the inability of obtaining enough proteins for molecular characterization.
- Differential display is highly reproducible.
- High-speed of analysis is possible for differential display.

VI. SUMMARY AND CONCLUSION

A variety of stresses including ischemia, hypoxia, heat shock, and oxidative stress have been found to stimulate the expression of early responsive genes such as c-fos and c-myc; genes of antioxidant enzymes including superoxide dismutase and catalase; as well as genes for heat shock proteins (HSP) namely, HSP 27, HSP 32, HSP 70, and HSP 89. To examine other, potentially more specific gene responses, we performed subtractive screening of genes after 20 minutes of global ischemia with no reperfusion in isolated rat hearts. We have also used PCR-based differential display technique to identify differentially expressed genes during ischemic preconditioning. Subtractive hybridization and differential display revealed that a short duration of ischemia can trigger the induction of several mitochondrial genes, including ATPase 6 and cyt b, as well as a gene encoding the ribosomal protein L23a, which are further enhanced during ischemic preconditioning. In addition, many other as yet unidentified genes are either up-regulated or down-regulated in response to ischemia or ischemic preconditioning. These results indicate that the initial gene response of the heart to energy deprivation is its attempt to increase the capacity of mitochondrial oxidative phosphorylation, which is followed by the induction of the expression of several stress-related genes, These responses suggest that reprogramming of gene expression is an essential adaptive response to stress.

ACKNOWLEDGMENTS

This study was supported by NIH grants HL-22559 and HL-34360, and by a Grant-in-Aid from the American Heart Association (No. 95015490).

REFERENCES

Andres, J., Sharma, H.S., Knoll, R., Stahl, J., Sassen, L.M.A., Verdouw, P.D., and Schaper W. (1993). Expression of heat shock proteins in the normal and stunned myocardium. Cardiovasc. Res. 27, 1421-1429.

Banerjee, A., Locke-Winter, C., Rogers, K.B., Mitchell, M.B., Brew, E.C., Cairns, C.B., Bensard, D.D., and Harken, A.H. (1993). Preconditioning against myocardial dysfunction after ischemia and reperfusion by an alpha 1-adrenergic mechanism. Circ. Res. 73, 656-670.

Currie, R.W., and Tanguay, R.M. (1991). Analysis of RNA for transcripts for catalase and SP 71 in rat hearts after in vivo hyperthermia. Biochem. Cell. Biol. 69, 375-382.

Das, D.K. (1993). Ischemic preconditioning and myocardial adaptation to ischemia. Cardiovasc. Res. 27, 2077-2079.

Das, D.K., Engelman, R.M., and Kimura, Y. (1993). Molecular adaptation of cellular defenses following preconditioning of the heart by repeated ischemia. Cardiovasc. Res. 27, 578-584.

Das, D.K., Maulik, N., and Moraru, I.I. (1995). Gene expression in acute myocardial stress. Induction by hypoxia, ischemia, reperfusion, hyperthermia and oxidative stress. J. Mol. Cell. Cardiol. 27, 181-193.

Das, D.K., Maulik, N., Yoshida, T., Engelman, R.M., and Zu, Y-L. (1996).. Preconditioning potentiates molecular signaling for myocardial adaptation to ischemia. Ann. N.Y. Acad. Sci. 793, 191-209 .

Das, D.K., Moraru, I.I., Maulik, N., and Engelman, R.M. (1994). Gene expression during myocardial adaptation to ischemia and reperfusion. Ann. N.Y. Acad. Sci. 723 , 292-307.

Donnelly, T.J., Sievers, R.E., Vissern, F.L.J., Welch, W.J., and Wolfe, C.L. (1992). Heat shock protein induction in rat hearts: a role for improved salvage after ischemia and reperfusion? Circulation 85, 769-778.

Engelman, D.M., Watanabe, M., Engelman, R.M., Rousou, J.A., Kisin, E., Kagan, V.E., Maulik, N., and Das, D.K. (1995). Hypoxic preconditioning preserves antioxidant reserve in the working rat heart. Cardiovasc. Res. 29, 133-140.

Flack, J.E., Kimura, Y., Engelman, R.M., Rousou, J.A., Iyengar, J., Jones, R., and Das, D.K. (1991). Preconditioning the heart by repeated stunning improves myocardial salvage. Circulation 84 (suppl III), 369-374.

Frass, O., Sharma, H.S., Knoll, R., Duncker, D.J., McFalls, E.O., Verdouw, P.D., and Schaper, W. (1993). Enhanced gene expression of calcium regulatory proteins in stunned porcine myocardium. Cardiovasc. Res. 27, 2037-2043.

Gross, G.J., and Auchampach, J.A. (1992). Blockade of ATP-sensitive potassium channels prevents myocardial preconditioning in dogs. Circ. Res. 70, 223-233.

Haag, E., and Raman, V. (1994). Effects of primer choice and source of Taq DNA polymerase on the banding patterns of differential display RT-PCR. Biotechniques 17, 226-228.

Hendrikx, M., Toshima, Y., Mubagwa, K., and Flameng, W. (1993). Improved functional recovery after ischemic preconditioning in the globally ischemic rabbit heart is not mediated by adenosine A_1 receptor activation. Basic Res. Cardiol. 88, 576-593.

Hutter, M.M., Sievers, R.E., Barbosa, V., and Wolfe, C.L. (1994). Heat-shock protein induction in rat hearts. A direct correlation between the amount of heat-shock protein induced and the degree of myocardial protection. Circulation 89, 355-360.

Kimura, Y., Iyengar, J., Subramanian, R., Cordis, G.A., and Das, D.K. (1992). Preconditioning of the heart by repeated stunning: attenuation of post-ischemic dysfunction. Basic Res. Cardiol. 87, 128-138.

Knowlton, A.A., Brecher, P., and Apstein, C.S. (1991). Rapid expression of heat shock protein in the rabbit after brief cardiac ischemia. J. Clin. Invest. 87, 139-147.

Lawson, C.S., Coltart, D.J.O., and Hearse, D.J. (1992). Ischemic preconditioning and protection against reperfusion-induced arrhythmias, reduction in vulnerability or delay in onset? Studies in the isolated blood perfused rat heart. Eur. Heart J. 13, 2334.

Li, G.C., Vasques, J.A., Gallagher, K.P., and Lucchesi, B.R. (1990). Myocardial protection with preconditioning. Circulation 82, 609-619.

Liang, P., and Pardee, A.B. (1992). Differential display of eukaryotic messenger RNA by means of the polymerase chain reaction. Science, 257, 967-971

Liang, P., Zhu ,W., Zhang, X., Guo, Z.., O'Connell, P.O., Averboukh, L., Wang, F., and Pardee, A.B. (1994). Differential display using one base anchored oligo-dT primers. Nucleic Acid Res. 22, 5763-5764

Liu, G.S., Thornton, J., VanWinkle, D.M., Stanley, A.W.H., Olsson, R.A., and Downey, J.M. (1991). Protection against infarction afforded by preconditioning is mediated by A1 adenosine receptors in rabbit heart. Circulation 84, 350-256.

Liu, X., Engelman, R.M., Moraru, I.I., Rousou, J.A., Flack, J.E., Deaton, D.W., Maulik, N., and Das, D.K. (1992). Heat shock: a new approach for myocardial preservation in cardiac surgery. Circulation 86 (suppl II), 358-363.

Lu, D., Maulik, N., Moraru, I.I., Kreutzer, D.L., and Das, D.K. (1993). Molecular adaptation of vascular endothelial cells to oxidative stress. Am. J. Physiol. 264, C715-C722.

Maulik, N., Engelman, R.M., and Das, D.K. (1996a). Hunting for differentially expressed mRNA species in preconditioned hearts. Ann. N.Y. Acad. Sci. 793, 240-258.

Maulik, N., Engelman, R.M., Wei, Z., Liu, X., Rousou, J.A., Flack, J., Deaton, D., and Das, D.K. (1995a). Drug-induced heat shock improves post-ischemic ventricular recovery after cardiopulmonary bypass. Circulation 92 (suppl II), 381-388.

Maulik, N., Engelman, R.M., Wei, Z., Lu, D., Rousou, J.A., and Das, D.K. (1993). Interleukin-1a preconditioning reduces myocardial ischemia reperfusion injury. Circulation 88 (suppl II), 387-394.

Maulik, N., Sharma, H.S., and Das, D.K. (1996b). Induction of the heme oxygenase gene expression during the reperfusion of ischemic rat myocardium. J. Mol. Cell. Cardiol. 28, 1261-1270.

Maulik, N., Watanabe, M., Engelman, D., Engelman, R.M., and Das, D.K. (1995b). Oxidative stress adaptation improves postischemic ventricular recovery. Mol. Cell. Biochem. 144, 67-74.

Maulik, N., Watanabe, M., Engelman, D., Engelman, R.M., Kagan, V.E., Kisin, E., Tyurin, V., Cordis, G.A., and Das D.K. (1995c). Myocardial adaptation to ischemia by oxidative stress induced by endotoxin. Am. J. Physiol. 269, C907-C916.

Maulik, N., Watanabe, M., Zu,Y-L., Haung, C-K., Cordis, G.A., Schley, J.A., and Das, D.K. (1996c). Ischemic preconditioning triggers the activation of MAP kinases and MAPKAP kinase 2 in rat hearts. FEBS Lett. 396, 233-237.

Maulik, N., Wei, Z., Liu, X., Engelman, R.M., Rousou, J.A., and Das, D.K. (1994). Improved postischemic ventricular recovery by amphetamine is linked with its ability to induce heat shock. Mol. Cell. Biochem. 137, 17-24.

Mehta, H.B., Popovich, B.K., and Dillman, W.H. (1988). Ischemia induces changes in the level of mRNAs coding for stress protein 71 and certain kinase M. Circ. Res. 63, 512-517.

Moraru, I.I., Engelman, D.T., Engelman, R.M., Rousou, J.A., Flack, J.E., Deaton, D.W., and Das, D.K. (1994). Myocardial ischemia triggers rapid expression of mitochondrial genes. Surg. Forum 40, 315-317.

Murry, C.E., Richard, V.J., Reimer, K.A., and Jennings, R.B. (1990). Ischemic preconditioning slows energy metabolism and delays ultrastructural damage during a sustained ischemic episode. Circ. Res. 66, 913-931.

Ono, M., Kohda, H., Kawaguchi, T., Ohhira, M., Sekiya, C., Namiki, M., Takeyasu, A., and Taniguchi, N. (1992). Induction of Mn-superoxide dismutase by tumor necrosis factor, interleukin-1 and interleukin-6 in human hepatoma cells. Biochem. Biophys. Res. Commun. 182, 1100-1107.

Schott, R.J., Rohmann, S., Braun, E.R., and Schaper, W. (1990). Ischemic preconditioning reduces infarct size in swine myocardium. Circ. Res. 66, 1133-1142.

Sharma, H.S., Wunch, M., Kandolf, R., and Schaper, W. (1989). Angiogenesis by slow coronary artery occlusion in the pig heart: expression of different growth factors mRNAs. J. Mol. Cell. Cardiol. 21 (suppl III), 69.

Shizukuda, Y., Iwamoto, T., Mallet, R.T., and Downey, H.F. (1993). Hypoxic preconditioning attenuates stunning caused by repeated coronary artery occlusions in dog heart. Cardiovasc. Res. 27, 559-564.

Tosaki, A., Behjet, N.S., Engelman, D.T., Engelman, R.M., and Das, D.K. (1995). Alpha-1 adrenergic agonist-induced preconditioning in isolated working rat hearts. J. Pharmacol. Exp. Therapeu. 273, 689-694.

Tosaki, A., Cordis, G.A., Szerdahelyi, P., Engelman, R.M., and Das, D.K. (1994). Effects of preconditioning on reperfusion arrhythmias, myocardial functions, formation of free radicals, and ion shifts in isolated ischemic/reperfused rat hearts. J. Cardiovasc. Pharmacol. 23, 365-373, 1994.

Zhao, S., Ooi, S.L., and Pardee, A.B. (1995). New primer strategy improves precision of differential display. Biotechniques 18, 842-850.

MYOCARDIAL PRECONDITIONING VIA ATP-SENSITIVE POTASSIUM CHANNELS:
INTERACTIONS WITH ADENOSINE

Garrett J. Gross, Tsuneo Mizumura,
Kasem Nithipatikom, and David A. Mei

Advances in Organ Biology
Volume 6, pages 81-100.

ISBN: 0-7623-0391-3

I. INTRODUCTION

Numerous studies have shown an important role for the adenosine triphosphate–sensitive potassium (K_{ATP}) channel and adenosine in mediating the potent cardioprotective effect of a brief period of ischemia or hypoxia to protect the heart against a more prolonged period of ischemia (i.e., classical myocardial preconditioning [PC]; however, the mechanisms by which the K_{ATP} channel and adenosine interact to produce classical myocardial PC are still not clear. Several hypotheses exist that are supported by experimental data. One hypothesis suggests that adenosine enhances K_{ATP} channel opening via a G_i protein and/or protein kinase C (PKC)–linked mechanism. A second theory suggests that K_{ATP} channel opening results in activation and translocation of PKC, which then turns on ecto-5'-nucleotidase, an enzyme that is important in the formation of adenosine. Subsequently, activation of ecto-5'-nucleotidase results in enhanced adenosine concentrations in the interstitial fluid and coronary venous blood draining the ischemic–reperfused area. Adenosine then produces its cardioprotective effect by a variety of mechanisms, including enhanced K_{ATP} channel opening in the cardiac myocyte.

In this chapter, evidence will be presented to support or refute both theories, although the majority of evidence supports the first hypothesis, which suggests that an initial burst of adenosine during PC sensitizes the K_{ATP} channel so that it is activated to a greater extent during the subsequent prolonged ischemic period and that adenosine production and release is actually reduced in a preconditioned heart as compared to a nonpreconditioned heart as a result of a decrease in ATP breakdown. Recently, several pharmacological agents have been shown to produce a cardioprotective effect 24 to 48 hours following their administration. This phenomenon has been termed "late PC" or the "second window of protection" (SWOP). We will present evidence that the K_{ATP} channel and adenosine are also key players in this phenomenon. Finally, we will present clinical evidence to suggest a role for both the K_{ATP} channel and adenosine in mediating PC in humans and the potential clinical implications that may result from this adaptive mechanism.

II. MYOCARDIAL PRECONDITIONING

A. Classical Ischemic Preconditioning

In 1986, Murry and co-workers discovered that brief periods of ischemia and reperfusion produced a marked cardioprotective effect against myocardial infarction resulting from a subsequent prolonged period of ischemia and reperfusion in dogs. They termed this phenomenon "ischemic PC." Since its initial discovery, many laboratories have attempted to characterize the cardioprotective effect of ischemic PC. Ischemic PC has been observed in every species studied to date, including dogs (Murry et al., 1986), pigs (Schulz et al., 1994), rats (Qian et al., 1996), rabbits (Toombs et al., 1993), and humans (Tomai et al., 1994; Speechly-Dick et al., 1995). Ischemic PC was found to delay the development of myocardial infarction against a 60-minute ischemic insult; however, when the ischemic insult was increased to 90 minutes, no cardioprotection was observed (Nao et al., 1990). In addition, Li and co-workers (1990) demonstrated that the number of PC stimuli did not affect the magnitude of infarct size reduction produced by PC, indicating that it did not follow clear "dose" dependence. The threshold of the ischemic PC stimulus in the dog as well as other species has been demonstrated to be 2 to 3 minutes (Yao and Gross, 1994a). Furthermore, a reperfusion period of several minutes between the preconditioning stimulus and the prolonged occlusion period has been shown to be necessary for the cardioprotective effect of ischemic PC. However, one can precondition by using low-flow ischemia in the absence of a reperfusion period (Schulz et al., 1995) and by using hypoxia as the stimulus (Shizikuda et al., 1992).

The cardioprotective effect of classical myocardial PC has been shown to last as long as 90 minutes following the PC stimulus in dogs (Yao and Gross, 1996); however, this acute "memory" appears to be shorter in smaller mammals, such as rabbits and rats (Li et al., 1992). Since the duration of the potent cardioprotective effect of classical ischemic PC is rather brief (30 to 90 minutes), it is thought that this may limit its usefulness in the clinical situation where patients are often not diagnosed with a myocardial infarction until several hours after the onset of ischemia. For this reason, considerable effort has been expended in finding ways of prolonging the memory phase of classical PC.

B. Late Preconditioning or the Second Window of Protection

Several recent studies in rabbits (Marber et al., 1993) and dogs (Kuzuya et al., 1993) indicate that the cardioprotective effect of ischemic PC reappears approximately 24 hours after the PC stimulus. This second window of protection (SWOP) or late PC can be produced by many pathophysiological stressors, including brief periods of ischemia (Kuzuya et al., 1993), heat shock (Marber et al., 1993), free radicals (Zhou et al., 1996), and by treatment with the nontoxic endotoxin derivative, monophosphoryl lipid A (MLA) (Yao et al., 1993). The clinical potential of SWOP is quite attractive because of its wide time window of protection. However, little is known about the mechanism(s) involved in producing SWOP, although oxygen-derived free radicals (Zhou et al., 1996), adenosine (Baxter et al., 1994), heat shock proteins (Marber et al., 1993) the K_{ATP} channel (Mei et al., 1995) and nitric oxide (Qiu et al., 1997) have been implicated. In the remainder of this chapter, we will discuss the role of the K_{ATP} channel and adenosine in mediating the cardioprotective effect of both classical ischemic PC and late PC or SWOP.

III. ADENOSINE AND CLASSICAL PRECONDITIONING

ATP is broken down to adenosine diphosphate (ADP), and under aerobic conditions, ADP is rapidly reconverted to ATP. However, under ischemic, hypoxic, or metabolically challenged conditions, ADP is hydrolyzed to adenosine monophosphate (AMP). Under severe ischemic or hypoxic conditions, AMP is converted to inosine monophosphate (IMP) by adenylate deaminase; however, a portion of AMP is also hydrolyzed to adenosine via the activity of nucleotidases. Adenosine is rapidly metabolized to inosine, then both inosine and IMP are metabolized to hypoxanthine. Hypoxanthine is metabolized to xanthine, and xanthine to uric acid. In the ischemic myocardium, the metabolic products of ATP hydrolysis rapidly accumulate in the interstitial space, and it is thought that the accumulation of adenosine in the interstitial space may initiate myocardial PC.

The role of adenosine in initiating ischemic PC was first suggested by Liu and co-workers (1991), who demonstrated that a nonselective adenosine receptor antagonist abolished the cardioprotective effect of ischemic PC in rabbits. This work has been supported by studies demonstrating that nonselective adenosine receptor blockade abolishes preconditioning in dogs (Auchampach and Gross, 1993). In rats, however, adenosine receptor blockade has not been shown to prevent the cardioprotective effects of ischemic PC (Liu and Downey, 1992). A positive role for adenosine is supported by studies demonstrating that brief exposure to adenosine (Yao and Gross, 1994b) or selective adenosine A_1 receptor agonists (Liu et al., 1991; Grover et al., 1992) can induce a PC-like state. Furthermore, adenosine receptor antagonists have been shown to block ischemic PC when administered immediately prior to, and after, the ischemic PC stimulus in rabbits (Thornton et al., 1993) and prior to PC in dogs (Auchampach and Gross, 1993). Interestingly, selective adenosine A_1 receptor agonists have been shown to induce a preconditioned state in rabbits (Liu et al., 1991), although selective adenosine A_1 receptor blockers have not been shown to abolish PC in this species (Liu et al., 1994). Adenosine uptake inhibitors and an adenosine A_1 enhancer of receptor binding have been shown to decrease the threshold for preconditioning in dogs (Auchampach and Gross, 1993; Mizumura et al., 1996), which supports a role for adenosine in triggering or initiating PC.

Interstitial microdialysis studies have allowed investigators to directly sample interstitial fluid, without the confounding effects of enzymatic degradation, and have permitted investigators to construct profiles of interstitial adenosine and its metabolites in ischemic preconditioned hearts. Many studies have demonstrated that interstitial adenosine concentrations increase during the ischemic PC stimulus (Van Wylen, 1994; Lasley et al., 1995), and it has been proposed that this local increase in interstitial adenosine initiates or triggers a preconditioned state via the subsequent activation of A_1 or A_3 receptors.

Once the preconditioned state has been established, the role of adenosine is less clear. Some investigators have shown an increase in venous adenosine release during the reperfusion period following ischemia and have attributed this increase in local adenosine production to an increase in the activity of 5'-ectonucleotidase (Kitakaze et al., 1993), an enzyme responsible for the conversion of AMP to adenosine. Increases in 5'-ectonucleotidase activity have been shown to occur following brief periods of ischemia and reperfusion (Minamino et al., 1996). Furthermore, Minamino and co-workers (1996) have recently demonstrated a strong correlation between the time course of the increase in 5'-ectonucleotidase activity and the duration or memory of myocardial protection produced by ischemic PC. However, many other laboratories have demonstrated a decrease in interstitial adenosine concentrations during the prolonged ischemic insult (Van Wylen, 1994; Lasley et al., 1995) and have attributed this decrease in local adenosine release to a decrease in ATP hydrolysis. These results are supported by studies in which ischemic PC has been shown to result in a decreased rate of ATP consumption during the prolonged period of ischemia (Jennings et al., 1991).

Thus, the role of adenosine in myocardial PC is confusing at best. There appear to be considerable species differences in the role adenosine plays in PC. These species differences may be due to the fact that redundant pathways exist for the production of PC. Alternatively, adenosine may play a modulatory role in the production of PC, simply augmenting or paralleling other cardioprotective pathways, and thus may not be absolutely necessary for producing PC in all species or in the presence of all stimuli used to precondition the heart.

IV. K_{ATP} CHANNELS AND CLASSICAL PRECONDITIONING

The role of K_{ATP} channels in mediating myocardial PC has recently been defined. The K_{ATP} channel was first demonstrated to be involved in classical ischemic PC in dogs by Gross and Auchampach in 1992. They found that a sulfonylurea, glibenclamide, could abolish the cardioprotective effect of ischemic PC. They also showed that a K_{ATP} channel opener, aprikalim, could mimic the cardioprotective effect of ischemic PC. The role of K_{ATP} channels in mediating PC in dogs was further confirmed by using 5-hydroxydecanoate (Auchampach et al., 1992), a K_{ATP} channel antagonist structurally distinct from the sulfonylureas, to block ischemic PC. Subsequent work has demonstrated the involvement of K_{ATP} channels in mediating the cardioprotective effect of ischemic PC in rats (Qian et al., 1996), rabbits (Toombs et al., 1993), pigs (Schulz et al., 1994), and humans (Tomai et al., 1994; Speechly-Dick et al., 1995). It appears that K_{ATP} channels are involved in triggering as well as maintaining a preconditioned state since glibenclamide can block the infarct size–limiting effect of ischemic PC when given either before or after the PC stimulus (Gross and Auchampach, 1992). Furthermore, the blockade of ischemic PC by glibenclamide is not due to its systemic actions since intracoronary administration (Auchampach et al., 1992) blocks ischemic PC and glibenclamide can block PC in isolated heart muscle preparations (Speechly-Dick et al., 1995). Recent studies have shown that glibenclamide and 5-hydroxydecanoate can block PC in isolated rabbit and chick cardiac myocytes (Armstrong et al., 1995; Liang, 1996). Taken together, these data indicate that functional K_{ATP} channels are necessary at the level of the cardiomyocyte to produce and maintain a preconditioned state, although the mechanism(s) by which K_{ATP} channel opening produces its cardioprotective effect is still not well understood.

V. INTERACTION BETWEEN ADENOSINE AND THE K_{ATP} CHANNEL IN MEDIATING CLASSICAL PRECONDITIONING

Soon after the discovery of K_{ATP} channels in ventricular myocytes, it was shown that these channels could be regulated by G proteins (Parent and Coronade, 1989). Initial evidence came from insulin-secreting RINm 5f cells, where it was found that guanoosine triphosphate (GTP) could activate this channel, yet ATP sensitivity remained (Dunne and Peterson, 1986). Further evidence of a G protein interaction came from T-tubular membranes isolated from rabbit skeletal muscle (Parent and Coronade, 1989) and subsequently cardiomyocytes (Kirsch et al., 1990; Ito et al., 1994).

As relates to preconditioning, the most important demonstration of a G protein link to the K_{ATP} channel was demonstrated in 1990 by Kirsch and co-workers, who showed that adenosine A_1 receptors were coupled to K_{ATP} channels via G_i proteins in rat neonatal cardiomyocytes. Following this discovery, other laboratories (Ito et al., 1994) were able to confirm these results in freshly isolated myocytes from adult guinea pigs. This link provides important information on the regulation of K_{ATP} channels since it has been shown that during ischemic PC, intracellular concentrations of ATP do not drop below those known to be inhibitory to K_{ATP} channel opening in in vitro conditions. It is possible that adenosine acts in conjunction with ADP, intracellular acidosis, and lactate to decrease the sensitivity of K_{ATP} channels to ATP via a G_i protein–linked mechanism, thus allowing the channels to open during myocardial ischemia at much higher levels of ATP than have been shown to inhibit channel activity in vitro.

Many laboratories have been able to demonstrate that the cardioprotection of exogenous adenosine or adenosine A_1 receptor agonists could be abolished by K_{ATP} channel blockade in vivo (Grover et al., 1992, Auchampach and Gross, 1993; Van Winkle et al., 1994). Further support for the role of this pathway in myocardial PC came with the demonstration that PC with adenosine could be blocked by pertussis toxin (Lasley and Mentzer, 1993). Pertussis toxin catalyzes the ADP-ribosylation of the α subunit of G_i and G_o proteins, rendering them inactive. Recently, a novel compound, PD 81,723, has been developed which specifically increases the coupling of adenosine A_1 receptors to G_i proteins (Bhattacharya and Linden, 1995). This compound has been shown to reduce the threshold for ischemic PC (Mizumura et al., 1996), and its effects can be abolished by K_{ATP} channel blockade. This interaction has been further supported in the vascular system, where it has been demonstrated that adenosine receptor stimulation contributes to K_{ATP} channel–induced hypoxic vasodilation (Nakhostine and Lamontagne, 1993). Furthermore, hypoxia-induced increases in adenosine have been demonstrated to increase the potency of a K_{ATP} channel opener to induce vasodilation (Randall et al., 1994).

In spite of these numerous studies which suggest that there is an important link between adenosine and the K_{ATP} channel in mediating classical PC, it is still unclear as to how these two mediators or effectors interact to produce the potent cardioprotective effect observed. One hypothesis (Figure 1) suggests that the adenosine formed during ischemia acts on its A_1 receptor on the cardiac myocyte and via stimulation of a G_i protein enhances K_{ATP} channel opening during a subsequent prolonged ischemic period by reducing its sensitivity to blockade by ATP. In contrast, a second hypothesis (Figure 2) presented by Kitakaze et al., 1996a, 1996b) suggests that PC produced by ischemia or pharmacologically by several K_{ATP} channel openers activates ectosolic 5′-nucleotidase via a PKC-linked pathway and that activation

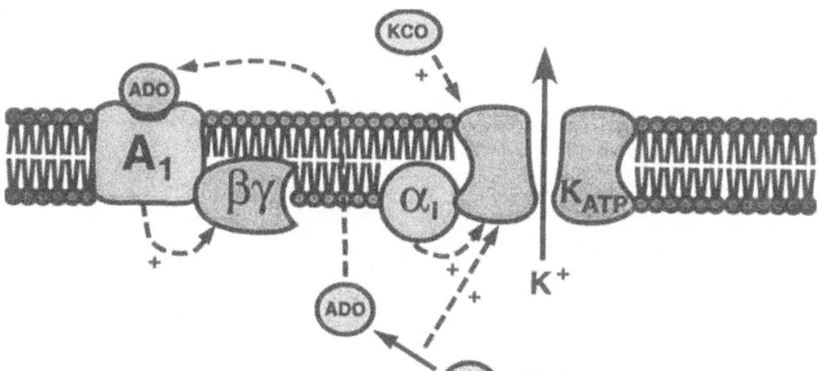

Figure 1. Schematic diagram of one hypothesis (1), which suggests a possible mechanism by which adenosine (ADO) and the adenosine triphosphate–sensitive potassium (K_{ATP}) channel interact to produce ischemic preconditioning (PC). This theory suggests that the ADO formed during ischemia or hypoxia diffuses out of the cardiac myocyte and attaches to its A_1 receptor. A_1 receptor activation turns on a G_i protein α_i subunit, which reduces the K_{ATP} channel's sensitivity to block by ATP and enhances further channel opening. Increased K_{ATP} opening during prolonged ischemia would reduce energy demand and calcium overload, thus producing a cardioprotective effect.

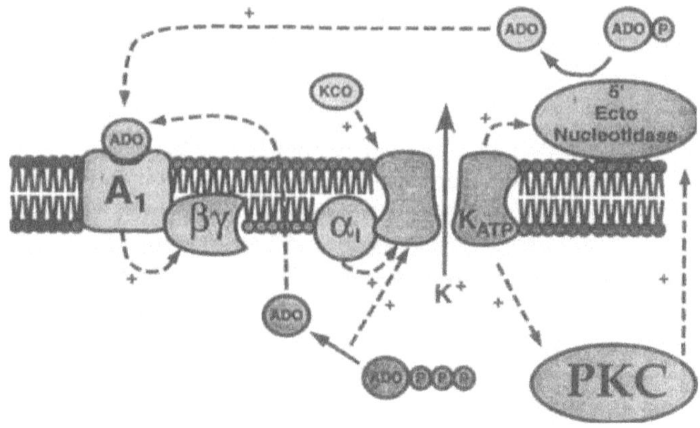

Figure 2. Schematic diagram of another hypothesis (2), which suggests a possible mechanism by which ADO and the K_{ATP} channel interact to produce ischemic PC. This theory suggests that K_{ATP} channel opening initially activates protein kinase C (PKC), which subsequently phosphorylates and activates ecto-5'-nucleotidase to increase the formation of ADO. The ADO formed will then further enhance K_{ATP} channel opening and by a positive feedback mechanism maintain this cycle, which is then transmitted into the cardioprotective effect of PC.

of ectosolic 5'-nucleotidase results in enhanced interstitial and/or coronary venous adenosine release during ischemia or reperfusion, which confers the cardioprotective effect observed. In the current discussion, we will present data in which we have measured the effects of ischemic PC and PC produced by K_{ATP} channel openers on interstitial and coronary venous adenosine release in the canine heart and will compare these results to those previously published and discuss how they relate to the two hypotheses presented above.

A. Comparative Effects of Ischemic Preconditioning and K_{ATP} Opener–induced Preconditioning on Coronary Venous Adenosine Release

Kitakaze and co-workers (1993) were the first to study the role of ecto-5'-nucleotidase and adenosine in mediating the cardioprotective effect of ischemic PC in the canine heart. These investigators produced PC by subjecting the heart to four 5-minute periods of coronary artery occlusion interspersed with 5-minute periods of reperfusion prior to a prolonged 40-minute period of occlusion and 120 minutes of reperfusion. They found that ecto- and cytosolic-5'-nucleotidase activities were significantly increased prior to and during the prolonged occlusion period in preconditioned dogs and that the release of adenosine from the ischemic area was markedly elevated in coronary venous blood during the first 30 minutes of reperfu-siogn.

More recently, this same group of investigators (Kitakaze et al., 1996b) showed that activation of PKC contributed to the infarct size–limiting effect of ischemic PC by activating ecto-5'-nucleotidase in canine hearts. These investigators reached this conclusion based on the observation that two inhibitors of PKC attenuated the increase in ecto-5'-nucleotidase and the reduction in infarct size produced by ischemic PC. Finally, these same investigators (Kitakaze et al., 1996a) have recently demonstrated that four 5-minute intracoronary infusions of two K_{ATP} channel openers, nicorandil and cromakalim, reduced infarct size, increased coronary venous adenosine concentrations, and increased ectosolic and cytosolic 5'-nucleotidase activity similar to that of ischemic PC; and that glibenclamide blunted these beneficial actions. It was suggested that K_{ATP} openers activate ectosolic-5'-nucleotidase via PKC and by this mechanism mimic the effect of ischemic PC.

In support of this hypothesis, Armstrong et al. (1995) demonstrated that the protective effects of pinacidil, a K_{ATP} channel opener, was blocked by an adenosine A_1 receptor antagonist in isolated rabbit cardiac myocytes. Furthermore, Armstrong et al. (1995) demonstrated that a PKC inhibitor or adenosine deaminase blocked the protective effect of pinacidil. These authors concluded that pinacidil activated PKC, which was responsible for enhanced adenosine production and release and subsequently to the cardioprotective effect observed.

Two recent studies from our laboratory (Mizumura et al., 1995a, 1995b) have further addressed the hypothesis that K_{ATP} channel opening which results from ischemic PC or from the administration of K_{ATP} channel openers, such as nicorandil or bimakalim, increases the formation and release of adenosine from the ischemic–reperfused area. Local coronary venous concentrations of adenosine were measured before and during the prolonged ischemic period and following reperfusion in anesthetized dogs subjected to one 5-minute PC stimulus 10 minutes prior to a prolonged 60-minute occlusion period and 3 hours of reperfusion or to an intravenous infusion of one of the K_{ATP} channel openers 15 minutes prior to coronary artery occlusion and continued until the time of reperfusion. Infarct size (IS) expressed as a percent of the area at risk (AAR) was used as an index of irreversible cellular injury in all dogs. Ischemic PC (9.8 ± 3.0%), bimakalim (14.3 ± 3.4%), and nicorandil (7.8 ± 1.6%) all produced a significant reduction in IS/AAR as compared to the control group (31.0 ± 5.6%). Similarly, all three interventions resulted in a marked reduction in adenosine release into coronary venous blood (Figure 3) draining the ischemic–reperfused area (data shown for ischemic PC and bimakalim) as compared to the control group. Interestingly, when bimakalim was administered just prior to reperfusion in another group of dogs, infarct size was reduced, although to a lesser extent than when administered as a pretreatment, and there was no effect on adenosine release.

These data suggest that adenosine release is a reflection of the severity of ischemia, which agrees with previous studies of Bardenheuer and Schrader (1986). The results of these experiments also suggest that it is unlikely that ischemic PC or K_{ATP} opener–induced cardioprotection is mediated via an increase in adenosine release

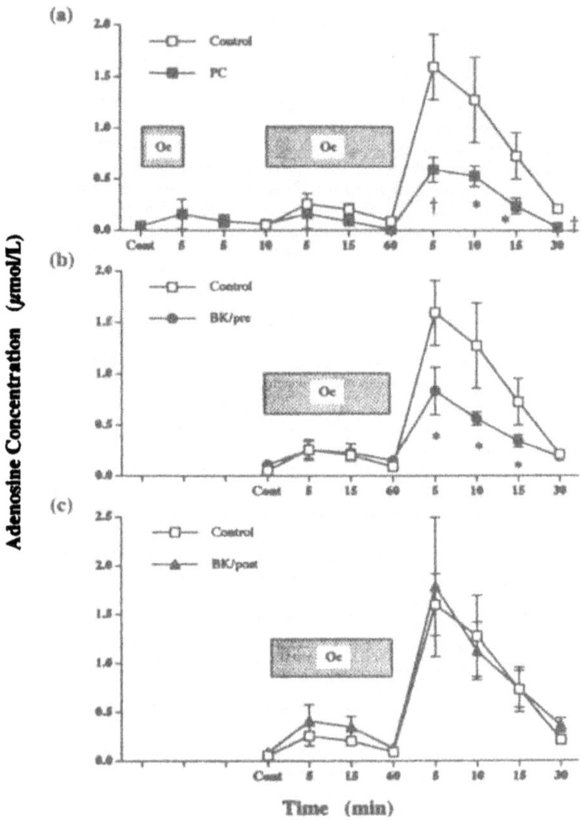

Figure 3. Plots illustrating coronary venous adenosine concentrations from the ischemic–reperfused region at various times during ischemia and following reperfusion. Ischemic PC (**a**) resulted in a significant reduction in adenosine release during reperfusion as compared to the control group. Similarly, bimakalim (BK) pretreatment (pre) also resulted in a significant reduction in adenosine release during reperfusion as compared to the control group. There were no differences in adenosine release when BK was administered at reperfusion as compared to the control group. All values are the mean ± SEM (n = 8/group). **P*<.05; †*P*<.01 versus the control group. From Mizumura et al. (1995a). Reprinted by permission of the American Heart Association.

from the ischemic–reperfused area during occlusion and/or reperfusion, which is in opposition to the second hypothesis mentioned above (see Figure 2). It is more likely that ischemic PC and K_{ATP} channel opening result in an anti-ischemic effect, which leads to a reduction in the breakdown of ATP and a decrease in adenosine release during reperfusion. Studies by Jennings et al. (1991) in canine hearts and by Grover et al. (1991) in rats and McPherson et al. (1993) in guinea pig myocardium all showed a reduction in ATP breakdown during ischemia in preconditioned animals or in those treated with K_{ATP} channel openers, which support the findings of Mizumura et al. (1995a, 1995b). However, since the endothelium has a marked influence on coronary venous adenosine concentrations, the results of the studies of Mizumura et al. (1995a, 1995b) may not be reflective of what is actually occurring in the interstitial space surrounding the myocyte; therefore, another group of experiments were performed in dogs in which ischemic PC and a K_{ATP} opener were compared for their effects on interstitial purine metabolism.

B. Comparative Effects of Ischemic Preconditioning and K_{ATP} Opener–Induced Preconditioning on Interstitial Adenosine

Van Wylen (1994) was the first investigator to study the effects of ischemic PC on interstitial purine metabolism using microdialysis probes inserted into the myocardium of anesthetized dogs. He found that dialysate concentrations of adenosine, inosine, hypoxanthine, and total purines all increased during and immediately after the PC stimulus, which consisted of two 5-minute periods of coronary artery occlusion followed by 10 minutes of reperfusion. However, PC resulted in a reduced rate of interstitial adenosine and total purine formation during the prolonged 60-minute occlusion period when compared to nonpreconditioned control animals. These data suggested that PC reduces the rate of adenine nucleotide degradation and release of purine metabolites during ischemia, which is most likely responsible for the delay in infarct size development observed in PC animals subjected to this same protocol. The increase in interstitial adenosine observed during PC is also consistent with the hypothesis that adenosine triggers PC by acting on its extracellular A_1 adenosine receptor.

Since Kitakaze et al. (1996) showed that two K_{ATP} openers, nicorandil and cromakalim, increased coronary venous adenosine concentrations similar to that of ischemic PC, we undertook a recent study to compare the effects of a K_{ATP} opener, bimakalim, to that of ischemic PC on interstitial purine metabolism by using the microdialysis technique in anesthetized dogs. PC was produced by one 5-minute coronary artery occlusion or by the intracoronary infusion of bimakalim (1 μg/min) prior to a 60-minute occlusion period and 3 hours of reperfusion. The profile demonstrating the dialysate concentrations of adenosine in the two groups is shown in Figure 4. Baseline concentrations did not differ between groups. During the ischemic PC period and the immediate reperfusion period following PC, there was a significant increase in adenosine. In contrast, intracoronary bimakalim resulted in a significant decrease in dialysate adenosine, which was maintained during the drug-free period immediately preceding the 60-minute occlusion period. During the prolonged occlusion period, dialysate adenosine concentrations peaked at 10 to 20 minutes of occlusion in all three groups then slowly returned toward control values at 60 minutes of occlusion (see Figure 4). In the control group, the peak concentration was 3.68 ± 0.72 μmol/L, which was significantly higher than that observed in the ischemic preconditioned (1.77 ± 0.54 μmol/L) and bimakalim-treated (2.07 ±

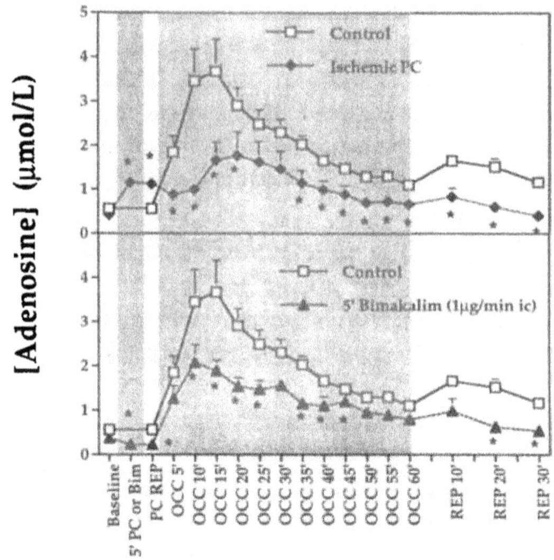

Figure 4. Dialysate adenosine concentrations (μM) in control, ischemic-preconditioned, and bimakalim-preconditioned hearts. Ischemic PC resulted in a significant increase in dialysate adenosine concentration during and immediately after the PC stimulus; however, bimakalim infusion resulted in a significant decrease. Both PC protocols resulted in a decrease in dialysate adenosine concentrations during the prolonged occlusion period. All values are the mean ± SEM (n = 7 to 8/group). *$P<.05$ versus the control group.

0.40 μmol/L) groups. Furthermore, at most time points sampled during the prolonged occlusion and initial 30-minute reperfusion period, the ischemic-preconditioned and bimakalim-treated groups had significantly lower dialysate concentrations of adenosine as compared to that observed in the nontreated group. Similar results were obtained with dialysate concentrations of inosine, hypoxanthine, and xanthine (data not shown).

These results obtained in ischemic preconditioned dogs agree with those previously obtained by Van Wylen (1994) using the microdialysis technique in dogs and those of Mizumura et al. (1995a, 1995b) using coronary venous adenosine as an index of interstitial adenosine in dogs and suggest that PC results in a decrease in adenine nucleotide metabolism during prolonged ischemia. That bimakalim also produced a similar profile in purine metabolism during the prolonged ischemic period suggests that opening of K_{ATP} channels results in a similar energy-sparing effect to that of ischemic PC and suggests that these two interventions may work through a similar pathway to produce their cardioprotective effects (i.e., opening K_{ATP} channels). In contrast to the findings of Kitakaze et al. (1996), bimakalim did not increase interstitial levels of adenosine but actually produced a marked reduction. Thus, based on these results and on those of Mizumura et al. (1995a, 1995b), it is unlikely that the cardioprotective effect exerted by K_{ATP} channel opening in PC or by K_{ATP} openers is the result of an enhanced production of adenosine during the prolonged occlusion or reperfusion period.

To further support these findings, we have recently performed experiments in which we compared the cardioprotective efficacy of bimakalim to reduce infarct size in the absence or presence of an adenosine A_1 receptor antagonist, DPCPX (1 mg/kg). Bimakalim significantly reduced infarct size from 27 ± 6% to 11 ± 3% in the absence of DPCPX and to 12 ± 4% in the presence of DPCPX. This dose of DPCPX completely abolished the cardioprotective effect of ischemic PC in dogs (Gross et al., 1997).

VI. PHARMACOLOGICALLY INDUCED LATE PRECONDITIONING OR SECOND WINDOW OF PROTECTION

Recently, there have been increasing reports that support the concept that brief periods of coronary artery occlusion result in a secondary, delayed phase of protection that is manifest 24 to 48 hours later and is thought to be the result of some as yet undefined molecular adaptation (Yellon and Baxter, 1995). This phenomenon termed "late PC" or the "second window of protection" (SWOP), has been shown to occur in both dogs (Kuzuya et al., 1993) and rabbits (Marber et al., 1993). The cellular mechanisms responsible for late PC are still not clearly understood; however, recent evidence suggests that the K_{ATP} channel and adenosine may also be important components of this phenomenon similar to that observed in classical ischemic PC. Experimental data to support this concept have been obtained by the use of two pharmacological tools, monophosphoryl lipid A (MLA) and the adenosine A_1 receptor agonist, 2-chloro-N^6-cyclopentyl-adenosine (CCPA).

A. Monophosphoryl Lipid A—Role of K_{ATP} Channels

MLA is a nontoxic endotoxin derivative that we have previously shown to reduce infarct size and attenuate stunning in anesthetized dogs (Yao et al. 1993; Yao et al., 1995) 24 hours after drug administration. Recently, we tested the hypothesis that the K_{ATP} channel may be an important effector involved in mediating the potent cardioprotective effect of MLA. In this regard, we administered either one of two K_{ATP} channel antagonists, glibenclamide (0.3 mg/kg, IV) or 5-hydroxydecanoic acid (7.5 mg/Hg, intracoronary) immediately prior to or during the first few minutes of a 60-minute occlusion period in anesthetized dogs that had been pretreated with MLA (35 μg/kg, IV) 24 hours earlier. Both agents completely blocked its cardioprotective effect, which suggests that the K_{ATP} channel is an important mediator of this late PC-like effect of MLA (Mei et al., 1995).

More recently, we have obtained preliminary data in which we compared the effect of MLA pretreatment to that of ischemic and K_{ATP} opener–induced PC on interstitial adenosine concentrations prior to and during a prolonged 60-minute ischemic period. Interestingly, baseline interstitial adenosine concentrations were enhanced in MLA-pretreated dogs (Figure 5). In addition, interstitial adenosine concentrations were reduced in MLA-pretreated dogs during the prolonged occlusion period and coronary venous adenosine release following reperfusion was decreased similar to that previously seen in preconditioned or bimakalim-treated dogs. Taken together, these data suggest that adenosine and the K_{ATP} channel may be interacting in a similar manner in late PC produced by MLA to that observed in classical ischemic PC, although more studies are needed to determine the mechanism by which adenosine and the K_{ATP} channel interact to produce the cardioprotective effect of MLA.

B. Adenosine and Late Preconditioning

Although a clear role for adenosine appears to be established in classical ischemic PC, there is only one study that has addressed the involvement of adenosine in late PC. Baxter et al. (1994) demonstrated a reduction in infarct size (53.6 ± 5.7% to 32.9 ± 4.6%) in rabbits subjected to four 5-minute periods of PC 24 hours prior to a prolonged 30-minute ischemic period and 120 minutes of reperfusion. This cardioprotective effect of ischemic PC was abolished when the rabbits were pretreated with the nonselective adenosine receptor antagonist 8-(p-sulfophenyl)-theophylline (SPT) prior to the four PC stimuli, which suggests that adenosine receptor activation is necessary as a trigger for the cardioprotective effect of late PC. Simi-

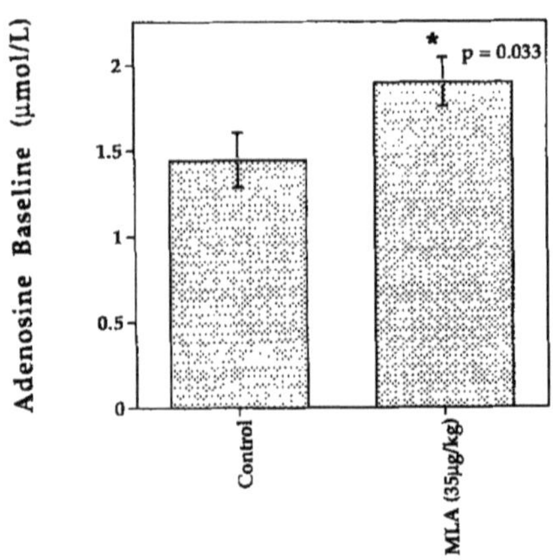

Figure 5. Baseline dislysate adenosine concentrations in vehicle-treated control animals and those treated with monophosphoryl lipid A (MLA) 24 hours previously. MLA-treated dogs had significantly higher baseline adenosine concentrations (*$P<.03$). All values are the mean ± SEM (n = 6 to 8/group).

larly, if conscious rabbits were pretreated with several intravenous bolus doses of the adenosine A_1 receptor agonist, CCPA, a reduction in infarct size similar to that observed following ischemic PC was noted (26.3 ± 5.7%) at a high dose (100 μg/kg). Whether this protective effect of adenosine A_1 receptor stimulation is mediated via K_{ATP} channel activation as is the case in classical PC, is still unknown and deserves further study.

VII. CLINICAL EVIDENCE TO SUPPORT A ROLE FOR K_{ATP} CHANNELS AND ADENOSINE IN PRECONDITIONING IN HUMANS

Clinical evidence that PC occurs in humans was first presented by Deutsch et al. (1990) in which less evidence of myocardial ischemia was found as indicated by attenuated ST-segment shifts, less lactate production, and smaller hemodynamic changes following the second of two periods of coronary artery occlusion during percutaneous transluminal coronary angioplasty (PTCA). Several recent clinical studies (Kloner et al., 1995; Ottani et al., 1995) showed that previous angina or new-onset prodromal angina may confer a beneficial effect on patient outcome following an acute myocardial infarction, which might be explained by a preconditioning-like effect of the preceding angina.

The first evidence that the K_{ATP} channel may be involved in PC in humans was provided by Tomai et al. (1994). This group found that PC occurs following the second balloon inflation in patients undergoing PTCA and that this beneficial effect of PC was blocked by a 10-mg oral dose of glibenclamide given 60 minutes prior to PTCA, which suggests that the beneficial effect of PC in humans is primarily mediated via the K_{ATP} channel. Using a human experimental model of simulated ischemia in which right atrial trabeculae were preconditioned by a brief 3-minute period of hypoxia, a PKC activator, 1,2-dioctanoyl-sn-glycerol (DOG) or the K_{ATP} opener cromakalim, Speechly-Dick et al. (1995) showed that all three PC stimuli were blocked by glibenclamide. These data were the first to suggest that the K_{ATP} channel is the end effector of PC produced in an experimental human model.

In a similar model of isolated superfused human right atrial trabeculae, Walker et al. (1995) found that the beneficial effect of hypoxic PC was blocked by pretreatment with SPT, a nonselective adenosine antagonist, and mimicked by administration of R-phenyl-isopropyladenosine (R-PIA), an adenosine A_1 receptor agonist. These data provided the first direct evidence for the involvement of adenosine in PC in human myocardium. In support of these results, Kerensky et al. (1995) showed that intracoronary adenosine can precondition patients when administered prior to coronary angioplasty. Yao and Gross (1994b) also showed that intracoronary infusion of adenosine mimicked the infarct size–reducing effect of ischemic PC in dogs. Thus, taken together, these data suggest that both adenosine and the K_{ATP} channel are important components of the PC response in humans similar to that observed in most experimental animal models. However, future studies are necessary to determine if adenosine and the K_{ATP} channel are linked to a common signalling pathway in humans as appears to be the case in experimental animals.

VIII. CONCLUSIONS AND FUTURE DIRECTIONS

In conclusion, the data presented support an important role for adenosine and the K_{ATP} channel in mediating the powerful cardioprotective effect of PC produced by several pathophysiological stimuli such as ischemia or hypoxia or by several pharmacological agents such as a K_{ATP} channel opener or an adenosine A_1 receptor agonist. These two mediators of the PC response appear to be important in classical PC as well as late PC in most experimental animal models, including humans. A future goal will be to exploit these systems pharmacologically for therapeutic benefit in the clinical setting. Future studies are also needed at the molecular and cellular level to determine the mechanisms and signalling pathways involved in transducing the adenosine receptor–K_{ATP} channel interaction into the potent cardioprotective effect that occurs when both of these systems are activated simultaneously. More studies similar to a recent one reported by Liu et al. (1996) in isolated rabbit cardiac myocytes—in which a synergistic modulation of the K_{ATP} channel by PKC and adenosine was described—are needed to better understand cellular mechanisms responsible for PC. In this regard, the recent identification of a full-length K_{ATP} channel clone from pancreatic β cells (Inagaki et al., 1995) suggests that we may be able to study the molecular regulation of K_{ATP} channel function in the near future. Furthermore, the importance of the recently identified mitochondrial K_{ATP} channel in mediating ischemic PC will also be an intense area of investigation in the future (Garlid et al., 1997; Liu et al., 1998).

ACKNOWLEDGMENTS

The authors wish to thank Jeannine Moore and Anna Hsu for their excellent technical assistance in carrying out the experiments described and Carol Knapp for her outstanding help in the preparation of this chapter. The experiments described were supported by funds from NIH Grant HL 08311 and contracts from Merck KGa, Darmstadt, Germany, and RIBI ImmunoChem Research, Inc., Hamilton, Montana.

REFERENCES

Armstrong, S.C., Liu, G.S., Downey, J.M., and Ganote, C.E. (1995). Potassium channels and preconditioning of isolated rabbit cardiomyocytes: effects of glyburide and pinacidil. J. Mol. Cell. Cardiol. 27, 1765-1774.

Auchampach, J.A., and Gross, G.J. (1993). Adenosine A_1 receptors, K_{ATP} channels and ischemic preconditioning in dogs. Am. J. Physiol. 264, H1327-H1336.

Auchampach, J.A., Grover, G.J., and Gross, G.J. (1992). Blockade of ischemic preconditioning in dogs by the novel ATP-dependent potassium channel antagonist sodium 5-hydroxydecanoate. Cardiovasc. Res. 26, 1054-1062.

Bardenheuer, H.J., and Schrader, J. (1986). Supply-to-demand ratio for oxygen determines formation of adenosine by the heart. Am. J. Physiol. 250, H173-H180.

Baxter, G.F., Marber, M.S., Patel, V.C., and Yellon, D.M. (1994). Adenosine receptor involvement in a delayed phase of myocardial protection 24 hours after ischemic preconditioning. Circulation 90, 2993-3000.

Bhattacharya, S., and Linden, J. (1995). The allosteric enhancer, PD 81,723, stabilizes human A_1 adenosine receptor coupling to G proteins. Biochim. Biophys. Acta. 1265, 15-21.

Deutsch, E., Berger, M., Kussmaul, W.G., Hirshfeld, J.W. Jr., Herrmann, H.C., and Laskey, W.K. (1990). Adaptation to ischemia during percutaneous transmural coronary angioplasty: clinical, hemodynamic and metabolic features. Circulation 82, 2044-2051.

Dunne, M.J., and Peterson, O.H. (1986). GTP and GDP activation of K^+ channels that are inhibited by ATP. Pfugers Arch. 407, 564-565.

Garlid, K.D., Paucek, P., Yarovoy, V., Murray, H.H., Darbenzio, R.B., D'Alonzo, A.J., Lodge, N.J., Smith, M.A., and Grover, G.J. (1997). Cardioprotective effect of diazoxide and its interaction with mitochondrial ATP-sensitive K^+ channels: Possible mechanism of cardio-protection. Circ. Res. 81, 1072-1082

Gross, G.J., Mei, D.A., Sleph, P., and Grover, G.J. (1997). Adenosine A_1 receptor blockade does not abolish the cardioprotective effects of the adenosine triphosphate-sensitive potassium channel opener bimakalim. J. Pharmacol. Exp. Ther. 280, 533-540.

Gross, G.J., and Auchampach, J.A. (1992). Blockade of ATP-sensitive potassium channels prevents myocardial preconditioning in dogs. Circ. Res. 70, 223-233.

Grover, G.J., Newburger, J., Sleph, P.G., Dzwonczyk, S., Taylor, S.C., Ahmed, A.Z., and Atwal, K.S. (1991). Cardioprotective effects of the potassium channel opener cromakalim: stereoselectivity and effects on myocardial adenine nucleotides. J. Pharmacol. Exp. Ther. 257, 156-162.

Grover, G.J., Sleph, P.G., and Dzwondzyk, S. (1992). Role of myocardial ATP-sensitive potassium channels in mediating preconditioning in the dog heart and their possible interaction with adenosine A_1-receptors. Circulation 86, 1310-1316.

Inagaki, N., Gonoi, T., Clement J.P. IV, Namba, N., Inazawa, J., Gonzalez, G., Aguilar-Bryan, L., Seino, S., and Bryan, J. (1995). Reconstitution of I_{KATP}: an inward rectifier subunit plus the sulfonylurea receptor. Science 270, 1166-1170.

Ito, H., Vereecke, J., and Carmeliet, E. (1994). Mode of regulation by G protein of the ATP-sensitive K^+ channel in guinea-pig ventricular cell membrane. J. Physiol. 478.1, 101-108.

Jennings, R.B., Murry, C.E., and Reimer, K.A. (1991). Energy metabolism in preconditioning and control myocardium: effect of total ischemia. J. Mol. Cell. Cardiol. 23, 1449-1458.

Kerensky, R.A., Kutcher, M.A., Braden, G.A., Applegate, R.J., Solis, G.A., and Little, W.C. (1995). The effects of intracoronary adenosine on preconditioning during coronary angioplasty. Clin. Cardiol. 18, 91-96.

Kirsch, G.E., Codina, J., Birnbaumer, L., and Brown, A.M. (1990). Coupling of ATP-sensitive K^+ channels to A_1 receptors by G proteins in rat ventricular myocytes. Am. J. Physiol. 259, H820-H826.

Kitakaze, M., Hori, M., Takashima, S., Sato, H., Inoue, M., and Kamada, T. (1993). Ischemic preconditioning increases adenosine release and 5'-nucleotidase activity during myocardial ischemia and reperfusion in dogs. Implications for myocardial salvage. Circulation 87, 208-215.

Kitakaze, M., Minamino, T., Node, K., Komamura, K., Shinozaki, Y., Chujo, M., Mori, H., Inoue, M., Hori, M., and Kamada, T. (1996a). Role of activation of ectosolic 5'-nucleotidase in the cardioprotection mediated by opening of K^+ channels. Am. J. Physiol. 270, H1744-H1756.

Kitakaze, M., Node, K., Minamino, T., Komamura, K., Funaya, H., Shinozaki, Y., Chujo, M., Mori, H., Inoue, M., Hori, M., and Kamada, T. (1996b). Role of activation of protein kinase C in the infarct size-limiting effect of ischemic preconditioning through activation of ecto-5'-nucleotidase. Circulation 93, 781-791.

Kloner, R.A., Shook, T., Przyklenk, K., Davis, V.G., Junio, L., Matthews, R.V., Burstein, S., Gibson, C.M., Poole, W.K., Cannon, C.P., McCabe, C.H., and Braunwald, E. (1995). Previous angina alters in-hospital outcome in TIMI 4: a clinical correlate to preconditioning? Circulation 91, 37-47.

Kuzuya, T., Hoshida, S., Yamashita, N., Fuji, H., Oe, H., Hori, M., Kamada, T., and Tada, M. (1993). Delayed effects of sublethal ischemia on the acquisition of tolerance to ischemia. Circ. Res. 72, 1293-1299.

Lasley, R.D., Konyn, P.J., Hegge, J.O., and Mentzer, R.M. Jr. (1995). Effects of ischemic and adenosine preconditioning on interstitial fluid adenosine and myocardial infarct size. Am. J. Physiol. 269, H1460-H1466.

Lasley, R.D., and Mentzer, R.M. Jr. (1993). Pertussis toxin blocks adenosine A_1 receptor–mediated protection of the ischemic rat heart. J. Mol. Cell. Cardiol. 25, 815-821.

Li, G.C., Vasquez, J.A., Gallagher, K.P., and Lucchesi, B.R. (1990). Myocardial protection with preconditioning. Circulation 82, 609-619.

Li, Y.W., Whittaker, P., and Kloner, R.A. (1992). The transient nature of the effect of ischemic preconditioning on myocardial infarct size and ventricular arrhythmia. Am. Heart J. 123, 346-353.

Liang, B. (1996). Direct preconditioning of cardiac ventricular myocytes via adenosine A_1 receptors and K_{ATP} channels. Am. J. Physiol. 271, H1769-H1777.

Liu, G.S., Richards, S.C., Olsson, R.A., Mullane, K., Walsh, R.S., and Downey, J.M. (1994). Evidence that the adenosine A_3 receptor may mediate the protection afforded by preconditioning in the isolated rabbit heart. Cardiovasc. Res. 28, 1057-1061.

Liu, G.S., Thornton, J., Van Winkle, D.M., Stanley, A.W.H., Olsson, R.A., and Downey, J.M. (1991). Protection against infarction afforded by preconditioning is mediated by A_1 adenosine receptors in rabbit heart. Circulation 84, 350-356.

Liu, Y., and Downey, J.M. (1992). Ischemic preconditioning protects against infarction in rat heart. Am. J. Physiol. 263, H1107-H1112.

Liu, Y., Gao, W.D., O'Rourke, B., and Marban, E. (1996). Synergistic modulation of ATP-sensitive K^+ currents by protein kinase C and adenosine: implications for ischemic preconditioning. Circ. Res. 78, 443-454.

Liu, Y., Sato, T., O'Rourke, B., and Marban, E. (1998). Mitochrondrial ATP-dependent potassium channels: Novel effectors of cardioprotection? Circulation 97, 2463-2469.

Marber, M.S., Latchman, D.S., Walker, J.M., and Yellon, D.M. (1993). Cardiac stress protein elevation 24 hours after brief ischemia or heat stress is associated with resistance to myocardial infarction. Circulation 88, 1264-1272.

McPherson, C.D., Pierce, G.N., and Cole, W.C. (1993). Ischemic cardioprotection by ATP-sensitive K^+ channels involves high-energy phosphate preservation. Am. J. Physiol. 265, H1809-H1818.

Mei, D.A., Elliott, G.T., and Gross, G.J. (1995). ATP-sensitive K^+ channels mediate the cardioprotective effect of monophosphoryl lipid A. Circulation 92, I-388.

Minamino, T., Kitakaze, M., Morioka, T., Node, K., Komamura, K., Takeda, H., Inoue, M., Hori, M., and Kamada, T. (1996). Cardioprotection due to preconditioning correlates with increased ecto-5'-nucleotidase activity. Am. J. Physiol. 270, H238-H244.

Mizumura, T., Auchampach, J.A., Linden, J., Bruns, R.F., and Gross, G.J. (1996). PD 81,723, an allosteric enhancer of the A_1 adenosine receptor, lowers the threshold for ischemic preconditioning in dogs. Circ. Res. 79, 415-423.

Mizumura, T., Nithipatikom, K., and Gross, G.J. (1995a). Bimakalim, an ATP-sensitive potassium channel opener, mimics the effects of ischemic preconditioning to reduce infarct size, adenosine release, and neutrophil function in dogs. Circulation 92, 1236-1245.

Mizumura, T., Nithipatikom, K., and Gross, G.J. (1995b). Effects of nicorandil and glyceryl trinitrate on infarct size, adenosine release, and neutrophil infiltration in the dog. Cardiovasc. Res. 29, 482-489.

Murry, C.E., Jennings, R.B., and Reimer, K.A. (1986). Preconditioning with ischemia: a delay of lethal cell injury in ischemic myocardium. Circulation 74, 1124-1136.

Nakhostine, N., and Lamontagne, D. (1993). Adenosine contributes to hypoxia-induced vasodilation through ATP-sensitive K^+ channel activation. Am. J. Physiol. 265, H1289-H1293.

Nao, B.S., McClanahan, T.B., Groh, M.A., Schott, R.J., and Gallagher, K.P. (1990). The time of effective ischemic preconditioning in dogs. Circulation 88, III-271 [Abstr.].

Ottani, F., Galvani, M., Ferrini, D., Sorbello, F., Limonetti, P., Pantoli, D., and Rusticali, F. (1995). Prodromal angina limits infarct size: a role for ischemic preconditioning. Circulation 91, 291-297.

Parent, L., and Coronade, R. (1989). Reconstitution of the ATP-sensitive potassium channel of skeletal muscle. Activation by a G protein-dependent process. J. Gen. Physiol. 94, 445-463.

Qian, Y., Levasseur, J.E., Yoshida, K., and Kukreja, R.C. (1996). K_{ATP} channels in rat heart: blockade of ischemic and acetylcholine-mediated preconditioning by glibenclamide. Am. J. Physiol. 271, H23-H28.

Qiu, Y., Risvi, A., tang, X.L., Manchikalapudi, S., Takano, H., Jadoon, A.K., Wu, W.J., and Bolli, R. (1997). Nitric oxide triggers late preconditioning against myocardial infarction in conscious rabbits. Am. J. Physiol. 273, H2931-H2926.

Randall, M.D., Ujie, H., and Griffith, T.M. (1994). Modulation of vasodilation to levcromakalim by adenosine analogues in the rabbit ear: an explanation for hypoxic augmentation. Br. J. Pharmacol. 112, 49-54.

Schulz, R., Post, H., Sakka, S., Wallbridge, D.R., and Heusch, G. (1995). Intraischemic preconditioning. Circ. Res. 76, 942-950.

Schulz, R., Rose, J., and Heusch, G. (1994). Involvement of activation of ATP-dependent potassium channels in ischemic preconditioning. Am. J. Physiol. 36: H1341-H1352.

Shizikuda, Y., Mallet, R.T., Lee, S.C., and Downey, H.F. (1992). Hypoxic preconditioning of the ischemic canine myocardium. Cardiovasc. Res. 26, 534-542.

Speechly-Dick, M.E., Grover, G.J., and Yellon, D.M. (1995). Does ischemic preconditioning in the human involve protein kinase C and the ATP-dependent K^+ channel? Studies of contractile function after simulated ischemia in an atrial in vitro model. Circ. Res. 77, 1030-1035.

Thornton, J.D., Thornton, C.S., and Downey, J.M. (1993). Effect of adenosine receptor blockade: preventing protective preconditioning depends on the time of initiation. Am. J. Physiol. 265, H504-H508.

Tomai, F., Crea, F., Gaspardone, A., Versaci, F., De Paulis, R., Penta de Peppo, A., Chiariello, L., and Gioffre, P.A. (1994). Ischemic preconditioning during coronary angioplasty is prevented by glibenclamide, a selective ATP-sensitive K^+ channel blocker. Circulation 90, 700-705.

Toombs, C.F., Moore, T.L., and Shebuski, R.J. (1993). Limitation of infarct size in rabbits by ischaemic preconditioning is reversible with glibenclamide. Cardiovasc. Res. 27, 617-622.

Van Winkle, D.M., Chien, G.L., Wolff, R.A., Soifer, B.E., Kuzume, K., and Davis, R.F. (1994). Cardioprotection provided by adenosine receptor activation is abolished by blockade of the K_{ATP} channel. Am. J. Physiol. 259, H820-H826.

Van Wylen, D.G.L. (1994). Effect of ischemic preconditioning on interstitial purine metabolite and lactate accumulation during myocardial ischemia. Circulation 89, 2283-2289.

Walker, D.M., Walker, J.M., Pugsley, W.B., Pattison, C.W., and Yellon, D.M. (1995). Preconditioning in isolated superfused human muscle. J. Mol. Cell. Cardiol. 27, 1349-1357.

Yao, Z., Auchampach, J.A., Pieper, G.M., and Gross, G.J. (1993). Cardioprotective effects of monophosphoryl lipid A, a novel endotoxin analogue, in the dog. Cardiovasc. Res. 27, 832-838.

Yao, Z., Elliott, G.T., and Gross, G.J. (1995). Monophosphoryl lipid A preserves myocardial contractile function following multiple, brief periods of coronary artery occlusion in dogs. Pharmacology 51, 152-159.

Yao, Z., and Gross, G.J. (1994a). Activation of ATP-sensitive potassium channels lowers threshold for ischemic preconditioning in dogs. Am. J. Physiol. 267, H1888-H1894.

Yao, Z., and Gross, G.J. (1994b). A comparison of adenosine-induced cardioprotection and ischemic preconditioning in dogs: efficacy, time course and the role of K_{ATP} channels. Circulation 89, 1229-1236.

Yao, Z., and Gross, G.J. (1996). Role of K_{ATP} channels in memory associated with myocardial preconditioning. Drug News and Perspectives 9, 13-18.

Yao, Z., Rasmussen, J.L., Hirt, J.L., Mei, D.A., Pieper, G.M., and Gross, G.J. (1993). Effects of monophosphoryl lipid A on myocardial ischemia/reperfusion injury in dogs. J. Cardiovasc. Pharmacol. 22, 653-663.

Yellon, D.M., and Baxter, G.F. (1995). A "second window of protection" or delayed preconditioning phenomenon: future horizons for myocardial protection? J. Mol. Cell. Cardiol. 27, 1023-1034.

Zhou, X., Zhai, X., and Ashraf, M. (1996). Direct evidence that initial oxidative stress triggered by preconditioning contributes to second window of protection by endogenous antioxidant enzyme in myocytes. Circulation 93, 1177-1184.

ISCHEMIC PRECONDITIONING:
ROLE OF MULTIPLE KINASES IN SIGNAL AMPLIFICATION AND MODULATION

Dipak K. Das

Advances in Organ Biology
Volume 6, pages 101-124.

ISBN: 0-7623-0391-3

I. INTRODUCTION

Myocardial adaptation to ischemic stress, which is the manifestation of the earlier stress response that occurs during repeated episodes of brief ischemia and reperfusion, can render the myocardium more tolerant to a subsequent potential lethal ischemic injury (Flack et al., 1991). This transient adaptive response has been demonstrated to be associated with decreased reperfusion-induced arrhythmias (Tosaki et al., 1994; Liu and Downey, 1992; Hagar et al., 1991), increased recovery of postischemic contractile functions (Kimura et al., 1992; Maulik et al., 1993; Tosaki et al., 1995; Engelman et al., 1995; Asimakis et al., 1992; Cave et al., 1993; Steenbergen et al., 1993; Banerjee et al., 1993), and reduction of the infarct size (Murry et al., 1986; Schott et al., 1990; Liu et al., 1991; Li et al., 1990; Lawson et al., 1993; Sandhu et al., 1993). The adaptive protection has been found to be mediated by gene expression and their transcriptional regulation (Das et al., 1995; Das et al., 1994; Maulik and Das, 1996).

Ischemic preconditioning provides a powerful anti-ischemic protection of the heart. For example, ischemic preconditioning induced by repeated short-term ischemia and reperfusion has been found to reduce postischemic left ventricular functional abnormalities, ventricular arrhythmias, infarct size, and cell damage (Flack et al., 1991; Tosaki et al., 1994; Liu and Downey, 1992; Hagar et al., 1991), and to promote increased recovery of postischemic contractile functions (Kimura et al., 1992; Maulik et al., 1993; Tosaki et al., 1995; Engelman et al., 1995; Asimakis et al., 1992; Cave et al., 1993; Steenbergen et al., 1993; Banerjee et al., 1993; Murry et al., 1986; Schott et al., 1990; Liu et al., 1991; Li et al., 1990; Lawson et al., 1993; Sandhu et al., 1993). Although preconditioning is a powerful tool in protecting the myocardium from ischemic–reperfusion injury, considerable debate remains regarding its mechanism of action. Most of the existing studies have been focused on the role of adenosine A_1 receptors in preconditioning (Murry et al., 1990). In addition, α_1-adrenergic receptors (Tosaki et al., 1995), muscarinic receptors (Kida et al., 1991), adenosine triphosphate (ATP)-dependent potassium channels (Gross and Auchampach, 1992), multiple receptors including bradykinin and angiotensin II receptors (Liu et al., 1995) and G protein (Lawson et al., 1993) have been implicated in ischemic preconditioning. Irrespective of the signaling pathways involved, it is more or less universally accepted that the intracellular signaling leads to the translocation and activation of protein kinase C (PKC) (Ytrehus et al., 1994). Involvement of PKC has been demonstrated directly by biochemical and immunological studies and indirectly by experiments demonstrating that PKC inhibitors block the beneficial effects of preconditioning (Prasad and Jones, 1992; Liu et al., 1995).

II. BACKGROUND

Murry et al. (1986) was the first investigator to report the phenomenon of ischemic preconditioning. Subsequently, the same investigators measured glucose and energy metabolism in preconditioned hearts and determined that a reduction of glycolysis in conjunction with the preservation of ATP led to the conservation of ultrastructure during sustained ischemia (Murry et al., 1990). Subsequently, Kida et al. (1991) confirmed these results and further demonstrated that in addition to ATP and ultrastructure, preconditioning preserved creatine phosphate and pH during sustained ischemia in pig hearts. In the same study, ischemic preconditioning was found to attenuate acidosis.

Of many hypotheses, the most popular has been the adenosine hypothesis. This receives support from the observation that ischemic preconditioning can be blocked by adenosine A_1 antagonists, and that adenosine A_1 agonists can limit infarct size (Liu et al., 1991). Although there is general agreement regarding the beneficial role of adenosine on the ischemic tissue, the adenosine hypothesis remains controversial for at least three reasons: (1) the half-life of extracellular adenosine is extremely short (seconds to minutes), whereas the preconditioning effects lasts up to 2 hours; (2) the amount of adenosine necessary for preconditioning is unlikely to accumulate after two to four brief (usually 5-minute) repeated periods of ischemia, which do not result in cumulative depletion of ATP; and (3) the adenosine hypothesis has received limited experimental support in some species (e.g., rats).

To reconcile the adenosine hypothesis, an argument has been made that adenosine could trigger a secondary mechanism such as the activation of G_i protein, which in turn could open the ATP-sensitive potassium channel (K_{ATP}). However, a recent study did not support the role of G_i protein, because of the inability of the pertussis toxin to influence the antiarrhythmic protection afforded by preconditioning (Lawson et al., 1993). The K_{ATP} hypothesis has also come under criticism, principally because many K_{ATP} openers exert spontaneous proarrhythmic action in the face of ischemia. Moreover, after a short duration of ischemia, the concentration of ATP in the heart is usually still high enough to inhibit the K_{ATP}.

Another intriguing recent hypothesis is the concept of stimulation of an endogenous protective mechanism by way of myocardial adaptation to ischemia. Repeated short durations of ischemia have been found to modify lipid bilayer and membrane fluidity as well as to lead to the induction of the expression of several stress-related and antioxidant genes and proteins (Das et al., 1995; Das et al., 1994). These include induction of the expression of proto-oncogenes such as c-fos, c-myc, c-jun, and so forth; induction of mRNAs for heat shock proteins (HSPs) such as HSP 27, HSP 65, HSP 70, HSP 89, and for antioxidants such as manganese superoxide dismutase (Mn-SOD) and catalase; and stimulation of several antioxidant enzyme activities such as SOD, catalase, glutathione peroxidase, and glutathione reductase. Nevertheless, the list represents only a fraction of the entire list of genes and proteins actually induced by ischemia.

It should be clear from the above discussion that the precise mechanism of ischemic preconditioning is far from clear. However, it seems quite reasonable to speculate that ischemic preconditioning occurs in two different steps: (1) early effect (short-term adaptation) triggered between seconds to minutes, which is likely to be mediated by the release of some endogenous compound(s) such as catecholamines, adenosine, nitric oxide, acetylcholine, and so forth, and may last up to 1 to 2 hours (*ischemic preconditioning*); and (2) late effect (long-term adaptation), which may occur after several hours and may last up to days, and which is likely to be mediated by the transcription of genes and their subsequent translation into proteins (*myocardial adaptation to ischemia*).

III. SIGNAL TRANSDUCTION

A. Role of Protein Kinase C

As mentioned earlier, it has been demonstrated that cellular PKC activation is an important step in the mechanism of adaptive protection of the heart (Ytrehus et al., 1994; Bugge and Ytrehus, 1995). The PKC hypothesis received further support from the observations that any agent that can activate PKC can also precondition the heart. For example, phenylephrine, an α_1 agonist, angiotensin AT_1, and bradykinin B_2 receptors can activate PKC (Dixon et al., 1994). Phenylephrine, angiotensin AT_1, and bradykinin B_2 have been shown to precondition the heart when infused prior to ischemia (Tosaki et al., 1995; Liu et al., 1995; Goto et al., 1995).

B. Protein Kinase C Isoforms

A variety of stress signals have been found to translocate and activate PKC. For example, mechanical stress induced by stretching can activate PKC in cultured myocytes (Yazaki et al., 1993). Immediately after stretching, activation of phosphatidyl inositol turnover was observed, suggesting a role of phospholipase C in PKC activation. Even short-term ischemia or ischemia followed by reperfusion were previously shown to translocate and activate PKC (Prasad and Jones, 1992). Furthermore, both α_1-receptor stimulation and Ca^{2+} ion can translocate and activate PKC (Henrich and Simpson, 1988; Fearon and Tashzian, 1985). Given the fact that both α_1-receptor activation and intracellular Ca^{2+} overloading are the manifestations of ischemia–reperfusion injury, it was not surprising when ischemic preconditioning consisting of repeated ischemia and reperfusion was also found to translocate and activate PKC.

IV. TYROSINE KINASES

It is universally accepted that protein phosphorylation plays a crucial role in a wide variety of cellular processes that control signal transduction. It must be clearly understood that protein phosphorylation is a rapidly reversible process that regulates intracellular signaling in response to a specific stress, for example, environmental changes (Hunter et al., 1985). Protein phosphorylation is mediated by a number of protein kinases that can be grouped into two major classes: (1) phosphorylates serine/threonine residues, such as PKC and casein kinases; and (2) phosphorylate proteins on their tyrosine residues, such as tyrosine kinases. Tyrosine kinases can activate a number of different intracellular signaling pathways, including tyrosine phosphorylation in the case of phospholipase Cγ and phospholipase D (Sadowski et al., 1993), conformational changes induced by binding of the Src homology 2 (SH2) domain to phosphotyrosine for p13 kinases (Carpenter et al., 1993) as well as translocation to the plasma membrane for stimulation of Ras guanine nucleotide exchange by Sos (Quilliam et al., 1994).

A. Signaling Pathways

Among these signaling pathways, activation of the ERK family of mitogen-activated protein (MAP) kinases appear to be the most important event for signal transduction in response to a stress. This pathway is controlled by Ras, Raf, MEK, and ERK kinases, and a small guanine nucleotide binding protein links receptor tyrosine kinase activation to a cytosolic protein kinase cascade. This Ras/Raf/MEK/ERK pathway is generally known as the MAP kinases pathway for stress signal. In this pathway, the signal is sequentially transmitted through three kinases: serine/threonine protein kinase (mitogen-activated protein kinase kinase kinase [MAPKKK]), which phosphorylates and activates (mitogen-activated kinase kinase [MAPKK]), which in turn phosphorylates and activates another serine/threonine protein kinase (mitogen-activated protein kinase [MAPK]). In this pathway, Raf corresponds to MAPKKK, MEK corresponds to MAPKK, and ERK corresponds to MAPK. Current evidence indicates that stress responses in mammalian cells are mediated by this signaling pathway (Rouse et al., 1994).

B. Mechanisms of Signal Transduction

Signal transduction is triggered when a ligand binds the receptor of protein tyrosine kinase at the cell surface. After binding, a ligand-induced conformational change occurs in the external domain, which causes dimerization of the receptor. This then leads to intermolecular autophosphorylation, followed by transphosphorylation, which seems to be essential for substrate phosphorylation. Such transphosphorylation causes the activation of dimeric receptor protein tyrosine kinases for phosphorylation of cytoplasmic substrates. The autophosphorylated dimer then recruits substrates having increased affinity for the receptor. Many substrates or targets for receptor tyrosine kinases contain SH2 domains that are about 100 residues in length. Activated receptors bind to the SH2 domain–containing proteins. However, for many signaling proteins an Src homology 3 (SH3) domain also exists. Signaling by an activited receptor protein tyrosine kinase is initiated by the phosphorylation of cytoplasmic proteins, which potentiates the intracellular signaling cascade. Generally speaking, signal transduction by receptor protein tyrosine kinases is dependent solely on the tyrosine kinase activity.

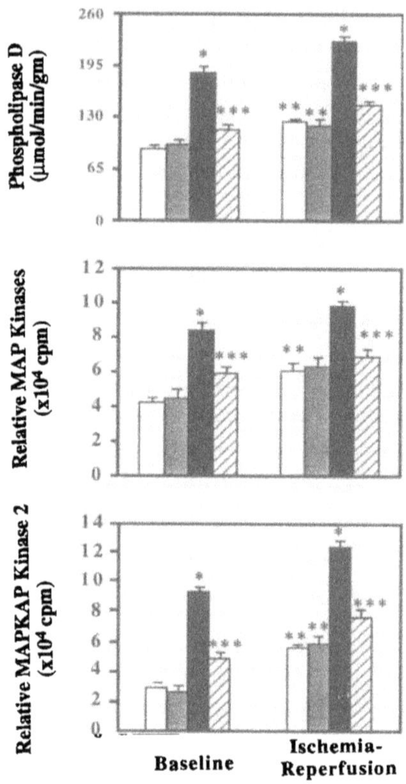

Figure 1. Effects of preconditioning and genistein on phospholipase D, mitogen-activated protein (MAP) kinases, and MAPKAP kinase 2 in rat heart. Isolated rat hearts were preperfused with 100 µM genistein for 10 minutes prior to preconditioning by repeated ischemia and reperfusion. The control group did not receive any genistein. After preconditioning, hearts were subjected to 30 minutes of ischemia followed by 30 minutes of reperfusion. Results are expressed as means ± SEM (n = 6 animals per group). Each assay was run in duplicate. *$p < .05$ compared to control; **$p < .05$ compared to baseline control; ***$p < .05$ compared to preconditioned group. [ ] control; [ ] control + genistein; [ ] preconditioned; [ ] preconditioned + genistein.

A recent study from our laboratory demonstrated that preconditioning of the rat hearts stimulated phospholipase D, PKC, MAP kinases, and MAPKAP kinase 2 activities which were inhibited by genistein, suggesting for the first time tyrosine kinase–phospholipase D as a potential signaling pathway for ischemic preconditioning (Maulik et al., 1996; Das et al., 1996) (Figure 1). The role of tyrosine kinases was further substantiated from the observations that phosphorylation of protein tyrosine kinases was enhanced after ischemic preconditioning (Figure 2).

V. MAP KINASES

Recently, many studies have indicated that MAP kinases, a novel serine/threonine protein kinase family, play an essential role in mediating intracellular signal transduction events (Seger and Krebs, 1995; Blumer and Johnson, 1994). In response to extracellular stimulation, MAP kinases are rapidly activated and in turn regulate cellular functions by inducing the phosphorylation of proteins, such as an oncogene product c-jun,

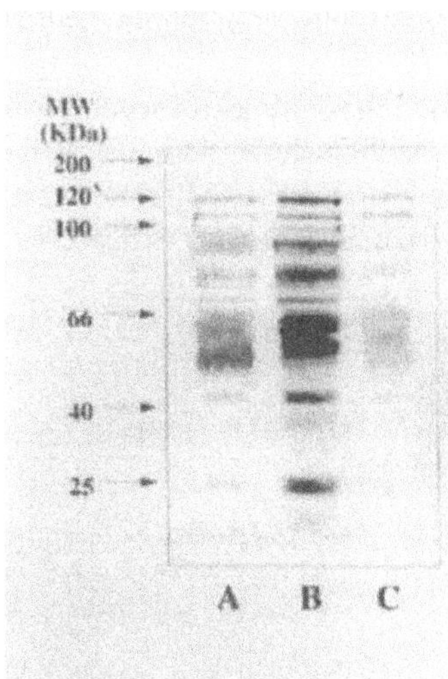

Figure 2. Effects of preconditioning on tyrosine phosphorylation and its inhibition by genistein. Isolated hearts were preconditioned (4 x PC) by repeated ischemia and reperfusion. Control hearts were perfused under identical conditions without preconditioning protocol. Another group of hearts were preperfused with genistein prior to preconditioning. Hearts were immediately frozen after the experiments. Phosphorylation of tyrosine kinase was examined by treating the samples with SDS-PAGE followed by Western blotting using a monoclonal antiphosphotyrosine antibody (1 μg/ml). Results are representative of four experiments per group. Each experiment was repeated at least three times with identical results. **A**, control; **B**, preconditioned; **C**, genistein followed by preconditioning.

S6 ribosomal protein kinase, and MAP kinase–activated protein kinase 2 (Novak-Hofer and Thomas, 1984; Anderson et al., 1990). MAPKAP kinase 2 has been implicated in a novel mammalian stress-activated signal transduction pathway initiated by a variety of mitogens, proinflammatory cytokines, or environmental stresses, where it regulates its substrate molecules by serine/threonine phosphorylation (Cobb and Goldsmith, 1995; Bogoyevitch et al., 1994). Stimulation of cultured cardiomyocytes with A_1 selective adrenergic analogs, endothelin 1, fibroblast growth factors, and mechanical stress activates MAP kinase signaling cascade (Yamazaki et al., 1996).

A. MAP Kinases Cascade

A growing body of evidence implicates that in response to an external stress, an intracellular kinase cascade becomes rapidly activated. This kinase cascade includes MAP kinases, MAPKK, and MAPKKK (Figure 3). Three distinct MAP kinases have been identified: ERK group, JNK/SAPK group, and p38 MAP kinase isoform, of which the latter can activate MAPKAP kinase 2 by phosphorylation (Davis, 1993).

B. Mechanisms of Signal Transduction

In the case of rat hearts, a MAP kinase cascade has already been identified (Lazou et al., 1994). These authors have demonstrated that MAPK isoforms p42 MAPK and p44 MAPK and two peaks of MEK were activated by more than 10-fold in perfused hearts or ventricular myocytes exposed to PMA for 5 minutes. In our

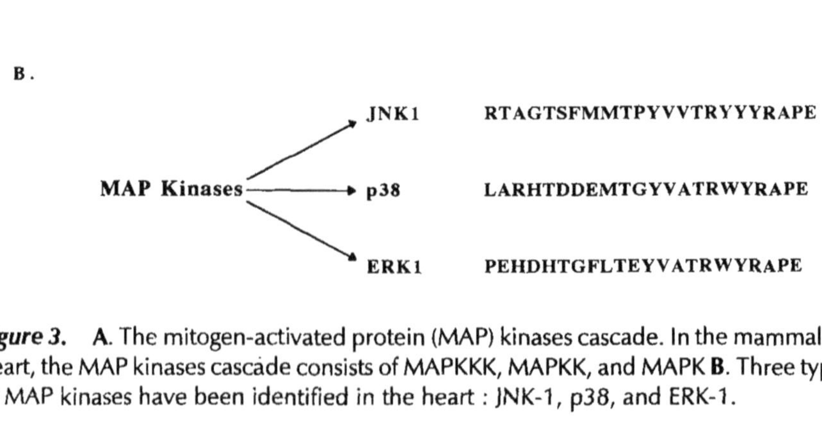

Figure 3. **A.** The mitogen-activated protein (MAP) kinases cascade. In the mammalian heart, the MAP kinases cascade consists of MAPKKK, MAPKK, and MAPK **B.** Three types of MAP kinases have been identified in the heart : JNK-1, p38, and ERK-1.

own study, we identified the participation of MAP kinase cascades in the ischemic preconditioning of rat hearts (Lazou et al., 1994). The results of our study demonstrated that a kinase cascade involving tyrosine kinase–phospholipase D–MAP kinases–MAPKAP kinase 2 is triggered after ischemic stress (see Figure 1).

In a recent study, MAP kinase activation was found to be essential during the bombesin-induced PKC-mediated sustained contraction in smooth muscle cells and the redistribution of MAP kinases were co-localized with the redistribution of HSP 27 in smooth muscle cells (Yamada et al., 1995). In view of the evidences that HSP 27 gene is induced after ischemic preconditioning (Das and Maulik, 1995), it seems likely that MAP kinases are involved in signal transduction leading to this gene expression. Indeed, in another related study, activation of cardiac gene expression during phenylephrine-induced hypertrophy seemed to require ERK activation (Thorburn et al., 1995).

Recently, a new member of the MAP kinase family, p38 MAP kinase, has been identified (Han et al., 1994). This MAP kinase seems to possess a dual phosphorylation motif Thr-Gly-Tyr in place of the Thr-Pro-Tyr motif present in Jnk and Thr-Glu-Tyr motif present in ERK (see Figure 3). A recent study from our laboratory has demonstrated that p38 MAP kinase is translocated and activated after ischemic preconditioning (Figure 4). Additionally, an inhibitor of p38 MAP kinase blocked the effects of ischemic preconditioning (Table 1).

Table 6.1. Effects of Genestein and SB 203580 on Myocardial Functions

		Heart rate (beats/min)	*Developed Pressure (mm Hg)*	*dp/dtmax (mm Hg/s)*	*Aortic Flow (ml/min)*	*Coronary Flow (ml/min)*	*Infarct Size (%)*
Control	Baseline	305±3.2	73±1.5	2983±74	43.5±0.9	24.1±0.6	—
	R30	312±7.9	67±6.5	2618±261	30.3±2.2†	23.2±1.5	
	R60	304±11.8	47.1±1.0†	2018±268	24.3±3.2†	23.2±1.4	
	R120	295±4.7	35.4±1.8†	1520±63†	14.4±0.7†	18.1±0.4†	35±3.2
Adapted	Baseline	294±4.9	74±3.1	3064±111	44.0±0.9	24.3±0.4	
	R30	299±6.9	77.1±2.0	2972±125	34.8±1.4	26.5±1.3	
	R60	299±7.5	74.6±2.5*	2803±99*	33.4±1.2*†	27.1±1.3	
	R120	285±9.0	45.4±1.4*†	2055±89*†	24.3±1.2*†	20.6±0.3†	17.5±1.7*
Genistein	Baseline	302±11.3	74±0.8	2875±134	43±0.8	23.4±0.5	—
	R30	299±8.2	62±2.0†**	2552±207	29.5±1.4†	22.5±1.2	
	R60	290±6.5	44±3.2†**	1912±77†**	25.2±1.5†**	20.3±0.9	
	R120	289±9.1	34.3±3.5**†	1650±120**†	15.3±1.6**†	18.8±0.5†	32.6±4**
SB 203580	Baseline	301±7.5	70±2.2	2912±69	42.8±1.2	25.0±1.1	—
	R30	297±5.2	65±3.1	2529±144	31.5±2.0†	23.6±0.8	
	R60	295±7.8	47±2.7†**	2066±97†**	21.3±2.1†**	20.0†**	
	R120	292±11.4	39.1±0.9**†	1720±72**†	17.3±1.5**†	18.6±0.9	29.8±2.8*

Notes: †<0.05 compared to base line
* $p < 0.05$ compared to control
** $p < 0.05$ compared to adapted
Results are shown as Means ± SEM of six rats per group.
R30, R60, and R120: hearts subjected to 30 min ischemia followed by 30, 60, and 120 min reperfusion.

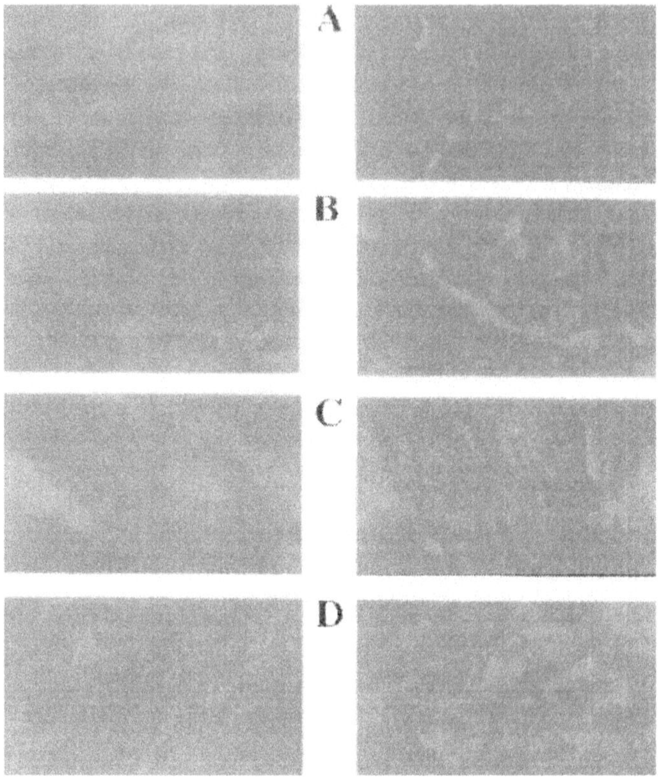

Figure 4. Translocation of p38 mitogen-activated protein (MAP) Kinases.

VI. INVOLVEMENT OF RAS AND RAF-1

The MAP kinases signal transduction pathway is likely to involve activation of Ras or Raf-1, which in turn induces MAPKK and MAP kinases. It is also known that Raf-1 kinase possesses MAPKKK activity and lies upstream from MAPKK and MAP kinases in various cell types (Force et al., 1994). A recent study has demonstrated that hypoxia and hypoxia/reoxygenation activated Raf-1, MAPKK, as well as MAP kinases in cultured rat cardiomyocytes (Seko et al., 1996). Raf-1 operates downstream from cell–surface associated tyrosine kinases and upstream from MAP kinases (Figure 5). Raf is not strictly a member of the MEKK family, but it is a function analog. Ras is part of the signal transduction chain extending from extracellular signals to transcriptional regulation in the nucleus. Upon activation, tyrosine kinase recruits a number of proteins including Ras-specific guanine nucleotide–releasing proteins, which then regulate the binding of Ras with guano-

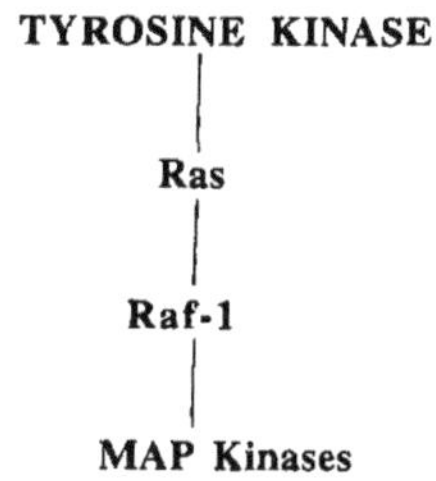

Figure 5. Position of Raf-1.

sine triphosphate (GTP), thereby potentiating the Ras signal. Ras proteins then interact with Raf kinases to induce downstream signals activating MAP kinases and other protein kinases. Once Raf is activated, Ras is no longer required.

The precise mechanism by which Ras controls Raf-1 is poorly understood. The binding of Raf-1 to Ras is largely GTP dependent and requires the effector region of Ras and the regulatory region of Raf-1.

A. Positive Regulation by Protein Kinase C

Activation of PKC results in an increase in Ras-GTP, suggesting that Ras can mediate PKC activation of Raf-1. Although phorbol esters can induce activation of Raf-1 and ERK-2, Ras is not essential for PKC-induced activation of Raf-1 and ERK-2 (deVries-Smiths et al., 1992). Raf-1 is a substrate for a variety of PKC isozymes, specifically for α, β, and γ PKC isozymes. Co-expression of PKCα and Raf-1 resulted in phosphorylation and enhanced autophosphorylation of Raf-1, but failed to affect the phosphorylation of Raf-1 and activate MEK (MacDonald, 1993), suggesting that direct phosphorylation of Raf-1 by PKC is not sufficient to activate MEK and ERK-2. Interestingly, Ras cannot activate Raf-1 by itself, instead, it requires PKC to fully activate Raf-1.

B. Role of Tyrosine Kinases in Raf-1 Activation

A recent report that tyrosine kinases can stimulate Raf-1 containing a mutation at the Ras interaction site raises the possibility that Raf-1 can be activated without Ras (Morrison, 1995). This author also demonstrated that Raf-1 can interact with members of the Src tyrosine kinase family and that such interaction is mediated in part by the SH2 domain of the tyrosine kinases (Cleghon and Morrison, 1994). The association between Raf-1 and the Src tyrosine kinase family was also identified by the same group when they found that Raf-1 interacted with the SH2 domain of Fyn (a member of the Src family), but not with the SH2 domains of phospholipase C_γ. Such interaction did not require Raf-1 tyrosine phosphorylation, but required serine phosphorylation of Raf-1. Taken together, these results suggest that an association with Fyn/Src family of tyrosine kinases may be responsible for the tyrosine phosphorylation of Raf-1.

VII. MAPKAP KINASE 2

MAPKAP kinase 2 is a downstream protein kinase in the stress-activated signal transduction pathway. Northern blot analysis indicated that this enzyme is highly expressed in heart tissues (Figure 6), suggesting that MAPKAP kinase 2 may function in myocardium to stress or mitogenic stimulation. In addition, MAPKAP kinase 2 activity of cardiomyocytes is stimulated by a myocardial hypertrophic factor (unpublished observation) and oxidative stress, as well as heat shock (Figure 7). Recombinant MAPKAP kinase 2 phosphorylates HSP 27 of cardiomyocytes dur-

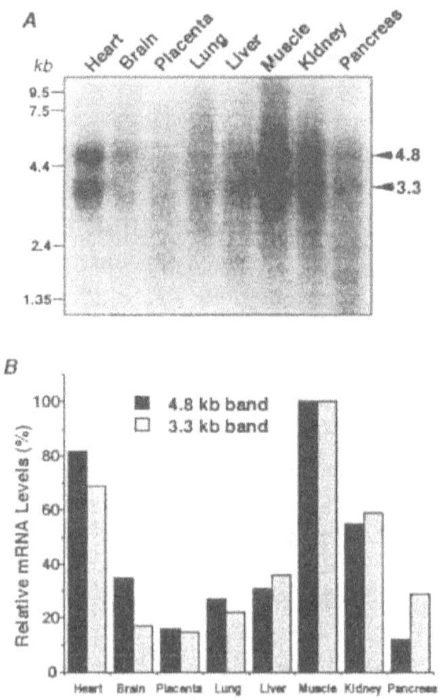

Figure 6. Amount of 3.3 kb and 4.8 kb mRNAs of MAPKAP kinase 2 in human tissues (relative to skeletal muscle). Human multiple tissue Northern blot (hMTN obtained from ClonTech Laboratories Inc,. NY) containing mRNAs from various organs were hybridized with a synthetic oligonucleotide probe corresponding to amino acid residues 344 to 353 in the C-terminal region of human MAPKAP kinase 2. Northern blot analysis revealed two mRNA species of MAPKAP kinase 2 with sizes of 4.8 and 3.3 kb. The highest amount was detected in the skeletal muscle with the second highest in the heart tissue. results are expressed as activities relative to skeletal muscle, which was set as 100%.

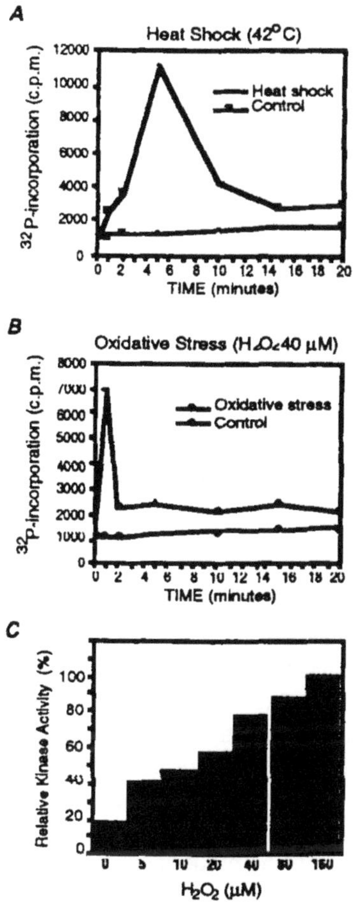

Figure 7. Ischemic preconditioning was achieved by inducing 5 minutes of ischemia followed by 10 minutes of reperfusion, repeated four times (4 x PC). Heat shock was produced by perfusing the hearts with krebs–hensseleit bicarbonate (KHB) buffer at 42.5°C for 10 minutes. Oxidative stress was induced by perfusing the heart with an OH-generating system for 10 minutes. $*p < .05$ compared to baseline.

ing in vitro phosphorylation assay. The known association between MAPKAP kinase 2 and heat shock proteins gives rise to the possibility that the kinase may be involved in a stress-related response of myocardium, which leads to cardioprotection. Our previous studies indicate that environmental stresses, including heat shock and oxidative stress, and phorbol ester treatment of cultured cardiac myoblast cells resulted in a rapid increase in cellular MAPKAP kinase 2 activity (Maulik et al., 1996; Das et al., 1996).

A. Regulation by Stress Signal

HSP 27 is an early target of phosphorylation upon stimulation by serum or a variety of mitogens, cytokines, inducers of differentiation, as well as by a variety of stress conditions (Ciocca et al., 1993; Stokoe et al., 1992). As mentioned earlier, a previous study from our laboratory demonstrated the induction for the expression of HSP 27 in rat hearts after ischemic preconditioning (Das et al., 1993). The induced phosphorylation regulates HSP 27 functions that may provide cardioprotection during myocardial ischemia and reperfusion (Das et al., 1994). It has been demonstrated that HSP 27 phosphorylation is regulated by the activation of MAP kinase–activated protein MAPKAP kinase2, a Ser/Thr protein kinase (Stokoe et al., 1992). MAPKAP kinase 2 was originally isolated as a substrate for MAP kinase (Stokoe et al., 1992). More recently, MAPKAP kinase 2 has been implicated in a novel mammalian stress-activated signal transduction pathway, which is activated by mitogens, cytokines, and a wide range of physical stresses, and in turn appears to modulate the function of HSP 25/27 by phosphorylation (Galchva-Gorgoba et al., 1994).

B. Abundance of MAPKAP Kinase 2 in Heart

In a recent study, we detected enzymatic activity during in vitro kinase assay by using MBP as a substrate resulted from the activation of the tissue ERK and/or p38 MAP kinases. In vitro studies both ERK and p38 MAP kinases can phosphorylate and activate MAPKAP kinase 2 (Zu et al., 1995; Stokoe et al., 1992). To detect tissue/cellular MAPKAP kinase 2 activity, the synthetic peptide derived from the *N*-terminus of glycogen synthase (Stokoe et al., 1992) is widely used as a specific substrate for in vitro kinase assays. In the present study, 20 μM H-7 was utilized in the kinase assay to inhibit enzyme activities mediated through other cellular kinases, including cyclic adenosine monophospate (cAMP)–dependent protein kinase (Ki = 3.0 μM), protein kinase C (Ki = 6.0 μM), and protein kinase G (Ki = 5.8 μM). The specificity of the kinase assay using whole tissue lysate and the peptide substrate has already been demonstrated from the observations that induced MAPKAP kinase 2 activity detected using in vitro kinase assay with whole tissue lysates and synthetic peptide as substrate, was inhibited by the presence of the competitive inhibitory peptide for MAPKAP kinase 2 (Zu et al., 1996). The MAP kinases activity, thus, truely reflects the total MAP kinases activity irrespective of whether they are derived from ERK, JNK, or p38 MAP kinases. It should be noted that a recent study demonstrated that p38 MAP kinases, and not ERK MAP kinase, leads to the activation of MAPKAP kinase 2 in vivo (Rouse et al., 1994).

C. Mechanisms of Signal Transduction

The precise physiological role of MAPKAP kinase 2 is not known, but this kinase has been implicated as a downstream molecule of the stress-activated protein kinase cascade. It has been shown that except for MAPKAP kinase 2, no other kinases, including PKC, are capable of inducing phosphorylation of HSP 27 directly. Induction of the expression of HSP 27 in response to diverse stresses has been demonstrated. For example, both ischemia–reperfusion and oxidative stress can induce the expression of HSP 27 in mammalian hearts (Das and Maulik, 1995) (Figure 8). The fact that phosphorylation of HSP 27 precedes its activation (Landry et al.,

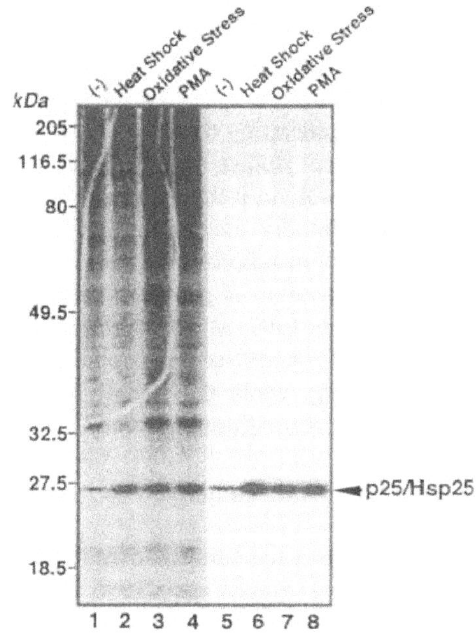

Figure 8. The induced hsp25 phosphorylation intact H9c2 cells. To study the myocardial protein phosphorylation in response to extracellular stimulation, H9c2 cells were prelabeled by $^{32}PO_4$, and treated with heat shock at 42°C, oxidative stress in the presence of 40 μM H_2O_2, 40 nM PMA, or nothing (—) at 37°C for 10 min. Treated cells were then lysed and cellular proteins were separated on 11% SDS-PAGE. The induced-protein phosphorylation in intact H9c2 cells was analyzed by autoradiography. The treatments of cells are indicated at the top of lanes 1–4. Molecular weights are shown on the left. The position of a myocardial 25-kDa phosphorprotein (p25), whose phosphorylation was markedly induced by stresses and PMA stimulation, is indicated by an arrow on the right. To examine the protein phosphorylation of cellular hsp25 in intact H9c2 cells, the cell lysates shown in lanes 1–4 were immunoprecipitated with commercial antibodies against rat hsp25. The immunoprecipitated proteins were analysed on an SDS-PAGE and the hsp25 phosphorylation was detected by autoradiography (lanes 5–8). The treatments are shown on the top of each lane. The position of the myocardial hsp25 is indicated by an arrow on the right. These results are representative of three similar experiments.

1992) and that HSP 27 is induced in response to stress, including ischemic preconditioning, and implicated in myocardial preservation (Das et al., 1993), suggests a role of MAPKAP kinase 2 in ischemic preconditioning and myocardial adaptation.

VIII. PHOSPHOLIPASE D

Phospholipid turnover plays a crucial role in the intracellular signaling process. At least three phospholipases are known to be regulated by the receptor protein tyrosine kinases: phospholipase A_2, phospholipase C_γ, and phospholipase D. The primary step of the signal transduction pathway for the activation of PKC involves the stimulation of phospholipase C generating the second messenger, diacylglycerol (Prasad et al., 1991). Several recent studies have demonstrated that activation of phospholipase D plays a crucial role in ischemic preconditioning (Cohen et al., 1996). Phospholipase D preferentially attacks phosphatidylcholine generating phosphatidic acid, which is readily metabolized by a phosphohydrolase present in the heart into diacylglycerol. Activation of phospholipase D was documented in ischemic–reperfused (Moraru et al., 1992) as well as in preconditioned hearts (Cohen et al., 1996; Trifan et al., 1996). Agonists of phospholipase D simulated the effects of ischemic preconditioning, while the inhibition of this phospholipase blocked the beneficial effects of preconditioning.

A. Phospholipase D in Ischemic Preconditioning

A recent study from our laboratory has confirmed the involvement of phospholipase D in ischemic preconditioning (Maulik et al., 1996; Das et al., 1996; Cohen et al., 1996; Trifan et al., 1996). Using specific polyclonal antibodies to phospholipase D, we found that these antiphospholipase D antibodies caused direct inhibition of phospholipase D, simultaneously reducing the amount of diacylglycerol and phosphatidic acid as well as significantly inhibiting the stimulation of PKC. In concert, these antiphospholipase D antibodies blocked the beneficial effects of ischemic preconditioning as evidenced by the increased incidence of ventricular arrhythmias. In a previous study, we found that the same antiphospholipase D antibodies blocked ischemia–reperfusion-mediated activation of phospholipase D (Moraru et al., 1993). In concert, preconditioning generated diacylglycerol and phosphatidic acid and led to the translocation and activation of PKC. Additionally, preconditioning reduced the incidence of ventricular arrhythmias, which was reversed by phospholipase D inhibition.

Phospholipase D catalyzes the hydrolysis of the terminal diester bond of phosphatidylcholine with the formation of choline and phosphatidic acid (Kanfer, 1980); the latter serves as a substrate for diacylglycerol biosynthesis by the action of phosphatidic acid phosphohydrolase. Diacylglycerol may serve as a second messenger leading to the activation of PKC. Phospholipase D also catalyzes a transphosphatidylation reaction in which the phosphatidyl moiety of the phospholipid is transferred to a nucleophilic alcohol–producing corresponding phosphatidyl alcohol (Billah, 1993). Ischemia–reperfusion was shown to activate phospholipase D, generating intracellular phosphatidic acid, part of which converted to diacylglycerol (Moraru et al., 1992). Additionally, activation of myocardial phospholipase D by sodium oleate resulted in a significant improvement of postischemic functional recovery and attenuation of cellular injury, suggesting that phospholipase D signaling in the ischemic myocardium is beneficial for the recovery of the heart.

In addition to phospholipase D, other phospholipases—including phospholipase A_2 and phospholipase C—also become activated in response to ischemia and reperfusion (Prasad et al., 1991). However, activation of phospholipase A_2 causes the generation of arachidonic acid and lysophosphoglycerides, which are detrimental to the heart (Das et al., 1986). Activation of phospholipase C also generates diacylglycerol, but it causes the production of inositol triphosphate, which mobilizes Ca^{2+} from the intracellular compartments (Moraru et al., 1995). Thus, unlike the detrimental effects of phospholipase A_2 and phospholipase C, activation of phospholipase D is beneficial to the heart. Activation of phospholipase D produces phosphatidic acid, while activation of phospholipase C produces diacylglycerol. However, these two products of hydrolysis are easily convertable; catalyzed by two enzymes: diacylglycerol kinase and phosphatidic acid phosphohydrolase. Inhibition of diacylglycerol kinase does not affect the amount of ischemia–reperfusion-induced generation of diacylglycerol or phosphatidic acid. To the contrary, phosphatidic acid phosphohydrolase inhibition leads to an enhancement of phosphatidic acid in concert with a reduction in diacylglycerol (Prasad et al., 1991), suggesting that phospholipase D activity is solely responsible for the generation of phosphatidic acid. Additionally, phospholipase D indirectly contributes to diacylglycerol generation since part of the phosphatidic acid is hydrolyzed to diacylglycerol by phosphatidic acid phosphohydrolase. It is interesting to note that phospholipase D–mediated stimulus-response couplings are known to exist in several cell types (Tosaki et al., 1996).

B. Regulation of Phospholipase D by Tyrosine Kinase

Signal transduction mediated by growth factors such as EGF, PDGF, and insulin occurs through protein tyrosine kinase found on the plasma membrane (Fantl et al., 1993). Earliest events for such signal transduction include activation of phospholipase C_γ and phospholipase D (Meisenhelder et al, 1989; Ha et al., 1993). Several agonists can also enhance phospholipase D activation through protein tyrosine phosphorylation. For example, in human embryonic kidney cells, expression of the m_3 muscarinic acetylcholine receptor increased tyrosine phosphorylation of various cellular proteins and phospholipase D in response to carbachol (Schmidt et al., 1994). The involvement of protein tyrosine phosphorylation in phospholipase D activation was supported from the observations that the tyrosine kinase inhibitors such as genistein and tyrphostin blocked the phospholipase D activation (Rivard et al, 1994) and that phospholipase D activation was increased upon treatment with the tyrosine phosphatase inhibitors such as vanadate or pervanadate (Bourgoin et al., 1992). Many oxidants such as H_2O_2 can increase protein tyrosine phosphorylation (Fialkow et al, 1993) in conjunction with phospholipase D activation (Natarajan et al., 1993). In neutrophils and HL-60 cells, oxygen free radicals generated by f-Met-Leu-Phe resulted in increased protein tyrosine phosphorylation and phospholipase D activation (Uings et al., 1992). Genistein not only blocked the oxidant-mediated protein tyrosine phosphorylation in the endothelial cells, but also a correlation between tyrosine kinase inhibition and phospholipase D activation was observed with genistein (Natarajan et al., in press).

Although evidence supports a role of tyrosine kinases in phospholipase D activation, the mechanisms of activation remain unknown. It is possible that phospholipase D activation is the direct manifestation of protein tyrosine kinase phosphorylation. It is also possible that tyrosine kinase generates other intermediate proteins that are instrumental for phospolipase D activation. A number of recent studies identified MAP kinases as targets for tyrosine phosphorylation, as described earlier.

In a recent study, ischemic preconditioning–mediated activation of phospholipase D was found to be inhibited by a tyrosine kinase blocker, genistein (Maulik et al., 1996). In concert, the preconditioning effect was almost abolished by the genistein treatment. Additionally, preconditioning of the rat hearts stimulated multiple protein kinases including PKC, MAP kinase, and MAPKAP kinase 2, which were inhibited by genistein (see Figure 1), suggesting the existence of a tyrosine kinase–coupled phospholipase D pathway for ischemic preconditioning and implicating the involvement of multiple protein kinases in myocardial adaptation to ischemia (Figure 9).

IX. SIGNAL AMPLIFICATION AND MODULATION

It should be clear from the preceding discussion that an external stress such as ischemia–reperfusion is likely to be transmitted by a sequential stimulation of the MAP kinases signaling cascade. At least three kinases, and perhaps up to six kinases, can contribute to the amplification and modulation of the transmitted signals (see Figure 3). Much evidence exists to support the notion that the MAP kinases cascade involves the sequential activation of tyrosine kinases-Ras-Raf-MAPKKK-MAPKK-MAPK. Different signaling relay mechanisms have been identified within this cascade, including Raf-1 (MAPKKK), MEKs-1 and -2 (MAPKKs), and ERK-1 and -2 (MAPKs). The involvement of multiple kinases suggests the intriguing possibility that they are involved in the signal amplification process in a relay-type mechanism. For example, if each kinase can phosphorylate

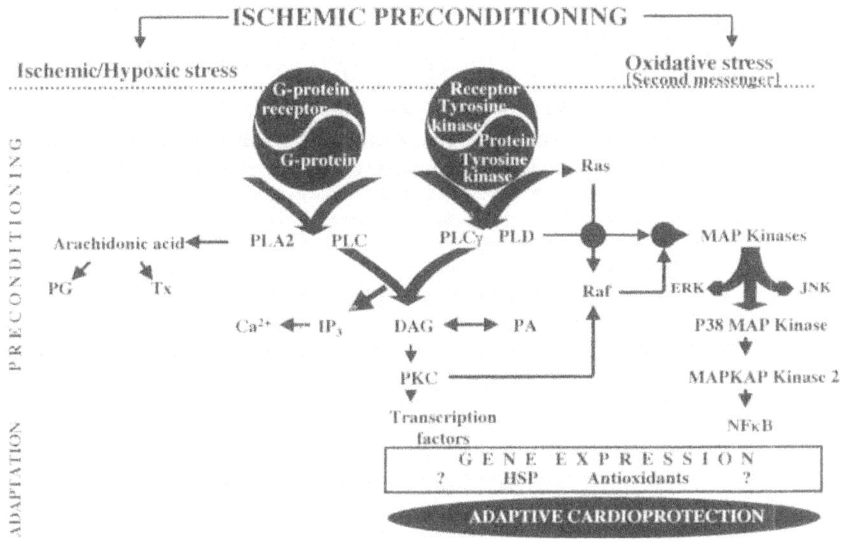

Figure 9. A model supported by our results. The figure indicates a role of *Tyrosine Kinase–Phospholipase D–p38 MAP Kinase–MAPKAP Kinase* 2 signaling pathway in *preconditioned* leading to the translocation and activation of NFκB (FEBS Lett. 396, 233-237, 1996; FEBS Lett. 429, 365-369, 1998). A role of *free radicals/oxidative stress* as *second messenger* and a *PKC-independent pathway* for signal transduction are also suggested (Mol. Cell. Biochem. Focused issue : Stress Adaptation, Prophylaxis, and Treatment, 1998).

100 target proteins, then the final amplication factor in the *three-kinase-relay system* will contribute to at least a 10^6-fold amplification. Recent studies from our laboratory also suggest a role of MAPKAP kinase 2 in the signaling process (Maulik et al., 1996; Das et al., 1996). However, the precise role of MAPKAP kinase 2 in signal amplification process remains unclear.

It is possible that such a high degree of amplification is not necessary for the signal transduction process. It is tempting to speculate that one or more of the kinases in the relay system is (are) responsible for the monitoring of the internal status of the cell and, accordingly, modulate the amplification process according to the need.

X. SUMMARY AND CONCLUSION

The heart possesses a remarkable ability to adapt itself against any stressful situation by increasing resistance to the adverse consequences. Stress induced by single or multiple brief periods of ischemia and reperfusion render the heart more tolerant to the subsequent lethal ischemic insult. This phenomenon has been termed ischemic preconditioning (a short-term phenomenon likely to be mediated by signal transduction) and adaptation to ischemia (a long-term process mediated by gene expression and their transcription regulation) (Das, 1993; Das and Maulik, 1997). The studies from this laboratory suggests that the initial ischemic-stress signal is triggered by the activation of protein tyrosine kinases, which then transmit the signals by activating phospholipase D through the MAP kinases and MAPKAP kinase 2 in a relay-type fashion.

ACKNOWLEDGMENTS

This study was supported by NIH grants HL-22559, HL-34360, and HL-33889, and by a Grant-in-Aid from the American Heart Association.

REFERENCES

Anderson, N.G., Maller, J.L., Tonks, N.K., and Sturgill T.W. (1990). Requirement for integration of signals from two distinct phosphorylation pathways for activation of Mapkinase. Nature (Lond) 343, 651-653.

Asimakis, G.K., Inners-McBride K., Medellin, G., and Conti, V.R. (1992). Ischemic preconditioning attenuates acidosis and postischemic dysfunction in isolated rat heart. Am. J. Physiol. 263, H887-H894.

Banerjee, A., Locke-Winter, C., Rogers, K.B., Mitchess, M.B., Brew, E.C., Cairns, C.B., Bensard, D.D., and Hartken A.H. (1993). Preconditioning against myocardial dysfunction after ischemia and reperfusion by an α_1-adrenergic mechanism. Circ. Res. 73, 656-670.

Billah, M.M. (1993). Phospholipase D and cell signaling. Curr. Opin. Immunol. 5, 114-123.

Blumer, K.J., and Johnson, G.L. (1994). Diversity in function and regulation of MAP kinase pathway. Trends Biochem. Sci. 19, 236-240.

Bogoyevitch, M.A., Glennon, P.E., Andersson, M.B., Clerk, A., Lazou, A., Marshall, C.J., Packer, P.J., and Sugden, P.H. (1994). Endothelin-1 and fibroblast growth factors stimulate the mitogen-activated protein kinase signaling cascade in cardiac myocytes. J. Biol Chem 269: 1110-1119.

Bourgoin, S., and Grinstein, S. (1992). Peroxides of vanadate induce activation of phospholipase D in HL-60 cells. Role of tyrosine phosphorylation. J. Biol. Chem. 267, 11908-11916.

Bugge, E., and Ytrehus, K (1995). Ischemic preconditioning is protein kinase C dependent but not through stimulation of α adrenergic or adenosine receptors in the isolated rat heart. Cardiovasc. Res. 29, 401-406.

Carpenter, C.L., Auger, K.R., Chanudhuri, M., Yoakim, M., Schaffhausen, B., Shoelson, S., and Cantley, L.C. (1993). Phosphoinositide 3-kinase is activated by phosphopeptides that bind to the SH2 domains of the 85-kDa subunit. J. Biol Chem. 268, 9478-9483.

Cave, A.C., Collis, C.S., Downey, J.M., and Hearse D.J. (1993). Improved functional recovery by ischemic preconditioning is not mediated by adenosine in the globally ischemic isolated rat heart. Cardiovasc. Res. 27, 663-668

Cleghon, V., and Morrison, D.K. (1994). Raf-1 interacts with Fyn and Src in a non-phosphotyrosine–dependent manner. J. Biol. Chem. 269, 17749-17755.

Ciocca, D.R., Oesterreich, S., Chamness, G.C., McGuire, W.L., and Fuqua, S.A.W. (1993). Biological and clinical implications of heat shock protein 27000 (HSP 27): a review. J. Nat. Cancer Inst. 85, 1558-1570.

Cobb, M.H., and Goldsmith, E.J. (1995). How MAP kinases are regulated? J. Biol. Chem. 270, 14843-14846.

Cohen, M.V., Liu,Y., Liu, G.S., Wang, P., Cordis, G.A., Das, D.K., and Downey J.M. (1996). Phospholipase D plays a major role in ischemic preconditioning in rabbit heart. Circulation 94, 1713-1718.

Das, D.K (1993). Ischemic preconditioning and myocardial adaptation to ischemia. Cardiovasc. Res. 27, 2077-2079.

Das, D.K., Engelman, R.M., and Kimura, Y. (1993). Molecular adaptation of cellular defenses following preconditioning of the heart by repeated ischemia. Cardiovasc. Res. 27, 578-584.

Das, D.K., Engelman, R.M., Rousou, J.A., Breyer, R.H., Otani, H., and Lemeshow, S. (1986). Role of membrane phospholipids in myocardial injury induced by ischemia and reperfusion. Am. J. Physiol. 251, H71-H79.

Das, D.K., and Maulik, N. (1995). Cross talk between heat shock and oxidative stress inducible genes during myocardial adaptation to ischemia. In: Cell Biology of Trauma (Lemasters, J.J., and Oliver, C., Eds.), pp. 193-211. CRC Press, Boca Raton, FL.

Das, D.K., Maulik, N., and Moraru, I.I. (1995). Gene expression in acute myocardial stress. J. Mol. Cell. Cardiol. 27, 181-193.

Das, D.K., and Maulik, N. (1997). Reprogramming of gene expression during myocardial adaptation to stress. In: Adaptation Biology and Medicine (Sharma, B.K., Takeda, N., Ganguly, P.K., and Singal, P.K. Eds.), pp. ___-___. Narosa Publishing House.

Das, D.K., Maulik, N., Yoshida, T., Engelman, R.M., and Zu, Y-L. (1996). Preconditioning potentiates molecular signaling for myocardial adaptation to ischemia. Ann. N.Y. Acad. Sci. 793, 191-209.

Das, D.K., Moraru, I.I., Maulik, N., and Engelman, R.M. (1994). Gene expression during myocardial adaptation to ischemia and reperfusion. Ann. N.Y. Acad. Sci. 723, 292-307.

Davis, R.J. (1993). The mitogen-activated protein kinase signal transduction pathway. J. Biol. Chem. 268, 14553-14556.

de Vries-Smits, A.M.M.M., Burgering, B.M.T., Leevers, S.J., Marshall, C.J., and Bos, J.L. (1992). Involvement of $P21^{ras}$ in activation of extracellular signal-regulated kinase 2. Nature 357, 602-604.

Dixon, B.S., Sharma, R.V., Dickerson, T., and Fortune, J. (1994). Bradykinin and angiotensin II: activation of protein kinase C in arterial muscle. Am J. Physiol. 266, C1406-C1420.

Engelman, D., Watanabe, M., Engelman , R.M., Rousou, J.A., Kisin, E., Kagan, V.E., Maulik, N., and Das, D.K. (1995). Hypoxic preconditioning preserves antioxidant reserve in the working rat heart. Cardiovasc. Res. 29, 133-140.

Fantl, W.J., Johnson, D.E., and Williams, L.T. (1993). Signalling by receptor tyrosine kinases. Ann. Rev. Biochem. 62, 453-481.

Fearon, C.W., and Tashjian, A.H. (1985). Thyrotropin-releasing hormone induces redistribution of protein kinase C in GH4C1 rat pituitary cells. J. Biol. Chem. 260, 8366-8371.

Fialkow, L., Chan, C.K., Chan, S., Grinstein, S., and Downey, G.P. (1993). Regulation of tyrosine phosphorylation in neutrophils by the NADPH oxidase. Role of reactive oxygen intermediates. J. Biol. Chem. 268, 17131-17137.

Flack, J., Kimura, Y., Engelman, R.M., and Das, D.K. (1991). Preconditioning the heart by repeated stunning improves myocardial salvage. Circulation 84, III369-374.

Force, T., Bonventrem J.V., Heidecker, G., Rapp, U., Avruch, J., and Kyriakis, L.M. (1994). Enzymatic characteristics of the Raf-1 protein kinase. Proc. Nat. Acad. Sci. U.S.A., 91, 1270-1274.

Galchva-Gorgova, Z., Deriard, B., Wu, I-H., and Davis, R.J. (1994). An osmosensing signal transduction pathway in mammalian cells. Science 265, 806-808.

Goto, M., Liu, Y., Yang, X-M., Ardell, J.L., Cohen, M.V., and Downey, J.M. (1995). Role of bradykinin in protection of ischemic preconditioning in rabbit hearts. Circ. Res. 77, 611-621

Gross, G.J., and Auchampach, J.A.. (1992). Blockade of ATP-sensitive potassium channels prevents myocardial preconditioning in dogs. Circ. Res. 73, 656-670.

Ha, K.S., and Exton, J.H. (1993). Differential translocation of protein kinase C isozymes by thrombin and platelet derived growth factor. A possible function for phosphatidylcholine-derived diacylglycerol. J. Biol. Chem. 268, 10534-10539.

Hagar, J.M., Hale S.L., and Kloner, R.A. (1991). Effect of preconditioning of ischemia on reperfusion arrhythmias after coronary artery occlusion and reperfusion in the rat. Circ. Res. 68, 61-68.

Han, J., Lee, J-D., Bibbs, L., and Ulevitch, R.J. (1994). A new MAP kinase targeted by endotoxin and hyperosmolarity in mammalian cells. Science 265, 808-811.

Henrich, C.J., and Simpson, P.C. (1988). Differential acute and chronic response of protein kinase C in cultured neonatal rat heart myocytes to α1-adrenergic and phosbol ester stimulation. J. Mol. Cell. Cardiol. 20, 1081-1085.

Hunter, T., Alexander, C.B., and Cooper, J.A. (1985): Protein tyrosine kinases. Ann. Rev. Biochem. 54, 897-930.

Kanfer, J.N. (1980). The base exchange enzymes and phospholipase D of mammalian tissue. Can. J. Biochem. 58, 1370-1380.

Kida, M., Fujiwara, H., Ishida, M., Kawai, C., Ohura, M., Miura, I., and Yabuuchi, Y. (1991). Ischemic preconditioning preserved creatine phosphate and intracellular pH. Circulation 84, 2495-2503.

Kimura, Y., Iyengar, J., Subramanian, R., Cordis, G.A., and Das, D.K. (1992). Myocardial adaptation by repeated short term ischemia reduces post-ischemic dysfunction. Basic Res. Cardiol. 87, 128-138.

Landry, J., Lambert, H., Zhou, M., Lavoie, J.N,, Hickey, E., Weber, L.A., and Anderson, C.W. (1992). Human HSP 27 is phosphorylated at serines 78 and 82 by heat shock and mitogen-activated kinases that recognize the same amino acid motif as S6 kinase II. J. Biol. Chem. 267, 794-803.

Lawson, C.S., Coltart, D.J., and Hearse, D.J. (1993). The antiarrhythmic action of ischemic preconditioning in rat hearts does not involve functional Gi proteins. Cardiovasc. Res. 27, 681-87.

Lazou, A., Bogoyevitch, M.A., Clerk, A., Fuller, S.J., Marshall, C.J., and Sugden, P.H. (1994). Regulation of mitogen-activated protein kinase cascade in adult rat heart preparations in vitro. Circ. Res. 75, 932-941.

Li, G.C., Vasquez, B.S., Gallagher, K.P., and Lucchesi, B.R. (1990). Myocardial protection with preconditioning. Circulation 82, 609-19.

Liu, G.S., Thornton, J., Vanwinkle, D.M., Stanley, A.W.H., Olsson, R.A., and Downey J.M. (1991). Protection against infarction afforded by preconditioning is mediated by A1 adenosine receptors in rabbit heart. Circulation 84, 350-356.

Liu, Y., and Downey, J.M. (1992). Ischemic preconditioning protects against infarction in rat heart. Am. J. Physiol. 253, H1107-H1112.

Liu, Y., Tsuchida, A., Cohen, M.V., and Downey, J.M. (1995). Pretreatment with angiotensin II activates protein kinase C and limits myocardial infarction in isolated rabbit hearts. J. Mol. Cell. Cardiol. 27, 883-892.

Macdonald, S.G., Mcrews, C., Wu, L., Driller, J., Clark, R., Erikson, R.L., and McCormick, F. (1993). Reconstruction of the Raf-1-MEK-ERK signal transduction pathway *in vitro*. Mol. Cell. Biol. 13, 6615-6620.

Maulik, N., and Das, D.K. (1996). Hunting for differentially expressed mRNA species in preconditioned myocardium. Ann. N.Y. Acad. Sci. 793, 240-258.

Maulik, N., Watanabe, M., Zu, Y-L., Huang, C-K., Cordis, G.A., Schley, J.A., and Das, D.K. (1996). Ischemic preconditioning triggers the activation of MAP kinases and MAPKAP kinase 2 in rat hearts. FEBS Lett. 396, 233-237.

Maulik, N., Wei, Z-J., Engelman, R.M., Lu, D., Moraru, I.I., Rousou, J.A., and Das, D.K. (1993). Interleukin-1α preconditioning reduces myocardial ischemic reperfusion injury. Circulation 88, II387-394.

Meisenhelder, J., Suh, P.G., Rhee, S.G., and Hunter, T. (1989). Phospholipase C gamma is substrate for the PDGF and EGF receptor protein tyrosine kinases in vivo and in vitro. Cell 57, 1109-1122.

Moraru, I.I., Jones, R.M., Popescu, L., Engelman, R., and Das, D.K. (1995). Prazosin reduces myocardial ischemia/reperfusion-induced Ca^{2+} overloading in rat heart by inhibiting phosphoinositide signaling. Biochim. Biophys. Acta 1268, 1-8.

Moraru, I.I., Popescu, L.M., Liu, X., Engelman, R.M., and Das, D.K. (1993). Role of phospholipase A_2, C, and D activities during myocardial ischemia and reperfusion. Ann. N.Y. Acad. Sci. 723, 328-332.

Moraru, I.I., Popescu, L., Maulik, N., Liu, X., and Das, D.K. (1992). Phospholipase D signaling in ischemic heart. Biochim. Biophys. Acta 1139, 148-154.

Morrison, D.K. (1995). Mechanisms regulating Raf-1 activity in signal transduction pathways. Mol. Reproduc. Develop. 42, 507-514.

Murry, C.E., Jennings R.B., and Reimer, K.A. (1986). Preconditioning with ischemia: a delay of lethal cell injury in ischemic myocardium. Circulation 74, 1124-1136.

Murry, C.E., Richard, V.J., Reimer, K.A., and Jennings, R.B. (1990). Ischemic preconditioning slows energy metabolism and delays ultrastructural damage during a sustained ischemic episode. Circ. Res. 66, 913-931.

Natarajan, V., Scribner, W.M. and Taher, M.M. (1993). 4-hydroxynonenal, a metabolite of lipid peroxidation,activates phospholipase D in vascular endothelial cells. Free Rad. Biol. Med. 15, 365-375.

Natarajan, V., Vepa, S., Verma, R.S., and Scribner, W.M. (in press). Inhibitors of tyrosine kinases and protein tyrosine phosphatases modulate hydrogen peroxide-induced activation of endothelial cell phospholipase D. Am. J. Physiol.

Novak-Hofer, I., and Thomas, G. (1984). An activated S6 kinase in extracts from serum- and epidermal growth factor–stimulated Swiss 3T3 cells. J. Biol. Chem. 259, 5995-6000.

Prasad, M.R., and Jones, R.M. (1992). Enhanced membrane protein kinase C activity in myocardial ischemia. Basic Res. Cardiol. 87, 19-26.

Prasad, M.R., Popescu, L., Moraru, I.I., Liu, X., Engelman, R.M., and Das, D.K. (1991). Role of phospholipase A_2 and phospholipase C in myocardial ischemic reperfusion injury. Am. J. Physiol. 260, H877-H883.

Quilliam, L.A., Huff, S.Y., Rabun, K.M., Wei, W., Park, W., Broek, D., and Der, C.J. (1994). Membrane-targeting potentiates guanine nucleotide exchange factor CDC25 and SOS1 activation of Ras transforming activity. Proc. Nat. Acad. Sci. U.S.A. 91, 8512-8516.

Rivard, N., Rydzewska, G., Lods, J.S., Martinex, L.J., and Morisset, J. (1994). Pancreas growth, tyrosine kinase, PtdIns 3-kinase, and PLD involve high-affinity CCK-receptor occupation. Am J. Physiol. 266: G62-G70.

Rouse, J., Cohen, P., Trigon, S., Morange, M., Alonso-Llamazares, A., Zamanillo, D., Hunt, T., and Nebreda, A.R. (1994). A novel kinase cascade triggered by stress and heat shock that stimulates MAPKAP kinase 2 and phosphorylation of the small heat shock proteins. Cell 78, 1027-1037.

Sadowski, H.B., Shuai,K., Darnell, J.E., and Gilman, M.Z. (1993). A common nuclear signal transduction pathway activated by growth factor and cytokine receptors. Science 261, 1739-1744.

Sandhu, R., Diaz, R.J., and Wilson, G.J. (1993). Comparison of ischemic preconditioning in blood perfused and buffer perfused isolated heart models. Cardiovasc. Res. 27, 602-607.

Schmidt, M., Huwe, S.M., Fasselt, B., Homann, D., Rumenapp, U., Sandmann, J., and Jakobs, K.H. (1994). Mechanisms of phospholipase D stimulation by m3 muscarinic acetylcholine receptors. Evidence for involvement of tyrosine phosphorylation. Eur. J. Biochem. 225, 667-675.

Schott, R.J., Rohmann, S., Braun, E.R., and Schaper, W. (1990). Ischemic preconditioning reduces infarct size in swine myocardium. Circ. Res. 66, 1133-1142.

Seger, R., and Krebs, E.G. (1995). The MAPK signaling cascade. FASEB J. 9, 726-735.

Seko, Y., Tobe, K., Ueki, K., Kadowaki, T., and Yazaki, Y. (1996). Hypoxia and hypoxia/reoxygenation activate raf-1, mitogen-activated protein kinase kinase, mitogen-activated protein kinases, and S6 kinase in cultured rat cardiac myocytes. Circ. Res. 78, 82-90.

Steenbergen, C., Perlman, M.E., London, R.E., and Murphy, E. (1993). Mechanism of preconditioning. ionic alterations. Circ. Res. 72, 112-125.

Stokoe, D., Campbell, D.G., Nakielny, S., Hidaka, H., Leevers, S.J., Marshall, C., and Cohen, P. (1992a). MAPKAP kinase 2: a novel protein kinase activated by mitogen-activated protein kinase. EMBO J. 11, 3983-3994.

Stokoe, D., Engel, K., Campbell, D.G., Cohen, P., and Gaeste, M. (1992b). Identification of MAPKAP kinase 2 as a major enzyme responsible for the phosphorylation of the small mammalian heat shock proteins. FEBS 313, 307-313.

Thorburn, J., Carlson, M., Mansour, S.J., Chien, K.R., Ahn, N.G., and Thorburn, A. (1995). Inhibition of a signaling pathway in cardiac muscle cells by active mitogen-activated protein kinase kinase. Mol. Biol. Cell. 6, 1479-1490.

Tosaki, A., Behjet, N.S., Engelman, D.T., Engelman, R.M., and Das, D.K. (1995). Alpha-1 adrenergic receptor agonist–induced preconditioning in isolated working rat hearts. J. Pharmacol. Exp. Therapeu. 273, 689-694.

Tosaki, A., Cordis, G.A., Szerdahelyi, P., Engelman, R.M., and Das, D.K. (1994). Effects of preconditioning on reperfusion arrhythmias, myocardial functions, formation of free radicals and ion shifts in isolated ischemic/reperfused rat hearts. J. Cardiovasc. Pharmacol. 23, 365-373.

Tosaki, A., Maulik, N., Engelman, D.T., Engelman, R.M., and Das, DK. (1996). The role of protein kinase C in ischemic/reperfused preconditioning isolated rat hearts. J. Cardiovasc. Pharmacol. 28, 723-731.

Trifan, O.C., Popescu, L.M., Tosaki, A., Cordis, G., and Das, D.K. (1996). Ischemic preconditioning involves phospholipase D. Ann. N.Y. Acad. Sci. 793, 485-488.

Uings, I.J., Thompson, N.T., Randall, R.W., Spacey, G.D., Bonser, R.W., Hudson, A.T., and Garland, L.G. (1992). Tyrosine phosphorylation is involved in receptor coupling to phospholipase D but not phospholipase C in the human neutrophil. Biochem. J. 281, 597-600.

Yamada, H., Strahler, J., Welsh, M.J., and Bitar, K.N. (1995). Activation of MAP kinase and translocation with HSP 27 in bombesin-induced contraction of rectosigmoid smooth muscle. Am. J. Physiol. 269, G683-G691.

Yamazaki, T., Komuro, I., Kudoh, S., Zou ,Y., Shiojima, I., Hiroi, Y., Mizuno, T., Maemura, K., Kurihara, H., Aikawa, R., Takano, H., and Yazaki, Y. (1996). Endothelin-1 is involved in mechanical stress-induced cardiomyocyte hypertrophy. J. Biol. Chem. 271, 3221-3228.

Yazaki, Y., Komuro, I., Yamazaki, T., Tobe, K., Maemura, K., Kadowaki, T., and Nagai, R. (1993). Role of protein kinase system in the signal transduction of stretch-mediated protooncogene expression and hypertrophy of cardiac myocytes. Mol. Cell. Biochem. 119, 11-16.

Ytrehus, K., Liu, Y., and Downey, J.M. (1994). Preconditioning protects ischemic rabbit heart by protein kinase C activation. Am. J. Physiol. 266, H1145-H1152.

Zu, Y-L., Ai, Y., Gilchrist, A., Labadia, M.E., Sha'afi, R.I., and Huang, C.K. (1996). Activation of MAP kinase-activated protein kinase 2 in human neutrophils after phorbol ester or fMLP peptide stimulation. Blood 87, 5287-5296.

Zu, Y-L., Ai, Y., and Huang, C.K. (1995). Characterization of an autoinhibitory domain in human mitogen activated protein kinase activated protein kinase 2. J. Biol. Chem. 270, 202-206.

EARLY AND LATE PRECONDITIONING AGAINST MYOCARDIAL STUNNING: PATHOGENESIS AND PATHOPHYSIOLOGY

John A. Auchampach, Xian-Liang Tang,
Yumin Qiu, Peipei Ping, and Roberto Bolli

Advances in Organ Biology
Volume 6, pages 125-138.

ISBN: 0-7623-0391-3

I. INTRODUCTION

Ischemic preconditioning is the phenomenon whereby brief periods of ischemia protect the myocardium during subsequent ischemic insults. This phenomenon, first discovered by Murry and co-workers (1986) over 10 years ago, was originally demonstrated to reduce myocardial infarct size. Subsequent studies, however, have shown that preconditioning also protects against contractile dysfunction (myocardial stunning) induced by brief periods of ischemia, although this effect is less clear and somewhat controversial. The purpose of this chapter is to review the current knowledge regarding the influence of ischemic preconditioning on myocardial stunning. Similar to what has been observed for preconditioning against myocardial infarction, it has become apparent that preconditioning against myocardial stunning includes an early phase (which becomes manifest within minutes) and a late phase (which becomes manifest approximately 12 hours after the preconditioning ischemia). The mechanisms of cardioprotection probably differ between the two phases. Accordingly, these two phases will be discussed separately.

II. THE EARLY PHASE OF PRECONDITIONING AGAINST MYOCARDIAL STUNNING

A. Pathophysiology

1. Stunning Induced by a Single Ischemic Episode, 10 to 15 minutes in Duration

Early attempts to observe a preconditioning effect against myocardial stunning were made in models of stunning induced by a single coronary artery occlusion. In these studies, a single 2.5- or 5-minute coronary occlusion (Ovize et al., 1992) or two 5-minute coronary occlusions (Miyamae et al., 1993) were found not to attenuate the myocardial stunning induced by a subsequent 15-minute coronary occlusion in open-chest dogs or pigs. We performed a systematic investigation of the effects of preconditioning on the myocardial stunning induced by 10 or 15 minutes of coronary occlusion followed by 4 hours of reperfusion in 62 open-chest, anesthetized dogs. We attempted to precondition dogs with the following protocols:

- 5-minute occlusion (O)/10-minute reperfusion (R) × 1
- 2-minute O/5-minute R × 4
- 1-minute O/5-minute R × 10, or
- 40-second O/10-minute R × 5.

As compared to a control group that did not receive prior ischemia, none of the preconditioning protocols enhanced the recovery of function after the 10- or 15-

minute occlusion. Thus, taking into consideration these studies and the studies of others (Ovize et al., 1992; Miyamae et al., 1993), there is general agreement that preconditioning with brief periods of ischemia does not attenuate the myocardial stunning induced by a single, reversible ischemic insult of 10 to15 minutes duration.

It is well established that pretreatment with adenosine or adenosine A_1 receptor agonists can induce preconditioning against infarction (Cohen and Downey, 1995). We also attempted to precondition against stunning by stimulating adenosine receptors prior to a 15-minute coronary occlusion in open-chest dogs. Dogs were pretreated with an intracoronary infusion of large doses of adenosine (Sekili et al., 1995) or the A_1-selective agonist 2-chloro-N^6-cyclopentyladenosine (CCPA; Jeroudi et al., 1994). The severity of stunning was not influenced by either treatment protocol, demonstrating that unlike the protection conferred against cell death (Cohen and Downey, 1995), adenosine receptor activation does not precondition against postischemic dysfunction.

2. *Stunning Induced by Repetitive, Brief (≤ 5 Minute) Ischemic Episodes*

For several years the issue of whether recurrent ischemic episodes (which are likely to occur in patients with coronary artery disease) have a cumulative effect on postischemic dysfunction (i.e., whether myocardial stunning becomes progressively worse with each subsequent ischemic insult) has been the focus of intense research (reviewed in Bolli et al., 1995). One unifying theme that has arisen from these studies is that although function deteriorates during repetitive ischemia, the decrement in function becomes progressively smaller with each recurrent episode, with the greatest decrease in function occurring following the first occlusion. Such a pattern could be interpreted as evidence, as first proposed by Cohen and Downey (1990), that the initial ischemic episode preconditions against the stunning induced by subsequent ischemic episodes.

We addressed this hypothesis in a comprehensive study in open-chest dogs (Bolli et al., 1995) that were subjected to a variable number of cycles of 5 minutes of occlusion and 10 minutes of reperfusion. In dogs that were subjected to 10 cycles of occlusion, we observed similar results as previous investigators, in that the greatest decrease in contractile function occurred immediately after the first occlusion and additional occlusions produced little further decrease in contractile function (Figure 1). Based on these data, one would conclude that the first occlusion preconditioned the myocardium at least up until the 10th occlusion; that is, that the subsequent nine occlusions had little additional effect on postischemic dysfunction. However, we used a different approach to examine this issue. Instead of only measuring the contractile dysfunction immediately after each subsequent occlusion, we measured the severity of stunning (estimated from the total deficit of systolic wall thickening) over the entire 4-hour reperfusion period after the last episode of ischemia. The total deficit of wall thickening, which is calculated by measuring the area comprised between the wall thickening-versus-time line and the baseline

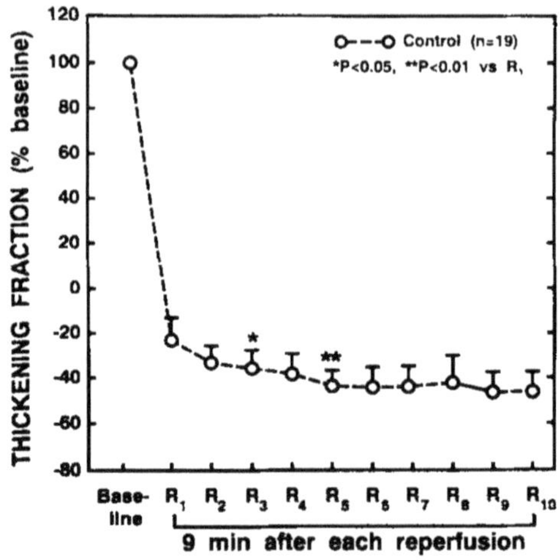

Figure 1. Systolic thickening fraction in the ischemic–reperfused region at baseline and 9 minutes into each of the 10 reperfusion (R) intervals after the 5-minutes occlusions. Open-chest dogs were subjected to a sequence of 10 5-minute coronary occlusion and 10-minute reperfusion cycles. Thickening fraction is expressed as a percentage of baseline levels. Data are the mean ± SEM. From Bolli et al., (1995), with permission.

(100% line) during the 4-hour period of observation after the last reperfusion period (see Figure 2), is an integrative measure of the overall severity of postischemic dysfunction (Bolli et al., 1995; Sun et al., 1995; Sun et al., 1996; Tang et al., 1996). We reasoned that the magnitude of injury caused by recurrent ischemic insults is more accurately reflected by the overall severity of postischemic dysfunction throughout the entire reperfusion phase than by the degree of dysfunction measured at a single time point after each ischemic insult (also because the early phase of reperfusion is a very unstable period). Strikingly different results were obtained when the issue was addressed in this manner. We found that the total deficit of wall thickening was similar in dogs that were subjected to one or three 5-minute occlusion/10-minute reperfusion cycles (Figure 2), indicating that the first occlusion did precondition against the stunning induced by the next two occlusions. However, the total deficit of wall thickening was significantly greater (approximately 2.5-fold) in dogs subjected to 10 occlusion/reperfusion cycles (see Figure 2). This indicates that the first three episodes failed to precondition against the stunning induced by the subsequent seven episodes; that is, the preconditioning effect was lost between the fourth and 10th occlusion and a cumulative effect occurred instead.

We have obtained similar results in conscious rabbits (Teschner et al., 1996). Specifically, we found that the total deficit of wall thickening was similar after one or two cycles of 4-minutes of coronary occlusion and 4-minutes of reperfusion, implying that the first cycle preconditioned against the second cycle. However, the

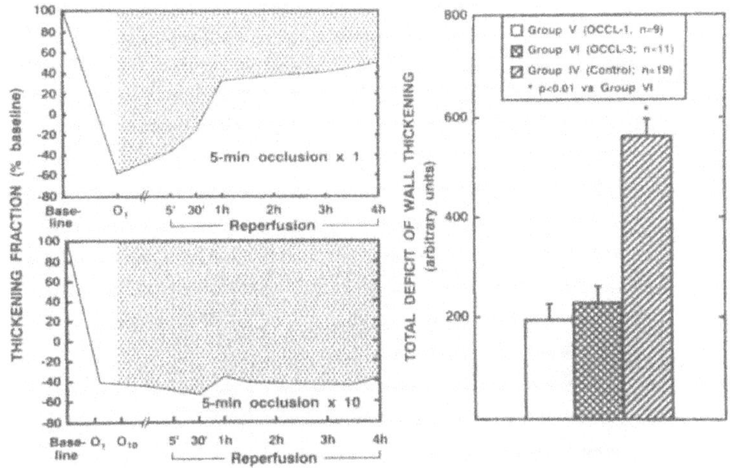

Figure 2. *Left*: Systolic thickening fraction in the ischemic-reperfused region at baseline, during the first left anterior descending (LAD) artery occlusion (O_1), during the 10th LAD occlusion (O_{10}), and at selected times during the final 4-hour reperfusion interval in a dog subjected to one 5-minute occlusion (top) and in a dog subjected to 10 5-minute occlusions (bottom). Thickening fraction is expressed as a percentage of baseline values. In both panels, the dotted area represents the area that is integrated to measure the total deficit of wall thickening after reperfusion. This measurement provides an estimate of the overall magnitude of postischemic dysfunction during the entire reperfusion interval. *Right:* The total deficit of wall thickening after reperfusion after one 5-minute LAD occlusion (open bar), three 5-min occlusions (crosshatched bar), and ten 5-minute occlusions (hatched bar). The total deficit of wall thickening is expressed as arbitrary units. Values are the mean ± SEM. From Bolli et al., (1995), with permission.

deficit increased after three cycles, indicating that the first two cycles failed to precondition against the third cycle. Furthermore, the deficit was similar after three or six cycles (indicating that the first three cycles preconditioned against the next three cycles) but increased after 12 cycles (indicating that the first six cycles failed to precondition against the next six). We conclude, therefore, that preconditioning against myocardial stunning does exist, but is limited, either in its efficacy (i.e., it protects only against a limited ischemic burden) or in its duration (i.e., it lasts less that the time period in which the subsequent cycles of ischemia occur).

B. Mechanisms

Little is known regarding the mechanism of early preconditioning against stunning. It is clear, however, that the protective effects cannot be explained by extrinsic factors such as changes in systemic hemodynamics (Cohen and Downey, 1990;

Bolli et al., 1995). Furthermore, based on measurements with radioactive microspheres, an increase in collateral blood flow to the ischemic region can also be excluded as a mechanism (Cohen and Downey, 1990; Bolli et al., 1995). In fact, we (Bolli et al., 1995) and others (Cohen and Downey, 1990) have observed in multiple coronary occlusion models that there is no correlation between the degree of collateral blood flow and the degree of stunning.

Bunch et al. (1992) have suggested that stimulation of adenosine receptors by adenosine released during the first ischemic episode may initiate early preconditioning against myocardial stunning. This hypothesis is based on their observation that a sequence of four 5-minute coronary occlusions, each followed by 10 minutes of reperfusion, resulted in a progressive decrease in segment shortening after each occlusion in anesthetized rabbits pretreated with the nonselective adenosine receptor antagonist, PD 115-199, whereas in control rabbits segment shortening decreased only after the first ischemic period. These results imply that the mechanism of early preconditioning against stunning may be similar to that of early preconditioning against infarction in which adenosine receptors activate a cascade of events thought to involve G_i proteins and other intracellular pathways (Cohen and Downey, 1995). However, using a canine model of stunning induced by six 5-minute occlusion/10-minute reperfusion cycles, Yao and Gross (1993) found that the selective A_1 adenosine receptor antagonist, 8-cyclopentyl-1,3-dipropylxanthine (CPX), decreased the recovery of segment shortening after the first coronary occlusion but caused no additional deterioration after the next five occlusions, suggesting that the adenosine released during the first ischemic episode was protective during that episode but did not precondition against the stunning induced by the subsequent episodes. The discrepancy between the results of this study and those of Cohen and Downey (1995) may be related to species differences or to the fact that additional adenosine receptor subtypes may be involved other than A_1 receptors. We have recently found that blockade of adenosine receptors with 8-*p*-sulfophenyltheophylline (SPT) fails to exacerbate the severity of myocardial stunning in conscious pigs (Sun et al., 1995) and rabbits (Maldonado et al., 1996) subjected to a sequence of brief coronary occlusion/reperfusion cycles, indicating that endogenous adenosine does not precondition against stunning during the early phase of protection in these conscious animal models. In conclusion, it seems unlikely that activation of adenosine receptors plays an important role in the early phase of preconditioning against myocardial stunning.

III. THE LATE PHASE OF PRECONDITIONING AGAINST MYOCARDIAL STUNNING

A. Pathophysiology

Kuzuya and co-workers (1993) and Marber et al. (1993) demonstrated that ischemic preconditioning protects against infarction not only immediately after the preconditioning ischemia, but also 24 hours later, suggesting that sublethal ischemia induces cellular adaptations that protect against the development of ischemic injury several hours later. We hypothesized that ischemic preconditioning would also produce delayed protection against myocardial stunning. To test this hypothesis, we subjected conscious pigs to a sequence of ten 2-minute coronary occlusions interspersed with 2-minute reperfusion intervals (Sun et al., 1995). We found that this sequence produced severe myocardial stunning, but when the same sequence was repeated 24 hours later, the severity of stunning was markedly reduced (approximately 50% reduction in the total deficit of wall thickening; Figure 3). The protection disappeared within 10 days after the last ischemic stress, but could be re-induced by another sequence of coronary occlusions. Using the same conscious pig model, we subsequently determined more accurately the time course of the protective effect (Tang et al., 1996). We found that preconditioning was not present 6 hours after the sequence of ischemic episodes, became manifest at 12 hours, peaked

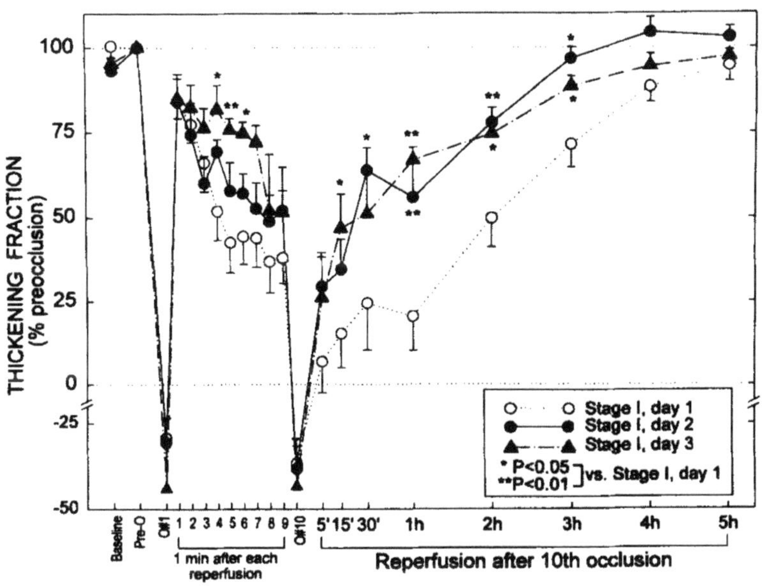

Figure 3. Systolic thickening fraction in the ischemic–reperfused region in conscious pigs subjected to a sequence of ten 2-minute left anterior descending (LAD) coronary artery occlusion/2-minute reperfusion cycles. Shown are the measurements of thickening fraction obtained at baseline, immediately before the first occlusion (preocclusion, Pre-O), 1 minute into the first LAD occlusion (O#1), 1 minute into each of the nine reperfusions, 1 minute into the 10th occlusion (O#10), and at selected times during the 5-hour reperfusion interval following the 10th coronary occlusion. Thickening fraction is expressed as a percentage of preocclusion values. Data are mean ± SEM (n = 8 to 10). From Sun et al., (1995), with permission.

at 1 and 3 days, and disappeared within 6 days. More recently, we have observed a similar preconditioning effect with a similar time-course in conscious rabbits subjected to a sequence of six 4-minute coronary occlusion/4 minute reperfusion cycles (Qiu et al., 1995). In this model, a marked protection against stunning (>50% reduction in the total deficit of wall thickening) was observed 24 hours after the ischemic challenge and disappeared 5 days later (Qiu et al., 1995). Taken together, these studies demonstrate that a brief ischemic stress induces a powerful, long-lasting adaptive response that renders the myocardium resistant to stunning. We have termed this response "late preconditioning" against myocardial stunning. Interestingly, we found that in the conscious rabbit model, three cycles of 4-minute coronary occlusions is the minimum number of occlusions capable of inducing a protective effect 24 hours later; additional bouts of ischemia (up to 12) did not increase the magnitude of the response, indicating that this is an all-or-none phenomenon (Teschner et al., 1996).

B. Mechanisms

1. *Do Adenosine Receptors Mediate Late Preconditioning Against Myocardial Stunning?*

It is well established that the mechanism of preconditioning against infarction involves activation of adenosine receptors as a result of the release of adenosine during the preconditioning ischemia (Cohen and Downey, 1995). Recent data by Baxter et al. (1994) indicate that this mechanism is involved in late preconditioning against infarction as well. In this study, it was shown that infarct size was not reduced 24 hours later if SPT was administered during the preconditioning ischemia, and that a preconditioning-like effect could be induced by administering CCPA 24 hours prior to ischemia. Accordingly, we tested the hypothesis that an adenosine-coupled pathway could also be responsible for the late preconditioning against stunning. Our results demonstrate this is not the case. In conscious pigs, we found that pretreating the animals with SPT during the preconditioning ischemia on day 1 did not prevent the late preconditioning against stunning on day 2 (Figure 4), despite the fact that the dose we used was sufficient to block the chronotropic and dromotropic effects of CCPA (Sun et al., 1995). We have subsequently obtained similar results in conscious rabbits (Maldonado et al., 1997). In this study, we could not block late preconditioning against stunning by treating rabbits with SPT during the preconditioning ischemia, nor could we induce late preconditioning against stunning by pretreating rabbits with large doses of CCPA 24 hours earlier. Thus, unlike the results reported by Baxter et al. (1994) for late preconditioning against infarction, it is clear that adenosine receptors do not contribute to late preconditioning against stunning, implying that there are distinct differences in the mechanism of late preconditioning against stunning and infarction.

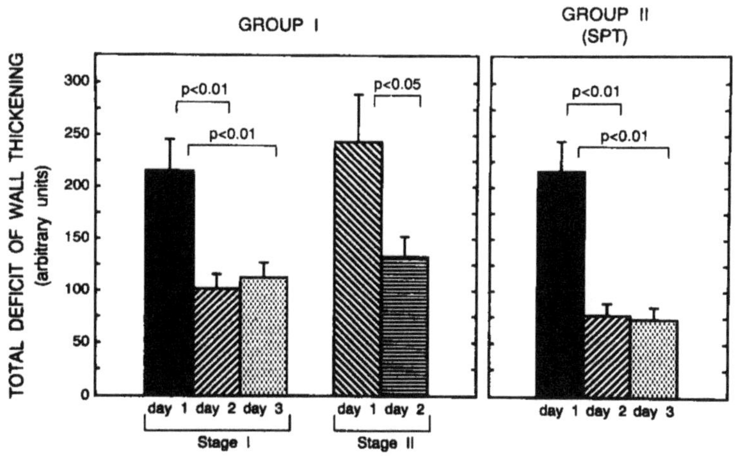

Figure 4. Total deficit of wall thickening after the 10th reperfusion in conscious pigs in group I, which did not receive 8-*p*-sulfophenyltheophylline (SPT), and in group II, which received SPT on day 1. Conscious pigs were subjected to a sequence of 10 2-minute LAD occlusion/2-minute reperfusion cycles and systolic thickening fraction was measured, as in Figure 3. The total deficit of wall thickening was calculated in a manner similar to that shown in Figure 2. Data are the mean ± SEM (n = 8 to 10). From Sun et al., (1995), with permission.

2. *Essential Role of Reactive Oxygen Species in Late Preconditioning Against Stunning*

We have recently demonstrated that administration of antioxidant therapy completely prevents the development of late preconditioning against stunning, indicating that reactive oxygen species (ROS) trigger the protective response (Sun et al., 1996). In conscious pigs subjected to a sequence of 10 2-minute occlusion/2-minute reperfusion cycles on three consecutive days (days 1, 2, and 3), we observed that administration of antioxidants (superoxide dismutase, catalase, and mercaptopropionyl glycine) on day 1 attenuated stunning on day 1 (indicating that ROS contribute to myocardial stunning), but resulted in a loss of preconditioning on day 2 (Figure 5). On day 3, antioxidant-treated pigs exhibited a marked attenuation of stunning, indicating that the ischemia produced on day 2 preconditioned the heart against stunning on day 3 (see Figure 5). The ability of the antioxidants to block late preconditioning is not a nonspecific result of the mitigation of postischemic dysfunction on day 1, since it was observed that administration of a calcium channel antagonist (nisoldipine) was as effective as antioxidants in attenuating myocardial stunning on day 1, yet it had no effect on the development of late preconditioning on day 2 (see Figure 5). Taken together, these results indicate that the oxidative stress

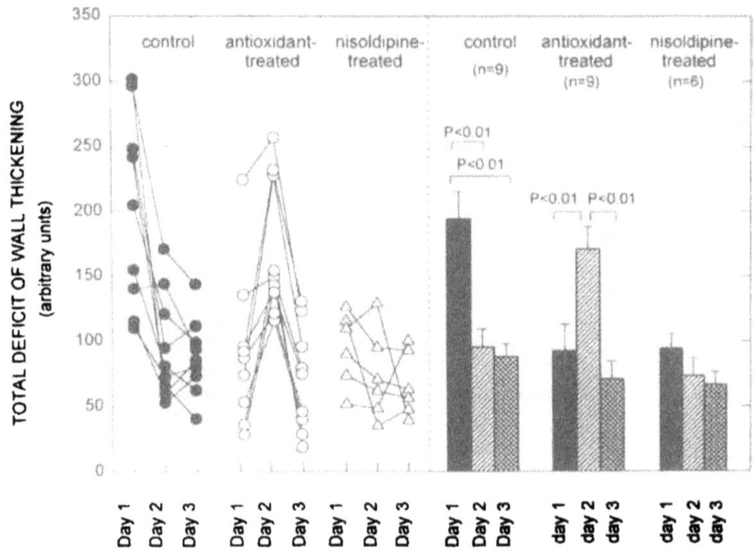

Figure 5. Total deficit of wall thickening after the 10th reperfusion in conscious pigs on days 1, 2, and 3 in the control group; in the group treated with antioxidants (superoxide dismutase plus catalase plus mercaptopropionyl glycine); and in the group treated with the calcium antagonist nisoldipine. The values of total deficit of wall thickening (calculated in a manner similar to that shown in Figure 2) in individual pigs are illustrated in the left panel; the mean (± SEM) values of total wall thickening in each group are depicted in the right panel. Note that antioxidant therapy on day 1 prevented preconditioning on day 2 in the treated group. Also note that antioxidant therapy attenuated stunning on day 1 in this group. Nisoldipine-treated pigs were protected on day 1, but also were preconditioned on day 2. From Sun et al., (1996), with permission.

incurred during brief ischemia and reperfusion triggers a protective response that makes the myocardium resistant to stunning 24 hours later.

These results support a new paradigm regarding the pathophysiological role of ROS in myocardial ischemia and reperfusion. Until now, generation of ROS has generally been viewed as a deleterious process. Our findings that ROS contribute to the genesis of myocardial stunning but, at the same time, trigger the development of late preconditioning against stunning suggest that the radical species generated after a brief ischemic episode are injurious in the short term (as mediators of the immediate injury), but at the same time, play a useful role in the long term (as triggers of a delayed, powerful, and long-lasting protective process). This dual function implies that ROS do not invariably have a detrimental effect: after a mild ischemic insult, ROS formation could serve as a warning signal that activates a protective response designed to minimize further injury through the up-regulation of redox-sensitive, cardioprotective genes. In this manner, the sublethal oxidative stress could act as a transduction pathway signaling an imminent threat and the need to develop defenses against it.

Further studies will be necessary to identify the precise mechanism whereby the exposure to ROS during the preconditioning ischemia leads to the development of late preconditioning against myocardial stunning. Although it has yet to be proven directly, there is circumstantial evidence that late preconditioning is caused by a sustained adaptive response such as the synthesis of cardioprotective proteins. Specifically, the fact that late preconditioning has a delayed onset (requiring > 6 hours to become apparent), peaks at 24 to 72 hours, and disappears by 6 days (Tang et al., 1996) is consistent with the notion that it is caused by the synthesis of new proteins. At first sight, the two most likely classes of proteins which may mediate the protection of preconditioning appear to be heat shock proteins (HSPs) and antioxidant enzymes. It is well-established that exposure to an oxidative stress can induce HSPs (Polla, 1988) and antioxidant enzymes (Stevens and Autor, 1977; Richter and Loewen, 1989; Das et al., 1995). In this regard, we have found that the presence of late preconditioning is associated with increased myocardial levels of HSP 70 in conscious pigs (Sun et al., 1995). We have not, however, detected increased activity of antioxidant enzymes in preconditioned myocardium. In conscious pigs subjected to a sequence of 10 2-minute occlusions 24 hours earlier, we observed that the levels of manganese superoxide dismutase (Mn-SOD), copper/zinc superoxide dismutase (Cu,Zn-SOD), catalase, glutathione peroxidase, and glutathione reductase in the ischemic–reperfused region were similar to the levels in the nonischemic region as well as to the levels in nonpreconditioned controls (Tang et al., 1996). It should be noted that an increase in HSPs, per se, does not prove causality, and may simply represent an epiphenomenon (Heads et al., 1995). Besides antioxidant enzymes and HSPs, it is possible that other proteins both known and unknown may also be induced by oxidative stress in late preconditioning, since a large number of genes have been identified which are regulated by reactive oxygen species (reviewed by Colburn, 1992; Das et al., 1995).

3. *Role of Protein Kinase C in Late Preconditioning Against Stunning*

More recently, we have explored potential mechanisms by which ROS may trigger gene expression in response to ischemic preconditioning. Specifically, we addressed the hypothesis that a protein kinase C–mediated pathway may be involved, based on previous studies which have shown that oxidative stress results in activation of protein kinase C (von Ruecker et al., 1989; Rao and Berk, 1992). We used both a direct and an indirect approach to test this hypothesis. First, in conscious rabbits subjected to six cycles of 4 minutes of ischemia and 4 minutes of reperfusion on three consecutive days, we administered the selective protein kinase C inhibitor, chelerythrine, on day 1 and observed its effect on the development on late preconditioning against stunning on day 2. Second, we attempted to induce a preconditioned state in conscious rabbits by administering phorbol-12-myristate-13-acetate

(PMA) without ischemia 24 hours prior to the initiation of the ischemic protocol (Qiu et al., 1998). Finally, to obtain direct evidence of protein kinase C activation in late preconditioning against stunning, we determined the subcellular distribution of all 11 known isoforms of protein kinase C in the preconditioned myocardium of conscious rabbits using immunoblotting techniques (Ping et al., 1997). We found that administration of chelerythrine effectively blocked the development of late preconditioning against stunning and, conversely, that administration of PMA 1 day prior to ischemia resulted in a reduction in stunning equivalent to that effected by ischemic preconditioning (Qiu et al., 1998). In addition, we observed that the ischemia–reperfusion cycles of preconditioning induced a selective translocation of the ε and η isoforms of protein kinase C with no significant change in the subcellular distribution of total PKC activity (Ping et al., 1997). Taken together, these results indicate that protein kinase C plays an important role in the signaling pathway involved in late preconditioning against stunning. It is plausible to speculate, therefore, that ROS generated during the preconditioning ischemia may activate protein kinase C, which in turn would result in the expression of cytoprotective proteins.

IV. CONCLUSIONS

Similar to the observations made for preconditioning against infarction, there are two phases of preconditioning against stunning: an early phase (which begins within minutes) and a late phase (which begins after ≥ 6 hours). These two phases have different pathophysiology and probably different mechanisms. The early phase is somewhat controversial, but most of the discrepancies can be explained by differences in the duration of the ischemic insult; that is, the available evidence indicates that a brief ischemic episode preconditions against the stunning induced by a second episode when the second episode lasts less than 5 minutes but not when it lasts more than 10 minutes. Furthermore, when early preconditioning occurs, the protection is limited to only a few ischemic episodes (probably three to five). Thus, the early preconditioning effect is rather weak. The mechanism of early preconditioning against stunning is unknown; a role of adenosine receptors seems unlikely.

In addition to this early phase of protection, brief ischemia induces a late phase of preconditioning against stunning that is long lasting (at least 72 hours) and powerful (resullting in approximately 50% reduction in total dysfunction). This late phase is triggered by the generation of ROS during the preconditioning ischemia–reperfusion cycles, and is mediated by a mechanism that involves a protein kinase C–mediated pathway. Based on the time-course of late preconditioning, it is likely that the synthesis of protective proteins contributes to the protection. In contrast to the results obtained for the late phase of preconditioning against infarction, activation of adenosine receptors is not involved in the mechanism of late preconditioning against stunning. Since the late phase of preconditioning offers pronounced, sustained protection, it is probably more important and more clinically relevant than the early phase. It is hoped that future studies aimed at unraveling the mechanism of preconditioning against stunning may lead to the development of therapeutic strategies capable of inducing a chronic state of preconditioning by pharmacological or genetic means.

ACKNOWLEDGMENTS

This study was supported in part by NIH RO1 grants HL-43151 (RB) and HL-55757 (RB); National American Heart Association grant 9630083N (JAA); and American Heart Association Kentucky Affiliate grants KY-96-GB-31 (XLT) and KY-96-GB-32 (YQ).

REFERENCES

Baxter G.F., Marber, M.S., Patel, V.C., and Yellon, D.M. (1994). Adenosine receptor involvement in a delayed phase of myocardial protection 24 hours after ischemic preconditioning. Circ. Res. 90, 2993-3000.

Bolli R., Zughaib, M., Li, X.-L., Tang, X.-L., Sun, J.-Z., Triana, J.F., and McCay, P.B. (1995). Recurrent ischemia in the canine heart causes recurrent bursts of free radical production that have a cumulative effect on contractile function: a pathophysiological basis for chronic myocardial "stunning." J. Clin. Invest. 96, 1066-1084.

Bunch F.T., Thornton, J., Cohen, M.V., and Downey, J.M. (1992). Adenosine is an endogenous protectant against stunning during repetitive ischemic episodes in the heart. Am. Heart J. 124, 1440-1445.

Cohen and M.V., Downey, J.M. (1990). Myocardial stunning in dogs: preconditioning effect and influence of coronary collateral flow. Am. Heart J. 120, 282-291.

Cohen and M.V., Downey, J.M. (1995). Preconditioning during ischemia: basic mechanisms and potential clinical applications. Cardiol. Rev. 3, 137-149.

Colburn, N.H. (1992). Gene regulation by active oxygen and other stress inducers: Role in tumor promotion and progression. In: Biological Consequences of Oxidative Stress. (Spatz, L., and Bloom, A.D., Eds.), pp. 121-137. Oxford University Press, New York.

Das, D.K., Maulik, N., and Moraru, I.I. (1995). Gene expression in acute myocardial stress. Induction by hypoxia, ischemia, reperfusion, hyperthermia and oxidative stress. J. Mol. Cell. Cardiol. 27, 181-193.

Heads, R.J., Baxter, G.F., Latchman, D.S., and Yellon, D.M. (1995). Delayed protection in rabbit heart following ischemic preconditioning is associated with modulation of HSP27 and superoxide dismutase at 24 hours. J. Mol. Cell. Cardiol. 27, A163 [Abstr.].

Jeroudi, M.O., Tang, X.-L., Abd-Elfattah, A., Sun, J.-Z., and Bolli, R. (1994). Effect of adenosine A1-receptor activation on myocardial stunning in intact dogs. Circulation 94, I-479 [Abstr.].

Kuzuya, T., Hoshida, S., Yamashita, N., Fuji, H., Oe, H., Hori, M., Kamada, T., and Tada, M. (1993). Delayed effects of sublethal ischemia on the acquisition of tolerance to ischemia. Circ. Res. 72, 1293-1299.

Maldonado, C., Qiu, Y., Tang, X.-L., Auchampach, J., and Bolli, R. (1997). Role of adenosine receptors in late preconditioning against myocardial stunning in conscious rabbits. Am. J. Physiol. 42, H1324-H1332.

Marbar M.S., Latchman, D.S., Waalker, J.M., Yellon, D.M. (1993). Cardiac stress protein elevation 24 hours after brief ischemia or heat stress is associated with resistance to myocardial infarction. Circulation 88, 1264-1272.

Miyamae, M.H., Fujiwara, H., Kida, M., Yokota, R., Tanaka, M., Katsuragawa, M., Hasegawa, K., Ohura, M., Koga, K., Yabuuchi, Y., and Sasayama, S. (1993). Preconditioning improves energy metabolism during reperfusion but does not attenuate myocardial stunning in porcine hearts. Circulation 88, 223-234.

Murry, C.E., Jennings, R.B., and Reimer, K.A. (1986). Preconditioning with ischemia: a delay of lethal cell injury in ischemia myocardium. Circulation 74, 1124-1136.

Ovize, M., Przyklenk, K., Hale, S., and Kloner, R.A. (1992). Preconditioning does not attenuate myocardial stunning. Circulation 85, 2247-2254.

Ping P., Zhang, J., Qiu, Y., Tang, X.-L., Manchikalapudi, S., Cao, X., and Bolli, R. (1997). Ischemic preconditioning induces selective translocation of protein kinase C isoforms ε and η in the heart of conscious rabbits without subcellular redistribution of total protein kinase C activity. Circ. Res. 81, 404-414.

Polla, B.S. (1988). A role for heat shock proteins in inflammation? Immunol. Today 9, 134-137.

Qiu, Y., Maldonado, C., Tang, X.-L., and Bolli, R. (1995). Late preconditioning against stunning in conscious rabbits. Circulation 92, I-715 [Abstr.].

Qiu, Y., Ping, P., Tang, X.-L., Manchikalapudi, S., Rizvi, A., Zhang, J., Takano, H., Wu, W-J., Teschner, S., and Bolli, R. (1998). Direct evidence that protein kinase C plays an essential role in the development of late preconditioning against myocardial stunning in conscious rabbits and that ε is the isoform involved. J. Clin. Invest. 101, 2182-2198.

Rao, G.N., and Berk, B. (1992). Activate oxygen species stimulate vascular smooth muscle cell growth and proto-oncogene expression. Circ. Res. 70, 593-599.

Richter, H.E., and Loewen, P.C. (1981). Induction of catalase in *Echerichia coli* by ascorbic acid involves hydrogen peroxide. Biochem. Biophys. Res. Commun. 109, 1039-1046.

Sekili, S., Jerudi, M.O., Tang, X.-L., Zughaib, M., Sun, J.-Z., and Bolli, R. (1995). Effect of adenosine on myocardial "stunning" in the dog. Circ. Res. 76, 82-94.

Stevens, J.B., and Autor, A.P. (1977). Induction of superoxide dismutase by oxygen in neonatal rat lung. J. Biol. Chem. 252, 3509-3514.

Sun, J.-Z., Tang, X.-L., Knowlton, A.A., Park, S.-W., Qiu, Y., and Bolli, R. (1995). Late preconditioning against myocardial stunning: an endogenous protective mechanism that confers resistance to postischemic dysfunction 24 h after brief ischemia in pigs. J. Clin. Invest. 95, 388-403.

Sun J.-Z., Tang, X.-L., Park, S.-W., Qiu, Y., Turrens, J.F., and Bolli, R. (1996). Evidence for an essential role of reactive oxygen species in the genesis of late preconditioning against myocardial stunning in conscious pigs. J. Clin. Invest. 97, 562-576.

Tang X.-L., Qiu, Y., Park, S.-W., Sun, J.-Z., Kalya, A., and Bolli, R. (1996). Time course of late preconditioning against myocardial stunning in conscious pigs. Circ. Res. 79, 424-434.

Tang, X.-L., Qiu, Y., Turrens, J.F., Sun, J.-Z., and Bolli, R. (1996). Is late preconditioning against myocardial stunning mediated by increased antioxidant defenses? Circulation 94, I-184 [Abstr.].

Teschner, S., Qiu, Y., Tang, X.-L., Maldonado, C., Rizvi, A., Manchikalpudi, S., Bagri, H., Jadoon, A., and Bolli, R. (1996). Late preconditioning against myocardial stunning in conscious rabbits: a dose-related or an all-or-none phenomenon? Circulation 94, I-423 [Abstr.].

Von Ruecker A.A., Han-Jeon, B.-G., Wild, M., and Bidlingmair, D. (1989). Protein kinase C involvement in lipid peroxidation and cell membrane damage induced by oxygen-based radicals in hepatocytes. Biochem. Biophys. Res. Commun. 163, 836-842.

Yao, Z., and Gross, G.J. (1993). Glibenclamide antagonizes adenosine A_1 receptor-mediated cardioprotection in stunned canine myocardium. Circulation 88, 235-244.

CHANGES IN CARDIAC ENERGETICS DURING PRECONDITIONING AND ADAPTATION

Nobuakira Takeda

I. INTRODUCTION

In myocardial ischemia, glycolysis is enhanced by hypoxia and hypocirculation induces the accumulation of H^+ and lactate produced by glycolysis; accumulation of these substances subsequently depresses glycolysis. The amount of adenosine triphosphate (ATP) produced by glycolysis is less than 10% of that produced under aerobic conditions, but it provides energy for Ca^{2+}-ATPase and Na^+-K^+-ATPase.

Advances in Organ Biology
Volume 6, pages 139-143.

ISBN: 0-7623-0391-3

ATP produced by glycolysis is an important energy source for the heart during ischemia. ATP produced by oxidative phosphorylation is utilized for myocardial contraction, while ATP produced by glycolysis is consumed for the ion transport system of the sarcolemma and sarcoplasmic reticulum. The influence of ischemic preconditioning on myocardial energy metabolism is important from the viewpoint of understanding myocardial protection by preconditioning. Myocardial protection against ischemia can be obtained either by increasing the oxygen supply to the ischemic area via the collateral circulation or by decreasing myocardial energy consumption. Murry et al. (1986, 1990) have suggested that decreased myocardial oxygen consumption is important and that the effect of preconditioning is to depress anaerobic glycolysis by reducing oxygen demand during myocardial ischemia.

II. GLYCOLYSIS

Ischemic preconditioning promotes anaerobic glycolysis (Janier et al., 1994; Schaefer et al., 1995; Finegan et al., 1995; Gota et al., 1995; Barbosa et al. 1996) and this results in a decrease in myocardial glycogen content. This decreases the accumulation of metabolic products of glycolysis, such as H^+ or lactate, and inhibits myocardial acidosis during ischemia. Ischemic preconditioning enhances glucose utilization after reperfusion (Fralix et al., 1992), leading to increased glycolysis. ATP produced by glycolysis becomes an energy source for ion transport by the sarcolemma and sarcoplasmic reticulum. Therefore, ATP produced by glycolysis is related to Ca^{2+} movement. Yabe et al. (1995) have reported that ischemic preconditioning inhibited the decrease of phosphofructokinase during myocardial ischemia and enhanced the recovery of this kinase after reperfusion. At the same time, the increase of myocardial glucose-6-phosphate (G-6-P) and fructose-6-phosphate (F-6-P) was lessened. This suggests that ischemic preconditioning improves glycolysis by influencing phosphofructokinase. See Figure 1.

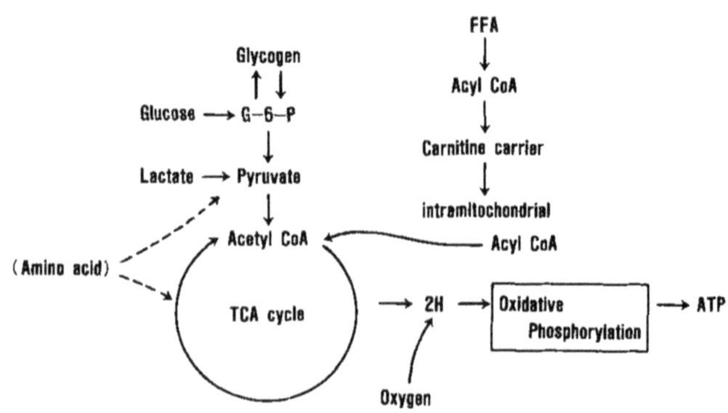

Figure 1. Myocardial energy metabolism. ATP, adenosine triphosphate; FFA, free fatty acids; G-6-P, glucose-6-phosphate; TCA, tricarboxylic acid.

III. MITOCHONDRIA

Myocardial ATP is consumed by mitochondrial F1F0-ATPase during ischemia. Ischemic preconditioning inhibits the activity of this enzyme (Jennings et al., 1991; Vuorinen et al., 1995) and decreases myocardial energy consumption. Ischemia damaged mitochondrial respiratory chains and increases H^+ (Borutaite et al., 1995). According to Yabe et al. (1995), the rate of mitochondrial oxygen consumption was decreased by ischemia, and ischemic preconditioning inhibited this decrease. Our laboratory found a high incidence of mitochondrial DNA deletion in the myocardium obtained at autopsy from patients with old myocardial infarctions (Takeda et al., 1993). Such deletion may be induced by free radicals produced by ischemia or reperfusion. If the proportion of mitochondrial DNA deletion exceeds a certain threshold, deterioration of energy production will become irreversible.

IV. MYOSIN

Ventricular myosin from rats is separated into three isozymes (V_1, V_2, and V_3) by pyrophosphate gel electrophoresis, and ATPase activity decreases in order from V_1 to V_3 (Hoh et al., 1978). In experimental myocardial infarction, there is an increase of isozyme V_3, which has the lowest ATPase activity (Mercadier et al., 1981; Geenen et al., 1989). The increase of V_3 in cardiac hypertrophy induced by pressure overload as well as in diabetes mellitus is thought to be an adaptive change to maintain myocardial force development with a lower oxygen consumption (Alpert and Mulieri, 1982; Kissling et al., 1982; Jacob et al., 1983; Holubarsch et al., 1985; Takeda et al., 1988). The increase of V_3 in infarcted myocardium may have the same significance. In human ventricular myocardium, however, myosin is only separated into two components (VM-A and -B) by pyrophosphate gel electrophoresis (Takeda et al., 1985). Our laboratory found that these ventricular myosin isozymes showed no significant alteration in autopsy samples of infarcted myocardium when compared with myocardium from patients with other diseases (Takeda et al., 1996). This suggest that the adaptation of ventricular myosin is less extensive in humans than in rats. See Figure 2.

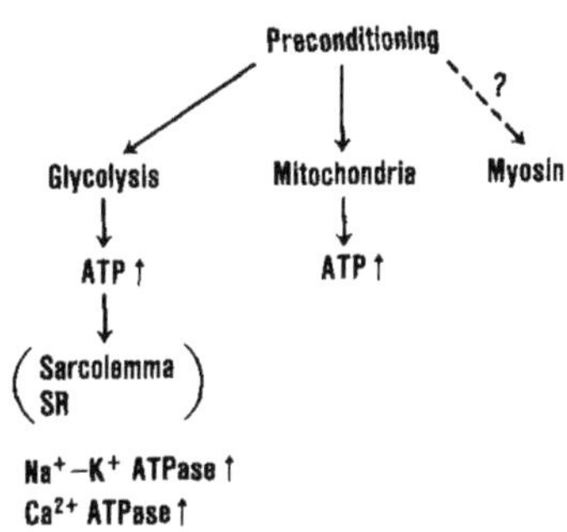

Figure 2. Effects of ischemic preconditioning on energetics. ATP, adenosine triphosphate.

V. SUMMARY AND CONCLUSION

Ischemic preconditioning protects the myocardium from ischemia–reperfusion injury by inhibiting the decrease of ATP production by glycolysis during ischemia and by oxidative phosphorylation after reperfusion. The cardioprotective effect of ischemic preconditioning is thought to be mediated via activation of adenosine receptors as well as by bradykinin. However, the contribution of alterations of myocardial energy metabolism to the cardioprotective effect of ischemic preconditioning remain unclear and should be investigated further.

REFERENCES

Alpert, N.R., and Mulieri, L.A. (1982). Increased myothermal economy of isometric force generation in compensated cardiac hypertrophy induced by pulmonary artery constriction in the rabbit. Circ. Res. 50, 491-500.

Barbosa, V., Sievers, R.E., Zaugg, C.E., and Wolfe, C.L. (1996). Preconditioning ischemia time determines the degree of glycogen depletion and infarct size reduction in rat hearts. Am. Heart. J. 131, 224-230.

Borutaite, V., Mildaziene, V., Brown, G.C., and Brand, M.D. (1995). Control and kinetic analysis of ischemia-damaged heart mitochondria:which parts of the oxidative phosphorylation system are affected by ischemia ? Biochim. Biophys. Acta 1272, 154-158.

Finegan, B.A., Lopaschuk, G.D., Gandhi, M., and Clanachan, A.S. (1995). Ischemic preconditioning inhibits glycolysis and proton production in isolated working rat hearts. Am. J. Physiol. 269, H1767-1775.

Fralix, T.A., Steenbergen, C., London, R.E., and Murphy, E. (1992). Metabolic substrates can alter postischemic recovery in preconditioned ischemic heart. Am. J. Physiol. 263, C17-23.

Geenen, D.L., Malhotra, A., and Sceuer, J. (1989). Regional variation in rat cardiac myosin isoenzymes and ATPase activity after infarction. Am. J. Physiol. 256, H745-750.

Goto, M., Tsuchida, A., liu, Y., Cohen, M.V., and Downey, J.M. (1995). Transient inhibition of glucose uptake mimics ischemic preconditioning by salvaging ischemic myocardium in the rabbit heart. J. Mol. Cell. Cardiol. 27, 1883-1894.

Hoh, J.Y., McGrath, P.A., and Hale, P.T. (1978). Electrophoretic analysis of multiple forms of rat cardiac myosin:effects of hypophysectomy and thyroxine replacement. J. Mol. Cell. Cardiol. 10, 1053-1076.

Holubarsch, C., Litten, R.Z., Mulieri, L.A., and Alpert, N.R. (1985). Energetic changes of myocardium as an adaptation to chronic hemodynamic overload and thyroid gland activity. Basic Res. Cardiol. 80, 582-593.

Jacob, R., Kissling, G., Ebrecht, G., Holubarsch, C., Medugorac, I., and Rupp, H. (1983). Adaptive and pathological alterations in experimental cardiac hypertrophy. In:Advances in Myocardiology. (Chazov, E., Saks, V., and Rona, G., Eds.), pp. 55-77. Plenum Press, New York.

Janier, M.F., Vanoverschelde, J.L., and Gergmann, S.R. (1994). Ischemic preconditioning stimulates anerobic glycolysis in the isolated rabbit heart. Am. J. Physiol. 267, H1353-1360.

Jennings, R.B., Reimer, K.A., and Steenbergen, C. (1991). Effect of inhibition of the mitochondrial ATPase on net myocardial ATP in total ischemia. J. Mol. Cell. Cardiol. 23, 1383-1395.

Kissling, G., Rupp, H., Malloy, L., and Jacob, R. (1982). Alteration in cardiac oxygen consumption under chronic pressure overload. Significance of the isoenzyme pattern of myosin. Basic Res. Cardiol. 77, 255-269.

Mercadier, J-J., Lompre, A-M., Wisnewsky, C., Samuel, J-L., Bercovici, J., Swynghedauw, B., and Schwartz, K. (1981). Myosin isozymic changes in several models of rat cardiac hypertrophy. Circ. Res. 49, 525-532.

Murry, C.E., Jennings, R.B., and Reimer, K.A. (1986). Preconditioning with ischemia: a delay of lethal cell injury in ischemic myocardium. Circulation 74, 1124-1136.

Murry, C.E., Richard, V.J., Reimer, K.A., and Jennings, R.B. (1990). Ischemic preconditioning slows energy metabolism and delays ultrastructural damage during a sustained ischemic episode. Circ. Res. 66, 913-931.

Schaefer, S., Carr, L.J., Prussel, E., Ramasamy, R. (1995). Effects of glycogen depletion on ischemic injury in isolated rat hearts: insight into preconditioning. Am. J. Physiol. 268, H935-944.

Takeda, N., Nakamura, I., Hatanaka, T., Ohkubo, T., and Nagano, M. (1988). Myocardial mechanical and myosin isoenzyme alterations in streptozotocin-diabetic rats. Japan. Heart J. 29, 455-463.

Takeda, N., Ota, Y., Tanaka, Y., Shikata, C., Hayashi, Y., Nemoto, S., Tanamura, A., Iwai, T., and Nakamura, I. (1996). Myocardial adaptive changes and damages in ischemic heart disease. Ann. N.Y. Acad. Sci. 793, 282-288.

Takeda, N., Rupp, H., Fenchel, G., Hoffmeister, H-E., and Jacob, R. (1985). Relationship between the myofibrillar ATPase activity of human biopsy material and hemodynamic parameters. Japan. Heart J. 26, 909-922.

Takeda, N., Tanamura, A., Iwai, T., Nakamura, I., Kato, M., Ohkubo, T., and Noma, K. (1993). Mitochondrial DNA deletion in human myocardium. Mol. Cell. Biochem. 119, 105-108.

Vuorinen, K., Ylitato, K., Peuhkurinen, K., Raatikainen, P., Ala-Rami, A., and Hassinen, I.E. (1995). Mechanisms of ischemic preconditioning in rat myocardium. Roles of adenosine, cellular energy state, and mitochondrial F1F0-ATPase. Circulation 91, 2810-2818.

Yabe, K., Nasa, Y., Sato, M., and Takeo, S. (1995). Enhancement of glycolysis during an early phase of reperfusion as a possible mechanism for preconditioning in isolated rat hearts. J. Mol. Cell. Cardiol. 27, A149.

MOLECULAR ADAPTATION OF THE TRANSCRIPTIONAL APPARATUS IN CARDIAC HYPERTROPHY AND EMBRYONIC DEVELOPMENT

Satish Ghatpande, Michael Wagner,
and M.A.Q. Siddiqui

I. INTRODUCTION

The heart occupies a central position in the ontogeny of all vertebrate organisms: as one of the first organs to arise during development, it becomes essential not only for development of the embryo and fetal life, but also to the viability of the vertebrate

Advances in Organ Biology
Volume 6, pages 145-153.

ISBN: 0-7623-0391-3

organism throughout its life. In children, as well as in adults, a significant percentage of cardiac morbidity can be ascribed to congenital defects of the heart. In other instances, disease states such as cardiac hypertrophy appear to have an "embryological nature" in that the organism's adaptive response to cardiac overload includes the reiteration of embryonic patterns of cardiac gene expression (Simpson et al., 1989; Yazaki et al., 1989).

Together, these observations provide a rationale for investigating the molecular events underlying cardiogenesis in that a thorough understanding of heart development may lead to new and useful insights into the cause as well as the treatment of heart disease. Some recent steps have already been taken along these lines: applying the modern techniques of molecular biology to the long-standing embryological paradigms of heart development in a number of different vertebrate (and nonvertebrate) species has provided a wealth of information on the molecules and mechanisms involved in heart development. The task ahead for clinicians and scientists alike remains to understand heart disease in light of this newly acquired information.

II. EMBRYOLOGICAL ASPECTS OF CARDIOGENESIS

Virtually all aspects of development and organogenesis in vertebrates rely on the interaction and inductive influences operating between the three primary germ cell layers of mesoderm, ectoderm, and endoderm. This is no less true for heart development. The earliest event in development of the heart is the establishment of the cardiogenic cell lineage from one of these cell layers, in this case, the mesoderm. Mesodermal cells that acquire the cardiogenic cell fate go on to contribute to the cardiac primordia or primitve heart tube. The heart in all vertebrates arises from the anterior lateral plate mesoderm within a horseshoe-shaped area bilateral to the node of the primitve streak (Rosenquist and DeHaan, 1966). Precardiac mesodermal cells form a cohesive cell sheet that migrates over the subjacent endoderm (Manasek, 1968; Stalsberg, 1969) toward the midline of the embryo and fuses to form the primitive heart tube. This heart tube loops towards the right side of the embryo, laying down the future "form" of the complete heart. Concomitant with heart tube formation and looping, heart cells start beating. The beginning of heart function is temporally related to myocardial differentiation and the expression of a number of cardiac-specific genes such as MHC, vMLC-2, and cardiac α-actin. Subsequently, cardiac neural crest cells migrate to the heart tube and begin to contribute to the building of the heart conduction system with the establishment of aortic arches, outflow tracts, endocardial cushions, and valves. Lastly, endocardial cells develop into the endothelial lining of the heart chambers, leading to the fomation of a fully functioning embryonic heart.

Underlying these processes is a complex network of both regulatory and structural genes that together interact to orchestrate these morphogenetic events. Some of these genes, such as those mentioned above which are expressed early in heart development and those expressed during later development (e.g., atrial natriuretic factor, atrial myosin light chain-1 [MLC-1]), have been characterized and their expression studied. Interestingly, these same genes are re-expressed in cardiomyocytes of adult hearts exhibiting hypertrophy. The reason and mechanism(s) for this remain at present unclear. What is clear is the attempt on the part of cardiomyocytes to adapt to the environmental stress of hypertrophy by re-expressing what might be considered cardiogenic genes. The following is a review of the developmental genetics of these genes in light of the mechanisms that may be operating in the adaptive response to hypertrophy.

The avian embryo has provided an excellent model system for the study of cardiac muscle development and heart formation, including analysis of the genes mentioned herein. Using sensitive assays based on RT-PCR and single avian blastoderms of defined stages, we demonstrated that the cardiac genes vMLC-2 and α-actin appear as early as stage 5 during development immediately after the commitment of mesodermal cells to the cardiogenic cell lineage at stage 4 (Rosenquist and DeHaan, 1965), but significantly earlier than the formation of the heart tube itself (Goswami et al., 1994). In situ hybridization analysis has shown that vMLC-2 is specifically expressed in ventricular progenitors of the developing cardiac primordia of stage 9 chicken embryos and continues to be expressed in the ventricular region of the heart tube of later stage embryos (unpublished observations). Similarly, cardiac α-actin is expressed in the early chicken embryo starting at stage 5 and continuing onward with expression in the early heart tube in which the α-actin contributes to the contractile properties of cardiomyocytes as evidenced by the contractile beating of the heart tube. Another contractile protein, myosin heavy chain (MHC), is temporally co-expressed with α-actin (i.e., around stage 7) with specific localization to the cardiogenic cresent. Expression of MHC in cardiomyocytes continues throughout embryonic development and into adulthood.

We have previously demonstrated that the temporal separation in the appearance of cardiac and skeletal muscle–specific gene markers is in agreement with the separation in the onset of the morphogenetic differentiation of the respective cell types (Goswami et al., 1994). The distinct temporal and spatial expression of early muscle-specific genes suggests that their expression is dependent upon the interaction of trans-acting, lineage-specific transcription factors with *cis*-elements in their promoters. For example, the mRNA for CMD-1, the chicken homologue of MyoD (Lin et al., 1989) appears at stage 5, at least two stages prior to the appearance of the first somite from which skeletal muscle progenitors are derived. Perhaps one of the best studied examples of activation of cardiac-specific gene promoters is the interaction of transcription factors with the promoter of the vMLC-2 gene. Our laboratory has focused on the MLC-2 gene promoter and has characterized a number of *cis* regulatory elements in the promoter of this gene including the myocyte enhancer factor, MEF-2, binding sequence (element B), a CArG sequence (element A) (Qasba et al., 1992; Zhou et al., 1993; Goswami et al., 1994), and a negative regulatory element, cardiac specific sequence (CSS) (Shen et al., 1993; Dhar et al., 1997). See Figure 1.

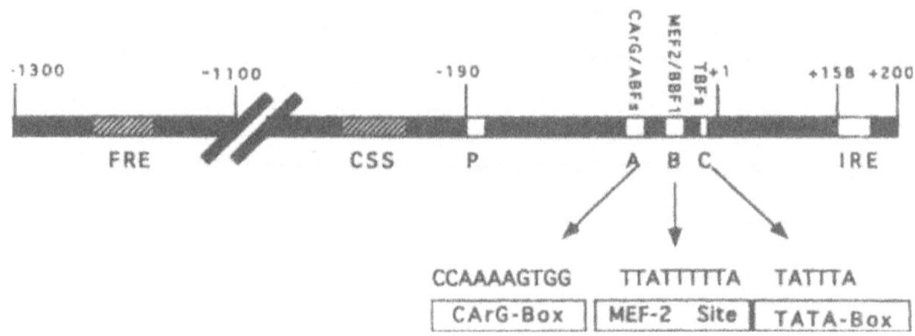

Figure 1. Functional sequence domains in the cardiac myosin light chain-2 (MLC-2) promoter. The sequences of elements A, B, and C are indicated. Both element A (CArG box) and element B (MEF-2–like sequence) are involved in up-regulation of the activity of the MLC-2 promoter.

Of the factors known to bind MLC-2 promoter elements, as well as the promoter elements of other muscle-specific genes, some of the best described are members of the MADS family of transcription factors (Olson et al., 1995). This includes the myocyte enhancer factors or MEF-2 proteins. In murine embryogenesis, MEF-2 expression occurs in cardiac and sketetal muscle as well as in neuronal cell lineages (Edmondson et al., 1994; Lyons et al., 1995). The early expression in muscle progenitor cells of MEF-2 members (of which there are four different isoforms) indicates that they may play a variety of roles in the early stages of myogenesis (Olson et al.,1995). In support of this idea, cMEF2 transcripts can be detected in early chicken embryos as early as stage 2 of development, prior to cardiogenic specification, and also in the stage 6 cardiogenic crescent with later restriction to the developing ventricle (Goswami et al., 1994 and unpublished data). Avian serum response factor (Croissant et al., 1996), a related member of the MEF-2 family, has been shown to play a role in cardiogenic differentiation and can be detected in developing heart primordia at stage 7 of chicken development.

Perhaps the most convincing evidence for the essential role of MEF proteins in cardiomyogenesis comes from recent genetic ablation studies in which the absence of MEF-2C in mice leads to developmental cardiac abnormalities and death in utero (Lyons, 1996). It is impotant to note from these studies that cardiac specification and differentiation are unaffected, suggesting that the MEF proteins may play important roles in later stages of cardiogenesis. The MEF-2 factors activate transcription through an AT-rich DNA sequence in control regions of various muscle specific genes (Olson et al., 1995).

In light of the complexity of the vMLC-2 gene promoter, it is likely that factors other than MEF-2 contribute to the complex regulation of this gene. In attempting to identify such factors, we have recently identified and characterized a new cardiac muscle transcription factor called BBF-1, which binds the B element of the vMLC-2 promoter. Despite binding the B element or MEF-2 binding site, BBF-1 is immunologically distinguishable from MEF-2 (Goswami et al., 1994). In avian embryos, BBF-1 and MEF-2 binding activities are of predominantly cardiac origin (Goswami et al., 1994) since they appear in chick embryos prior to skeletal muscle formation at stage 8.

For further investigation of the role of BBF-1 in cardiac development and differentiation, we have isolated a recombinant plasmid (pCLP-1) as a candidate for BBF-1 from a cDNA library constructed using mRNA from the heart-forming regions of stage 6 embryos (manuscript in preparation). CLP-1 is a novel transcription factor with binding sequence specificity similar to that of MEF-2 proteins. Analysis of CLP-1 expression by RT-PCR showed that CLP-1 transcripts were detected as early as stages 2 to 3, prior to the appearance of B element binding activity at stage 4. Whole mount in situ hybridization analysis shows that the pattern of CLP-1 expression in gastrulation stage embryos coincides with the heart forming regions (Rawles, 1943; DeHaan, 1965) as well as developing heart. Later expression is restricted to terminally differentiated cardiomyocytes in adult heart, (see Figure 2). The appearance of CLP-1 mRNA at stage 2 and cardiac muscle-specific transcripts at stage 4 support the idea that CLP-1 may be the earliest marker of cardiogenic differentiation yet described. Interestingly, the subcellular localization and nuclear import of CLP-1 appears to be controlled by phosphorylation (Vandromme et al., 1996). In order to ascertain the function of CLP-1, we attempted targeted inhibition of CLP-1 mRNA using modified antisense oligonucleotides in cultured embryos. These embryos upon development showed a clear anomaly in fusion of the cardiac primordia with a dramatic reduction in CLP-1 protein, substantiating the effectiveness of the antisense approach. Despite this finding, early heart development up until the point of cardiac primordia fusion appeared to be normal, based on both overt appearance of embryos and the apparently normal expression of cardiac MLC-2.

If CLP-1 is involved in early events such as cardiogenic specification by virtue of its expression at stage 2/3, a total absence of cardiac development would be expected if its function is disrupted by antisense oligonucleotides. Since this did not happen, the most likely explanation is that the antisense oligonucleotides were introduced too late in development(stage 4) to interfere with the early function of endogenous CLP-1 mRNA. Such a delay in manifestation of morphogenetic defects is not unexpected and has been observed previously regardless of the experimental technique used for introduction of the antisense oligonucleotides (Srivastava et al., 1995; Wei et al., 1996). Implicit in this, however, is the suggestion that CLP-1 may have two distinct roles: one that can be ascribed to its expression at stage 2/3, presumably in specification/commitment of mesodermal cells, and the other that is revealed via interference at a late stage by antisense oligonucleotides resulting in the arrest of fusion of cardiac primordia.

In hypertrophy, cardiac myocytes accumulate contractile proteins in a pattern of expression characteristic of cardiomyocytes during development of the cardiovascular system. In addition to the accumulation of MHC, expression of MLC-2 is also in-

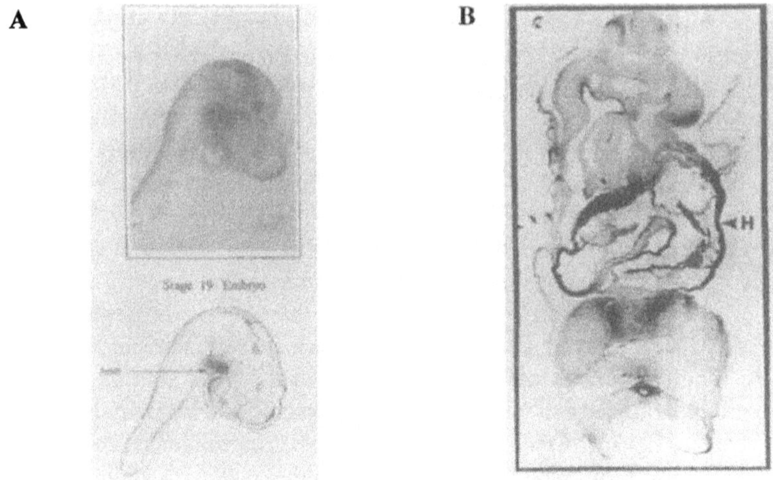

Figure 2. A. CLP-1 expression pattern by whole mount in situ hybridization. Dorsal view of stage 19 embryo hybridized with digoxigenin-labeled CLP-1 antisense riboprobe. **B**. Histological section passing through the heart region of the same embryo. Note that the CLP-1 transcripts are exclusively localized in heart.

creased (Lompre et al., 1991; Chassange et al., 1993). This increase is seen in cardiac muscle in which hypertrophy has been induced by hemodynamic overload, constriction of the aorta, or agonist-treated cardiomyocytes in culture and is present in cardiac muscle from the spontaneously hypertensive rat (SHR) (Swynghedauw, 1992). Together, these examples suggest that despite the particular etiology of the hypertrophic state, a common element in the adaptive response of cardiomyocytes to hypertrophy is the increase in MLC-2. Moreover, since the observed increase in MLC-2 protein results from an increase in MLC-2 mRNA levels, an important feature of the adaptive response to hypertrophy must be the re-expression and use of the same genetic mechanism(s) used to regulate MLC-2 gene expression during development (Chien, 1992).

III. REGULATION OF MYOSIN LIGHT CHAIN-2 GENE ACTIVITY IN HYPERTROPHY

As mentioned above, much is known about the transcription factors and promoter elements driving MLC-2 expression during cardiac progenitor differentiation and morphogenesis of the heart. The observed re-expression of MLC-2 during hypertrophy in adult cardiac muscle raises the question as to whether the same developmental mechanisms are being called upon to elicit MLC-2 gene expression or whether some alternative mechanism is operating. Recent experimentation in our laboratory has addressed this question. Using the SHR strain as a model of hypertrophy, we have assessed whether known elements of the transcriptional apparatus responsible for MLC-2 gene transcription undergo changes in response to this genetically manifested hypertrophic state.

As with our developmental studies, the electrophoretic mobility shift assay (EMSA) was used to assess the binding of transcription factors to known binding elements within the promoter of the MLC-2 gene. Use of EMSA in both developmental and hypertrophic studies allowed for a more direct comparison of the transcription factors involved (Doud et al., 1995). As previously seen in developmental studies, nuclear extracts from SHR hypertrophic cardiac muscle clearly exhibited enhanced binding to the B element (MEF-2 binding site). Three findings of these experiments are of note: (1) B-element binding is cardiac specific; (2) the B element appears to bind multiple complexes, indicating the presence of multiple B-element binding proteins; and perhaps most importantly, (3) binding activity is strictly correlated with the onset of hypertrophy. Additional EMSA analyses showed that another promoter element important for the transcription of the MLC-2 gene, the CArG box (or element A) also bound nuclear factors found in extracts of SHR cardiac muscle progressing toward hypertrophy. Hypertrophic cardiac muscle extracts exhibited normal levels of binding activity to basal promoter sequences suggesting that the up-regulation of the MLC-2 promoter in hypertrophic cardiomyocytes may be due solely to changes in binding activities of elements A and B. Since MEF-2 transcription factors bind element B, and serum response factors and perhaps cell-type–specific factors bind element A, the above findings suggest that these factors play a role in the expression of MLC-2 both during development and in the adaptive response of cardiomyocytes to hypertrophy.

One of the hallmarks of hypertrophy is the re-expression of contractile proteins in a way reminiscent of their expression during embryogenesis. We have used the vMLC-2 gene as a model to study the mechanisms controlling cardiac-specific gene transcription during both development and in the adaptive response of cardiomyocytes to hypertrophy. Our findings indicate that the adaptive response to hypertrophy may rely on the same factors and transcriptional elements used during development to bring about expression of the vMLC-2 gene. Given these observations, the overriding question now is how hypertrophic changes to the cell are transmitted via the cell's signal transduction pathways to bring about re-expression of contractile protein genes. In the case of vMLC-2, this would require redeployment of the transcription factors MEF-2, BBF-1, and perhaps other factors needed for the establishment of an active MLC-2 transcriptional gene complex. This, in turn, necessitates a more complete understanding of the regulation of the MEF-2 and BBF-1 gene family members than what is now presently known. Because the nature of the hypertrophic state and the signals it triggers may differ from that of a mesodermal cell undergoing committment and differentiation to the cardiogenic lineage, the regulatory mechanisms controlling MEF-2 and BBF-1 in hypertrophy may differ significantly from those acting during development. In this sense, control of

MEF-2 gene expression may be exerted at any one of a number of levels, for instance, via phosphorylation of transcription factors by different mitogen-activated protein kinase signal trandsduction pathways; G protein–mediated signal transduction; direct phosphorylation events mediated by protein kinases such as protein kinase C, heat shock, or stress-dependent signalling events; and so on.

Along these lines, one interesting possibility stemming from our studies of the transcription factor CLP-1 is the regulation of its activity by controlling its transport from cytoplasm to nucleus. The subcellular distribution of CLP-1 is dependent on treatment of cells with cyclic adenosine monophosphate (cAMP), which presumably stimulates phosphorylation by a cAMP-dependent kinase. In hypertrophy, this or a similar signalling mechanism may be used to translocate CLP-1 to the nucleus where it activates vMLC-2 gene expression and ultimately vMLC-2 protein accumulation. Given these possibilities, future studies focusing on the control and activation of transcription factors such as MEF-2 and BBF-1 may provide a more thorough understanding of the molecular events underlying the adaptive response of cardiomyocytes to hypertrophy.

REFERENCES

Chassange, C., Winewsky, C., and Schwartz, K. (1993). Antithetical accumulation of myosin heavy chain but not alpha-actin mRNA isoforms during early stages of pressure-overload-induced rat cardiac hypertrophy. Cir. Res. 72, 857-864.

Chien, K. (1992). Signaling mechanisms for the activation of an embryonic gene program during the hypertrophy of cardiac ventricular muscle. Basic Res. Cardiol. 87 (suppl. 2), 49-58.

Croissant, J.D., Kim, J.H., Eichele, G., Goering, L., Lough, J., Prywes, R., and Schwartz, R.J. (1996). Avian serum response factor expression restricted primarily to muscle cell lineages is required for α-actin gene transcription. Dev. Biol. 177, 250-264.

DeHaan, R. (1965). Morphogenesis of vertebrate heart. In: Organogenesis (DeHaan, R.L., and Ursprung, H., Eds.), p. 377. Holt, New York.

Dhar, M., Mascareno, E.M., and Siddiqui, M.A. (1997). Two distinct factor-binding DNA elements in cardiac myosin light chain 2 gene are essential for repression of its expression in skeletal muscle. Isolation of a cDNA clone for repressor protein Nished. J. Biol. Chem. 272, 18490-18497.

Doud, S.K., Pan, L., Carleton, C., Marmorstein, S., and Siddiqui, M.A.Q. (1995). Adaptational response in transcription factors during development of myocardial hypertrophy. J. Mol. Cell. Cardiol. 27, 2359-2372.

Edmondson, D.G., Lyons, G.E., Martin, J.F., and Olson, E.N. (1994). Mef2 gene expression marks the cardiac and skeletal muscle lineages during mouse development. Development 120, 1251-1263.

Goswami, S.K., Qasba, P., Ghatpande, S.K., Carlton, S., Deshpande, A.K., Baig, M.A., and Siddiqui, M.A.Q. (1994). Differential expression of the mef2 family of transcription factors in development: the cardiac factor BBF-1 is an early marker for cardiogenesis. Mol. Cell. Biol. 14, 5130-5138.

Lin, Z.Y., Dechesne, C.A., Eldridge, J., and Paterson, B.M. (1989). An avian muscle factor related to MyoD1 activates muscle specific promoters in nonmuscle cells of different germ layer origin and in Brdu-treated myoblasts. Genes Dev. 3, 986-996.

Lompre, A.M., Mercadier, J.J., and Schwartz, K. (1991). Changes in gene expression during cardiac growth. Int. Rev. Cytol. 124, 137-186.

Lyons, G. (1996). Vetebrate heart development. Curr. Opin. Genet. Dev. 6, 454-460.

Lyons, G.E., Micales, B.K., Schwarz, J., Martin, J.F., and Olson, E.N. (1995). Expression of mef2 genes in the mouse central nervous system suggests a role in neuronal maturation. J. Neuro. 15, 5727-5738.

Manasek, F.J. (1968). Embryonic development of the heart. I. A light and electron microscopic study of myocardiac development in the early chick embryo. J. Morphol. 125, 329-366.

Olson, E.N., Perry, M., and Schulz, R.A. (1995). Regulation of muscle differentiation by MEF2 family of MADS box transcription factors. Dev. Biol. 172, 2-14.

Qasba, P., Lin, E., Zhou, M.D. Kumar, A., and Siddiqui, M.A.Q. (1992). A single transcription factors binds to two divergent sequence elements with a common function in cardiac myosin light chain-2 promoter. Mol. Cell. Biol. 12, 1107-1116.

Rawles, M.E. (1943). The heart forming regions of the early chick blastoderm. Physiol. Zool. 16, 22-42.

Rosenquist, G.C., and DeHaan, R.L. (1966). Migration of precardiac cells in the chick blastoderm: a radioautographic study. Carnegie. Inst. Wash., Contrib. Embryol. 38, 111-121.

Shen, R., Goswami, S., Mascareno, E., Kumar, A., and Siddiqui, M.A.Q. (1992). Tissue-specific transcription of cardiac myosin light chain 2 gene is regulated by an upstream repressor element. Mol. Cell. Biol. 11, 1676-1685.

Simpson, P.C., Long, C.S., Waspe, L.E. Henrich, C.J., and Ordhal, C.P. (1989). Transcription of early developmental isogenes in cardiac myocyte hypertrophy. Mol. Cell. Cardiol. 21, 79-89.

Srivastava, D., Cserjesi, P., and Olson, E.N. (1995). A subclass of bHLH proteins required for cardiac morphogenesis. Science 270, 1995-1999.

Stalsberg, H. (1969). The origin of heart asymmetry: right and left contributions to the early chick embryo heart. Dev. Biol. 9, 109-127.

Swynghdauw, B. (1992). Chronic cardiac insufficiency, a disease of adaptation. C. R. Soc. Biol. (Paris) 186, 332-341.

Vandromme, M., Gauthier-Rouviere, C., Lamb, N., and Fernandez, A. (1996). Regulation of transcription factor localization: fine-tuning of gene expression. Trends Biochem. Sci. 21, 59-64.

Wei, Y., Bader, D., and Litvin, J. (1996). Identification of a novel cardiac-specific transcript critical for cardiac myocyte differentiation. Development. 122, 2779-2789.

Yazaki, T., Tsuchimochi, T., Kurabayashi, M., and Komuro, I. (1989). Molecular adaptation to pressure overload in human and rat hearts. Mol. Cell. Cardiol. 21, 91-101.

Zhou, M.D., Goswami, S.K., Martin, M.E., and Siddiqui, M.A.Q. (1993). A new serum responsive, cardiac tissue specific transcription factor that recognizes the MEF-2 site in the myosin light chain-2 promoter. Mol. Cell. Biol. 13, 1222-1231.

SIGNAL DIVERGENCE AND CONVERGENCE IN CARDIAC ADAPTATION

Anirban Banerjee, Alden H. Harken, Ernest E. Moore, Kyong Joo, Brian C. Cain, Daniel R. Meldrum, Fabia Gamboni Robertson, Charles B. Cairns, and Xianzhong Meng

Advances in Organ Biology
Volume 6, pages 155-179.

ISBN: 0-7623-0391-3

I. INTRODUCTION

Myocardial adaptation provides widespread opportunity for clinical application and presents new challenges for several branches of basic science endeavor. The authentication of this intrinsic myocardial property by Reimer and Jennings in 1986 (Murry et al., 1986) and the exploration of its mechanisms by Downey (Downey et al., 1994) and others (Reimer and Jennings, 1992; Przyklenk and Kloner, 1995; Yellon and Baxter, 1995; Banerjee et al., 1996) coincides fortuitously with advances in the study of signal transduction. Postreceptor events involving G proteins and protein kinases link hormones and extracellular receptors to the modifications of cardiac metabolic phenotype and resistance to injury. Although we are only beginning to understand the molecular mechanisms of cardioadaptation, a remarkable theme of efficiency and flexibility has begun to emerge. In essence, myocardial adaptation attests that specialized cells such as myocytes (and perhaps many others) remain sensitive to aspects of their past history and adapt their future response to subsequent stimuli accordingly (Banerjee et al., 1996; Meldrum et al., 1996a). Altered responses span the range from intermolecule to interorgan (e.g., receptor desensitization, neutrophil "priming" events, growth and apoptosis, post-traumatic multiple organ failure) (Meldrum et al., 1996a). Thus, sensitivity to previous conditions can be better understood due to an improved recognition of signaling mechanisms. In conjunction with an improved sense of how adaptation is initiated and how cardiac responses can be modified, dramatic advances in therapy are possible, capitalizing explicitly on the vast repertoire of intrinsic cellular adaptation mechanisms.

In this review, we survey the physiology and biochemistry underlying the signal-system interactions in cardiac adaptation. Acute and delayed adaptation are separately examined for evidence of recurring organizational themes. The physiological observations made about the heart are rephrased in the language of transmembrane signaling pathways that are interrelated by molecular homology. Lastly, the experimental insights from cardiac preconditioning and signaling are integrated with a thermodynamic hypothesis that postulates that reshuffling of resources optimizes postischemic outcome. Finding clear themes in signaling design within space constraints prevent us from reviewing many excellent studies on cardiac adaptation. The examples used are limited primarily to the developments made in our own laboratories. However, progress in signal transduction is still accelerating. Current concepts in specific areas are covered in much greater depth in recent reviews (Cohen, 1992; Inagaki et al., 1994; Clark and Brugge, 1995; Mochly-Rosen, 1995; Taniguchi, 1995; Alberts et al., 1996; Banerjee et al., 1996; Force et al., 1996; Iyengar, 1996).

Cardiac adaptation is an endogenous mechanism. However, the adaptation could be beneficial or not (Klug et al., 1993). Throughout the subsequent discussion, we define cardiac adaptation as an observable change in stimulated output. Experimentally, such an interrogative stimulus is not essential if physiological or biochemical markers of the adaptive state are known. It also follows that the adaptive state must itself be induced by a cellular stimulus preceding or concurrent with the interrogating stimulus.

II. EXPERIMENTALLY INDUCED ENDOGENOUS CARDIAC PROTECTION

A. Delayed Cardiac Adaptation

Our interest in inducing delayed cardiac adaptation began almost a decade ago with the finding that endotoxic stress induced cardiac protection in rats 24 hours after administration (Brown et al., 1989). After a single sublethal dose of endotoxin (0.5 mg/kg), both antioxidant defenses and functional recovery after ischemia–reperfusion (IR) injury were enhanced (Brown et al., 1989). These early successes in inducing protection in healthy animals with bioactive molecules have continued to propel our ongoing efforts to understand the therapeutic potentials and limitations of delayed cardiac adaptation (Brown et al., 1989; Brown et al., 1990; Nelson et al., 1991; Brown et al., 1992).

Endotoxin-induced cardiac adaptation likely includes a medley of mechanisms involving several cell types and different signaling pathways. The delayed protection induced by endotoxin occurs about 24 hours after administration and persists for approximately seven days (Meng et al., 1996b). The delayed cardiac adaptation seems to be protective against IR injury (Brown et al., 1989), as well as endotoxin-inflicted myocardial depression (Meng et al., 1996b). Due to the systemic and cellular toxicity of endotoxin (Ulevitch and Tobias, 1995), we searched for nontoxic dephosphorylated metabolites of endotoxin and discovered that monophosphoryl lipid A could induce similar cardiac protection (Nelson et al., 1991). In lieu of endotoxin, delayed cardiac adaptation could also be elicited by the cytokines tumor necrosis factor α (TNFα) and interleukin-1 (IL-1) that are produced systemically following exposure to endotoxin (Brown et al., 1990; Brown et al., 1992; Dinarello, 1996).

Endotoxin-induced cardiac adaptation involves a modest oxidant stress (Maulik et al., 1995), perhaps due to activation of neutrophils (Brown et al., 1992; Meldrum et al., 1996d). Bolli (1996) has established that delayed cardiac adaptation induced by cyclic ischemia in swine requires oxidant stress. Das and co-workers (1994) have found evidence for the induction and transcription of genes coding for mitochondrial proteins, antioxidant enzymes, and heat shock proteins by diverse cardioprotective stimuli ranging from cyclic ischemia to endotoxin and cytokines. Our results also indicate that endotoxin increases antioxidant enzyme activities (Brown et al., 1992; Meldrum et al., 1996d), and induces the expression of sarcomeric actin and myosin heavy chain isogenes (Meng et al., 1996c).

Endotoxin induces catecholamine release, and data from our laboratory and others suggest that an α_1-adrenergic receptor pathway is involved in acute preconditioning (Banerjee et al., 1993; Downey et al., 1994; Hu and Nattel, 1995; Przyklenk and Kloner, 1995; Yellon and Baxter, 1995; Cleveland et al., 1996b). We therefore investigated the role of norepinephrine in delayed cardiac adaptation (Meng et al., 1996a). Our results indicate that endotoxin-induced protection against ischemia and reperfusion injury does not occur if α_1-adrenergic receptors are blocked (unpublished data). Moreover, a single injection of norepinephrine produced protection in two phases (Meng et al., 1996a). The first acute phase appeared rapidly and faded by 2 hours. Delayed protection appeared at 4 hours after injection, and the adaptive phenotype was retained for several days. The mechanism of the delayed adaptation is clearly dependent on new protein synthesis since it can be inhibited by cycloheximide (Meng et al., 1996d). The delayed adaptation may also be dependent on altered gene transcription of c-fos and heat shock protein (HSP) 70 (Meng et al., 1996a). Thus, mRNA transcription and subsequent translation of proteins may result in constructive cardiac remodeling. Interestingly, Marber obtained direct evidence from a transgenic model that HSP 70 protects myocardium against IR injury (Marber et al., 1995). How HSP 70 and other unidentified proteins singly, or more likely in combination, might induce the ischemia-resistant phenotype has become an area of active investigation (Maulik et al., 1995; Yellon and Baxter, 1995; Bolli, 1996; Meng et al., 1996b).

B. Acute Cardiac Adaptation

Repeated bouts of brief ischemia are well tolerated and, remarkably, also produce a "preconditioning" effect that subsequently enables the myocardium to recover from a severe ischemic insult with cellular viability and contractile function relatively intact as compared to untreated myocardium. This acute adaptation occurs in all higher mammals examined (Przyklenk and Kloner, 1995; Yellon and Baxter, 1995; Cleveland et al., 1996b; Cohen and Downey, 1996). Researchers in this area have successfully identified a few proximal signaling components induced by ischemic stress. So far, these signals appear to involve neuroendocrine hormones and metabolites formed in the ischemically stressed myocardium (Reimer and Jennings, 1992; Yao and Gross, 1993; Goto et al., 1995; Banerjee et al., 1996; Cleveland et al., 1997a; Meldrum et al., 1996c). Many of these are agonists at receptors linked to protein kinase C (PKC).

Our initial efforts focused on the least ischemic stress that could improve postischemic ventricular functional recovery: a single 2-minute ischemic episode (Banerjee et al., 1993; Brew et al., 1995; Mitchell et al., 1995). In the isolated heart, this stimulus appeared to mobilize norepinephrine stores in adrenergic termini (Banerjee et al., 1993). Norepinephrine was found to act via α_1-adrenergic receptors in order to elicit cardiac protection against subsequent ischemia and reperfusion injury since α_1-adrenergic antagonists abolished protection induced by either norepinephrine or a 2-minute ischemic stress (Banerjee et al., 1993). Similar protection could also be obtained with direct stimulation of α_1-adrenergic receptors with phenylephrine (Banerjee et al., 1993). Based on the signaling of the α_1-adrenoreceptor, we anticipated that protein kinases might be involved (Banerjee et al., 1993). During reperfusion, energy metabolism was also beneficially modified (Banerjee et al., 1993), and indicated by simultaneously improved adenosine triphosphate (ATP) replenishment and augmented mechanical function.

Further investigations in several laboratories showed that the α_1-adrenergic receptor pathway to acute cardiac adaptation involved protein kinase C (PKC) (Downey et al., 1994; Mitchell et al., 1995; Przyklenk and Kloner, 1995; Yellon and Baxter, 1995). Direct PKC stimulation with an endogenous PKC activator diacylglycerol (DAG) induced cardiac protection against ischemia and reperfusion injury, and the cardiac protection afforded by this stimulus could be abolished with pan-PKC blockers. Downey therefore suggested that PKC-linked receptors in general might be cardioprotective and provided extensive supporting evidence in rabbit hearts (Downey et al., 1994; Goto et al., 1995). We evaluated other PKC-linked receptors in rat hearts and found that both bradykinin (Brew et al., 1995) and adenosine (Winter et al., 1996) could be cardioprotective. These two hormones also appeared to be partial mediators of the protection induced by a 2-minute ischemic stress. However, the protection induced by these two receptor pathways is not as prominent in rat hearts as that afforded by the α_1-adrenergic receptor pathway. In the case of adenosine, protection seems to require functioning (i.e., unblocked) α_1-adrenergic receptors (Winter et al., 1996).

Similarly, the robust protection afforded by several ischemia–reperfusion cycles also appears to involve a more complex interplay within the handful of known signaling contributors (Goto et al., 1995). The augmented protection against sustained IR injury suggests that for this stimulus, the combination of stress factors repeated in some special sequence may be crucial for optimal protection. It is not clear whether this is a purely dose or timing issue, or whether synergistic interactions among several factors modify the sense of the final signal qualitatively. Blockade experiments have shown that inhibition of a single signaling pathway failed to nullify the protection conferred by cyclic ischemic preconditioning, suggesting that complex stress stimuli may be more than the sum of their parts (Goto et al., 1995).

An obvious extension of pharmacologically induced acute and delayed protection is to maximize the cardiac protection by synergistic signal combination. Thus, it appears that the "genomic" protection that is evoked 24 hours after endotoxin injection can be further improved by a single ischemic preconditioning episode (Rowland et al., 1997). The combined protection is comparable to that obtained by several cycles of brief ischemia–reperfusion. Although offering superb and comprehensive protection, this protocol is lengthy and violently stressful. Therefore, we have investigated purely pharmacological combinations. It appears that acute adenosine receptor preconditioning can be added on beneficially to augment endotoxin-induced adaptation (Meldrum et al., 1997).

C. Summary and Implications

The neuroendocrine factors at play suggest that cardiac adaptation involves interactions among several myocardial structures. During ischemic stress, adrenergic and parasympathetic neurons may release norepinephrine (Banerjee et al., 1993) and acetylcholine respectively (Goto et al., 1995). Furthermore, the vascular endothelial system generates and releases bradykinin (Brew et al., 1995; Goto et al., 1995) while myocytes and other cell types produce the catabolite adenosine (Olsson and Pearson, 1990; Goto et al., 1995). Cardiac adaptation induced by endotoxemic stress might involve neuroendocrine factors and cytokines, as well as stimulated oxidant release from leukocytes. It should be noted that the final outcome of adaptation may be not only dependent on the timing of the production and release of these endogenous factors but also modified by the quantity, spatial distribution, and combination of these endogenous factors.

A casual overview of the blockade and pharmacological mimicry data might suggest that catecholamines, cytokines, oxidant stress, and cardiac remodeling are all linearly arranged in cause-and-effect sequences. However, such linearity is quite unlikely. Indeed, diverse intracellular signals converge at certain critical points. Moreover, there is a critical limitation inherent in gauging phenotypic adaptation on the basis of a single index: the use of postischemic ventricular recovery as an end point. Therefore, the mechanistic significance of inhibition or antagonism should be accepted as an indication that a signaling pathway is involved in undefined proportions (additive, synergistic, permissive, amplification, or inhibitory). The fact that each of these diverse putative mediators present at some optimal condition can induce functional protection suggests that (1) the mechanism of protection could be distinct (but indistinguishable by postischemic contractile measures), (2) additive or synergistic mechanisms may be at work, and/or (3) there are "convergence points" within the cell where diverse signals can be integrated to produce a common or similar phenotype.

It remains unclear how neurons, myocytes, and other cells sense complex stressors such as ischemia or endotoxin and how these mechanisms can detect and correctly interpret the hormone admixtures in the extracellular milieu. These issues are briefly discussed in the next section. Attention is drawn to the fact that cellular receptors receive diverse primary messages simultaneously. The stimulated receptor subtypes then generate several intracellular secondary messengers, which in turn go on to stimulate multiple other cascades simultaneously.

III. TRANSMEMBRANE SIGNALS, AMPLIFICATION, AND DIVERSIFICATION

In order to adapt, the myocardial cells must sense and interpret changes in the environment. Receptors, coupling elements, and effector proteins allow cells to communicate with and respond to the environment. Changes in the composition of the extracellular matrix can be detected by transmembrane hormone receptors and ion transporters (Alberts et al., 1996). The identity and character of the surrounding cells is sensed by integrins and other transmembrane adhesion complexes (e.g., cadherins and CAMs) (Alberts et al., 1996). Changes also involve strictly dynamic quantities such as rates of energy turnover or oscillating concentration gradients (Stucki, 1982; Katz, 1992). These variations are detected by sensory mechanisms which are themselves time and concentration sensitive, often based on dynamic equilibria formed by opposing kinases and phosphatases, or allosteric and feedback control of key enzymes (Cohen, 1992; Alberts et al., 1996).

A. G Protein–Linked Seven Transmembrane Helix Receptors

In ischemic preconditioning (Goto et al., 1995; Przyklenk and Kloner, 1995; Banerjee et al., 1996; Cohen and Downey, 1996), the best characterized class of myocellular signal sensing/transducing elements is the superfamily of seven transmembrane helix (TmHx) receptors that are linked to heterotrimeric G proteins (Hepler and Gilman, 1992; Watson and Girdlestone, 1996) (Figure 1). The ligands for the TmHx receptors may be hormones (epinephrine, angiotensin II, and endorphins) in the circulating or those produced locally. Local sources include sympathetic and parasympathetic termini (norepinephrine and acetylcholine) or endothelium (bradykinin and endothelin). The involvement of other receptors of this class, such as the ATP activated purinergic P2 receptors, in cardiac adaptation remains to be fully investigated. ATP is co-released with norepinephrine (Westfall et al., 1990) and purinergic P2 receptors are linked to phospholipase C (PLC). Thus, variations in extracellular ATP may indirectly modulate PKC activity (Olsson and Pearson, 1990).

Conduction of this extracellular signal from the sarcolemmal TmHx receptors into the cell typically involves the heterotrimeric G proteins (Hepler et al., 1992; Watson and Girdlestone, 1996) (see Figure 1). The TmHx receptors assist the α subunit to exchange guanosine trisphosphate (GTP) for bound guanosine diphosphate (GDP). This frees the G_α subunit from the $G_{\beta\gamma}$ subunits and causes G_α to acti-

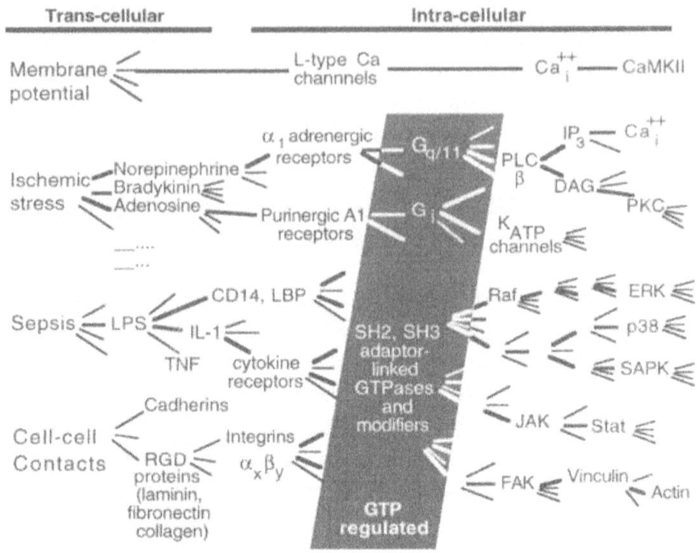

Figure 1. Signal divergence occurs at several levels during signal transduction. Downstream regulation (denoted by lines) can be either positive or negative, depending on species cell type and preceding signals.

vate powerful signal-amplifying elements such as adenylate cyclase (AC) or $PLC_{\beta 2}$. While an agonist remains bound to its TmHx receptor, active G_α units can be generated. However, the G_α subunits possess intrinsic GTPase activity. Thus, when the bound GTP is eventually hydrolyzed to GDP, the G_α subunit is inactivated and reforms the basal heterotrimeric complex with the $G_{\beta\gamma}$ units and the TmHx receptor (Hepler and Gilman, 1992; Ray et al., 1996). This terminates the hormone signal but not the downstream amplification. Additional proteins can bind, which either increases or decreases the intrinsic GTPase activity. This provides a means of fine tuning the duration of the G_α signal while ensuring its eventual termination (Iyengar, 1997). It should be recognized that signaling is not mediated solely by GTP-bound G_α units (Iyengar, 1997). Thus, signal divergence can occur immediately distal to the receptor, since the free $G_{\beta\gamma}$ units themselves can activate PLC, phospholipase A_2 (PLA_2), potassium channels, and the Ras signaling pathways (Ray et al., 1996).

The first step in this particular cascade illustrates three distinct processes: (1) sensing of the external signal (hormone) by a TmHx receptor, (2) transmission of switching inputs to the G_α, and $G_{\beta\gamma}$ units, and (3) disassembly of the TmHx–$G_{\alpha\beta\gamma}$ complex (see Figure 1). The amplification stage comprises many roving G_α and $G_{\beta\gamma}$ units. At this level, competition between specifically opposed G_α subunits, and their interactions with $G_{\beta\gamma}$ units provide an important mechanism for integrating postreceptor events with other parallel TmHx receptor pathways that may have been simultaneously stimulated. GTP dependence ensures that the G_α signal will be terminated if the receptor ligand is removed. This ensures that the system can respond to subsequent stimuli (Alberts et al., 1996). The third stage of signal processing initiated by extracellular agonists is the activation of major signal amplifiers by the G_α and $G_{\beta\gamma}$ subunits. As long as the G_α subunits remain active and bound to the signal amplifier, prodigious amounts of second messengers may be produced. For example, during its tenure in the GTP-bound active state (approximately 1 second) the G_α subunit might have stimulated the production of over 10^5 cyclic adenosine monophosphate (cAMP) molecules (the second messenger produced from ATP by AC), before it is inactivated by GTP hydrolysis (Alberts et al., 1996). Thus, a single hormone-receptor binding event can be impressively amplified.

The free energy (DG) driving the amplifier is obtained by catalyzing an energetically favored breakdown of stable high-energy compounds. These include the chemical potential of ATP (powering AC), GTP (powering GC and G_α subunits), membrane phospholipids (phosphatidyl inositol [PIP_2, powering $PLC_{\beta 2}$], as well as transmembrane electrochemical potentials. In order to economize on high energy usage, these amplifiers are designed to be active only while they remain stimulated. Some systems terminate the amplified signal by removing the second messengers. The cyclic nucleotide messengers can be efficiently degraded by phosphodiesterase enzymes. The products of PIP_2, inositol triphosphate (IP_3), and DAG, can be recombined, degraded, or metabolized to form other active messengers. Interestingly, the enzymes catalyzing elimination of second messengers are often themselves modulated by certain second messengers or their effector kinases (Cohen, 1992).

A second messenger can itself be considered as a conditioned signal that is an input to an amplifier (the kinase or phosphatase), similar to the design scheme for G_α units outlined earlier (see Figure 1). In most cases, the second messenger derepresses a kinase or phosphatase that temporarily modifies the phosphorylation status of cellular protein machinery. In this way, enzymatic affinity, turnover, and even location can be quickly and reversibly modified, without requiring synthesis of a structurally different kinase or phosphatase isoform.

Continuing the trend toward greater signal diversification, second messengers are not linked to any one amplifier (e.g., kinases) exclusively (see Figure 1). Thus, cAMP not only activates protein kinase A (PKA) but also cAMP-regulated phosphodiesterases and phosphatases (proximal and distal termination mechanisms) (Cohen, 1992). Similarly, one of the PLC products, IP_3 (co-generated with DAG), acts as a switch to release Ca^{2+} from internal stores (an amplifier powered by the gradient of stored Ca^{2+}). Among other switching effects, the released Ca^{2+} can synergize with the other PLC product, DAG, to augment PKC activity (Asaoka et al., 1992). Ca^{2+} also activates DAG metabolism (Asaoka et al., 1992), and facilitates Ca^{2+} reuptake into sarcoplasmic stores, illustrating distal self-termination by feedback (Katz, 1992). From these design trends, we can expect a dephosphorylating mechanism to reverse the phosphorylation conducted by stimulated kinases. This function is fulfilled by phosphatases that oppose the protein kinases (Cohen, 1992).

B. Ion Channels and Transporters

Ion channels and transporters act as both signal sensors and amplifiers (see Figure 1). The ion gradient across a cell membrane stores energy as chemical potential which can be harnessed by allowing influx down the concentration gradient, through various Ca^{2+} channels (Katz, 1992). This power source can be used as a switch to alter the intracellular ionic composition rapidly. However, the intracellular Ca^{2+} concentration is not raised solely by Ca^{2+} influx. In the heart, initial Ca^{2+} "sparks" from ion channels can trigger Ca^{2+} release from intracellular stores (Katz, 1992). This two step design is efficient in that all the influxing Ca^{2+} does not have to be pumped out against the vast concentration gradient. One mechanism of this energy conservation relies on sequestering the cytosolic Ca^{2+} within intracellular stores rather than on pumping it all out of the cell (Katz, 1992). Thus, the restorative energy expenditure is minimized. This design allows a small change in the open probability of a regulated channel to change the intracellular Ca^{2+} concentration by an order of magnitude. The Ca^{2+} ions within the cell act as an amplified second messenger signal that regulates downstream kinases and phosphatases. There are numerous schemes by which the flux through the transporter is coordinated with parallel signaling pathways. Examples include channel regulation by active G_{α} subunits, changes in the transmembrane potential due to fluxes of other cations or anions, and perhaps even membrane packing pressures.

C. Tyrosine Kinase–Linked Signaling

Many growth factors and cytokines can alter the profile of gene expression and metabolism in the cell, but these extracellular signals are not carried by the familiar TmHx receptors or G proteins (see Figure 1). The primary theme here is ligand binding leading to autophosphorylation of the receptor at tyrosine residues (Daum et al., 1994; Cano and Mahadevan, 1995; Alberts et al., 1996). Phosphotyrosine provides a sufficiently different binding site compared to the phenolic hydrogen on tyrosine itself. This altered affinity can be exploited for energetically favored binding or conformational changes. The Src homology 2 (SH2) domain is one such conserved phosphotyrosine binding domain found in numerous proteins (Daum et al., 1994; Cano and Mahadevan, 1995; Alberts et al., 1996). Many SH-2 containing proteins also contain Src homology 3 (SH3) motifs, which recognize proline rich sequences (PxxP). Hence, many such "adapter" proteins (Grb2, Shc) containing one or more of SH2 and SH3 domains help assemble various kinases with their protein substrates (Daum et al., 1994; Cano and Mahadevan, 1995; Alberts et al., 1996)

In contrast with the disassembly of the TmHx–$G_{\alpha\beta\gamma}$ complex, tyrosine phosphorylation signals the assembly of the preamplifier components at the cell membrane.

SH2 domains may also help recruit phosphatases that terminate the signals. Proteins assembled by the adapters could be effectors such as serine/threonine or tyrosine kinases, phosphatases, membrane-lipid kinases, or hydrolases. The recruited protein might also act as another adapter (such as the Grb2-recruited Sos, which localizes Ras, which in turn localizes the actual kinase Raf) (Daum et al., 1994; Cano and Madadeven, 1995). Signal conditioning by receptor tyrosine-kinases thus clusters proteins into kinetically favored reaction schemes. Interestingly, Sos, activates Ras, by facilitating GDP/GTP exchange (Daum et al., 1994; Cano and Mahadevan, 1995), similar to the TmHx receptors that activate G_α units by helping to bind GTP. The power for this pathway is provided by GTP-activated complexes and favorable steric alignment within and between specific proteins. The binding energy permits the adoption of inactive and active conformations that can conduct ATP-driven phosphorylation (Daum et al., 1994; Cano and Mahadevan, 1995).

In addition to growth factors and cytokine receptor tyrosine kinases, there are cascades of cytoplasmic threonine and tyrosine kinases (TxY) that can be initiated by UV oxidants or by TmHx-linked kinases (Cano and Mahadevan, 1995; Force et al., 1996). These effectors, which are triggered by physical and chemical stress (ERK-1 and -2, stress-activated kinase [SAPK], p38 kinase), are currently under intense study.

In addition to the three superfamilies of transmembrane signal sensors discussed above, advances in cell signaling have identified the superfamily of integrins, which sense and respond to the extracellular environment (Clark and Brugge, 1995) (see Figure 1). These transmembrane complexes detect arginine-glycine-aspartate (RGD) sequences on their extracellular matrix proteins, or on other cells. Integrin signaling produces large-scale modifications of transduction pathways by mechanisms that include the assembly of kinase complexes (often tethered by SH2 and SH3 domains to the cytoplasmic integrin tail). This leads to alterations of cytoskeletal structures and gene transcription (Clark and Brugge, 1995).

D. Summary and Implications

The recurrent theme of amplification of an initial stimulus followed by stimulation of diverse downstream agencies holds for both TmHx receptors and the phosphotyrosine pathways (Daum et al., 1994; Cano and Mahadeven, 1995; Alberts et al., 1996) (see Figure 1). For both, second messenger activities depend on GTP binding and hydrolysis. The notable difference is that whereas TmHx receptors disperse GTP bound active units throughout the cell, in the phosphotyrosine pathways the GTP-activated units are assembled and integrated to form signal initiating supercomplexes. The kinetic and regulatory implications of assembly versus disassembly are not completely clear. However, it appears that the use of phosphotyrosine pathways is prevalent among the growth factor, cytokine, and contact receptor pathways linking the extracellular milieu to gene transcriptional control. These pathways may be especially important in delayed and sustained changes of tissue phenotype.

As external stress stimuli are mechanistically traced through to their downstream components, we must be constantly aware that there are multiple interactions between stress factors and the previous stimulation history. These interactions hold true for both the experimentally manipulated conditions and the controls. Thus, experimentally derived mechanisms are only valid within a predefined context. Each extracellular agonist (i.e., catecholamines, peptides, purines, cytokines) often stimulates several transmembrane sensors in parallel. Moreover, downstream signaling intermediaries interact at several steps, both within and between pathways (Houslay, 1991; Liscovitch, 1992; Iyengar, 1996). Detection of these parallel, but subtle, contributions or cross-talk requires expanded experimental designs. These include, tangential cross-blocking experiments (usually deemed "unfocused") and broad but sensitive outcome indices (protein expression, function, viability, energetics, ionic fluxes, etc.). Lastly, detecting cross-talk requires suitable interrogative tools such as blockers, assays, or transgenic lines. Indeed, overlapping receptor stimulations and extensive cross-talk between parallel receptor pathways challenge the physiologist's experimental design where variables are often varied only singly. Similarly, identifying the steps that are not followed by indeterminate branching may help the perpetual pharmacological quest for greater selectivity.

Extensive interpathway relationships are anticipated during the induction of cardiac adaptation. The design scheme (see Figure 1) of receptor signaling implies that a single hormone that consistently leads to a well-defined result is likely to be the exception rather than the rule. Moreover, different species have probably evolved different dominant pathways that induce similar outcomes. The actual pathway by which a signal spreads is likely to become modified during growth, disease, or stress. Thus, an agent that provides a desired outcome, for example, in an obese adult male, may work poorly or even disastrously in a diabetic female. However, the interaction between signaling pathways also indicates that if a particular signaling pathway is interrupted, a desirable outcome may still be achieved by either bypassing this pathway, or by combining other receptors, G proteins, and second messengers.

IV. SIGNAL INTEGRATION

Clearly, signal diversification cannot proceed without end. The precise meanings denoted by combinations of different hormones, receptors, G proteins, protein kinases, or phosphorylated proteins and transcription factors must be interpreted. Thereafter, the appropriate response must be implemented. This *recombination* is conducted by enzymes and complexes that receive many inputs, but produce fewer outputs. In this respect, biological systems are wondrous in their ability to conduct parallel processing on a vast scale with remarkable speed and efficiency. This principle is illustrated with a few examples (Figure 2).

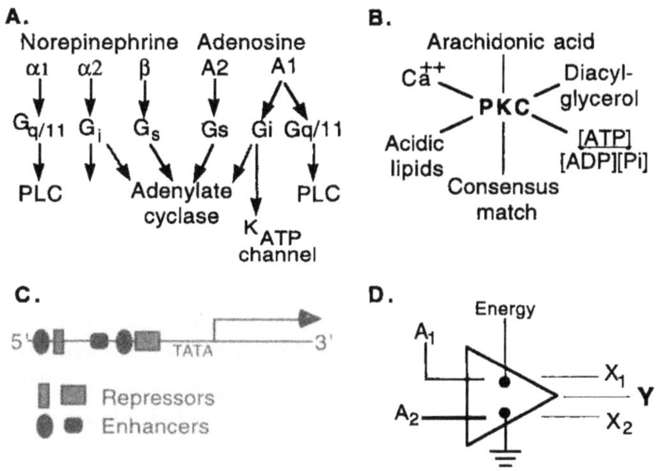

Figure 2. Signal recombination. (**A**) depicts presynaptic summation occurring at a secretory adrenergic nerve terminus. (**B**) shows how the activity of a regulated kinase such as protein kinase C is affected by numerous cellular conditions. (**C**) represents a collection of promoter elements regulating a gene. The correspondence between single or combined inputs and distinct outcomes across a signal recombining agency (∇) is delineated in (**D**) The illustrations are not intended to be complete and are for schematic understanding only.

A. Receptor Pathways

Recombination of messengers distal to their initiating receptors is one of the most proximal steps in signal processing. Many, if not most, neuroendocrine hormones typically stimulate several receptor types (Watson and Girdlestone, 1996), in addition to the receptor pathways that might be implicated in cardiac adaptation. For example, adenosine activates cardioprotective $A_{1a,b}$ receptors and also stimulates $A_{2a,b}$ and A_3 subtypes (Downey et al., 1994; Goto et al., 1995). Adrenergic receptors include the $\alpha_{1(a,b,c,d)}$ types, which are implicated in ischemic preconditioning (Banerjee et al., 1993; Hu and Nattel, 1995), and the distinct $\alpha_{2a,b}$ as well as β_{1-3} subtypes. Although the signal transduction pathways for these different receptors are quite diverse they all involve G proteins, amplifiers, and protein kinases (Watson and Girdlestone, 1996). Experimental inhibition–reconstitution strategies usually highlight a particular signaling pathway downstream of a receptor subtype. An example was the delineation of the noradrenergic α_1 receptors to PKC (Banerjee et al., 1993; Mitchell et al., 1995). However, the role of parallel signaling in the background was overlooked. In the case of norepinephrine, the β-adrenergic receptors, and α_2-adrenergic receptors mediated by $G_{\alpha s}$ and $G_{\alpha i}$ respectively, have opposing effects on AC (see Figure 2A). With the growing recognition of cross-talk between receptors and their pathways, this "unique transduction" hypothesis has been challenged profitably. Thus, a parallel role for the β-adrenergic receptors (provoking AC stimulation and Ca^{2+} influx) has been shown to provide a counterpoint pathway to the initially described α_1-adrenergic receptor–PKC pathway (Asimakis et al., 1994). Similarly, adenosine A_2 receptors have recently been shown to participate in preconditioning in coronary vascular endothelium (Zhou et al., 1996), expanding the known functions of adenosine receptors in cardiac adaptation.

In addition to engaging several receptors with a single agonist, complex stimuli, such as ischemic stress and endotoxin, engage a multitude of hormones and cytokines simultaneously. This feature compelled us to re-examine the α_1-adrenergic receptor pathway in acute ischemic preconditioning. Our previous study demonstrated α_1-adrenergic receptor signaling as a necessary and sufficient pathway for preconditioning (Banerjee et al., 1993). A partial role for bradykinin in preconditioning was also noted (Brew et al., 1995; Banerjee et al., 1996).

We therefore examined the adenosine mediation hypothesis of preconditioning that was demonstrated convincingly in rabbit hearts by Downey and others (Downey et al., 1994; Goto et al., 1995). Following Downey's initial observations that adenosine receptor preconditioning was a poorly developed signaling axis in the rat as compared to the rabbit (Downey et al., 1994), we found that the α_1-adrenergic receptor pathway appears indispensable for adenosine A_1 receptor–induced cardioprotection (Winter et al., 1996). Thus, rat myocardium requires "permissive activity" along the α_1-adrenergic pathway (Iyengar, 1996). Furthermore, in the rat, cardioprotection induced by modest episodes of ischemia is abolished by either α_1-adrenergic receptor antagonist or adenosine A_1 receptor blockade (Winter et al., 1996). It appears that the mechanism of cardiac adaptation after modest preconditioning stimuli involves a synergy between the adrenergic and purinergic pathways. In contrast, Downey's comprehensive cross-blocking study revealed that in rabbits, the adenosine A_1 pathway was indispensable (Downey et al., 1994). Indeed, preconditioning by either an α_1-adrenergic agonist or bradykinin required "permissive activity" from the purinergic signal transduction pathway (Goto et al., 1995).

For sublethal endotoxemia, the actual mediators of protection might include cytokines such as TNFα and IL-1β (Brown et al., 1990; Brown et al., 1992), produced by macrophages in response to the stimulation by endotoxin (Ulevitch and Tobias, 1995). Endotoxemia also elicits systemic catecholamine release (Jones and Yelich, 1987) and perhaps adenosine formation from hypoperfused tissues or decoupled mitochondrial respiration (Cairns et al., 1992). All of these putative secondary mediators may induce delayed cardiac adaptation. However, it is still unclear whether these processes are either necessary or sufficient in endotoxin-induced delayed cardiac adaptation.

B. Protein Phosphorylation

Protein kinases and phosphatases can recombine different signal branches and then "choose" the appropriate course of action. Two examples illustrating the principles involved are provided by examining PKC (Asaoka et al., 1992; Nishizuka, 1992; Mochly-Rosen, 1995) and Raf kinase (Daum et al., 1994; Alberts et al., 1996). Unlike the adenylate cyclase, which can add or subtract the strengths of the stimulatory and inhibitory G proteins (Hepler and Gilman, 1992), certain PKC isozymes appear to be able to perform more complex operations (Asaoka et al., 1992; Nishizuka, 1992; Mochly-Rosen, 1995) (see Figure 2B). Some PKC isozymes are activated by binding Ca^{2+} and/or DAG. However, the interactions are synergistic. Thus, DAG binding decreases the Ca^{2+} requirement by an order of magnitude (to about 0.1 μM, which is close to the physiological concentration). Similarly, Ca^{2+} binding also reduces DAG requirement (Asaoka et al., 1992; Nishizuka, 1992). Furthermore, these isoforms are also sensitive to the concentration of eicosanoids and the presence of acidic lipids (Asaoka et al., 1992; Nishizuka, 1992). Lastly, kinase activity is also affected by the ATP/(ADP × P_i) ratio in the cell. In this way, PLC-stimulated signaling (transient DAG elevation and IP_3-mediated increase in intracellular Ca^{2+}) is distinguished and/or combined with phospholipase D (PLD) activity (prolonged DAG elevation) (Asaoka et al., 1992; Nishizuka, 1992). The downstream products of PLA_2, and the numerous enzymes involved in DAG synthesis and degradation are thus automatically included in the expression of PKC catalytic activity. Moreover the final "message" takes into account the energy status of the cell. We have recently exploited the implicit synergy between DAG and Ca^{2+} to induce preconditioning (via PKC) with exogenous Ca^{2+} alone (Meldrum et al., 1996b).

The consensus site phosphorylated by PKC is rather general. Therefore, target specific activity requires PKC isozymes to be translocated into the immediate vicinity of the target protein (Inagaki et al., 1994; Mochly-Rosen, 1995). We and others have documented the stimulated translocation of the cardiac PKC isozymes to particular subcellular compartments of the rat myocyte (sarcolemma, nucleus, myofibrils, etc.) (Disatnik et al., 1994; Mitchell et al., 1995; Banerjee et al., 1996; Meldrum et al., 1996b). The isozyme/compartment translocation profile is rather specific to the extracellular stimulus (e.g., brief ischemia vs. adrenergic agonist vs. bradykinin) (Banerjee et al., 1996). Such localized integrators have evolved complex features to ensure that phosphorylation activity occurs only at the correct location, and to ensure signal termination, including a return to the basal cellular location. As signal integrators that sum the conditions of the cellular milieu and then provide an output (phosphorylation of compartmentalized target proteins), these kinases can be regarded as signal producing switches (Ferrell, 1996).

Raf kinases act on serines and threonines. These kinases illustrate the ability to combine inputs from tyrosine kinase-linked pathways and the TmHx receptor pathways (Daum et al., 1994; Cano and Mahadevan, 1995; Alberts et al., 1996). The initial assembly of GTP-dependent multiprotein assemblies provides anchoring sites for Raf. Raf is then activated by phosphorylation by local kinases such as PKC isozymes or the GTP-dependent Ras. Activated Raf engages serial phosphorylations in the triple-tiered mitogen-activated protein kinase (MAPK) pathways, which then provides activated transcription factors to activate or suppress specific genes (Cano and Mahadevan, 1995). Although the precise details are still being elucidated, it appears that a variety of environmental mitogenic factors and different types of stressors are funneled into these multitiered kinase cascades. This leads to transcriptional activation of gene programs that characterize the stress-induced adaptive phenotype (Cano and Mahadevan, 1995; Force et al., 1996).

Interestingly, the members of the tiered kinase cascades are activated by polyphosphorylation of specific residues (Cano and Mahadevan, 1995; Force et al., 1996). The step that is most understood is the lowest tier, where activation requires dual phosphorylation on certain threonine (T) and tyrosine (Y) residues. The precise sequence differs among the effectors of these pathways. The p38 MAPK is believed to transduce signals of endotoxin stress and osmolar shock (Ulevitch and Tobias, 1995; Force et al., 1996). It is phosphorylated at a TGY sequence. The stress-activated protein kinase (SAPK), which mediates UV- and oxidant-stimulated adaptation, is phosphorylated at a TPY sequence while the ERK-1 and ERK-2 of the classic mitogen-stimulated pathway at a TEY sequence (Cano and Mahadevan, 1995; Force et al., 1996). Thus, full activation relies on a sequence of phosphorylation events. It should be noted that dual phosphorylation implicitly incorporates the effects of distinct threonine and tyrosine phosphatases. In some cases, the order in which residues are phosphorylated can determine activation or inhibition (Cohen, 1992).

C. Transcription Factors

The tiered kinase cascades and the downstream kinases of the TmHx receptors can remodel the pattern of genes expressed by the existing phenotype. Mammalian genes contain transcription factor–binding elements, usually 20 to 5000 bases upstream of the initiation point (TATA box). Transcription factors can be activated by direct phosphorylation or by removal of an inhibitory portion via phosphorylation (Karin and Smeal, 1992; Daum et al., 1994; Force et al., 1996). Different factors can modulate the kinetics of RNA transcription positively, negatively, or conditionally (i.e., depending on the presence or absence of other factors) (see Figure 2C). Some transcription factors are necessary for the assembly of the full RNA polymerase II complex (comprised of more than a dozen different proteins), whereas others hinder or facilitate some steps (Karin and Smeal, 1992). Still other factors facilitate the release of the mRNA, and there are even factors that promote the initiation of protein translation by engaging the spliced mRNA to the ribosome (elongation factors). In many cases, the active transcription factor promotes the de novo synthesis of a second generation of transcription factors (e.g., Fos, Jun) (Force et al., 1996). Thus, these transcription cascades are examples of mechanisms capable of both delayed and sustained cardiac adaptation, as well as signal termination (synthesis of repressive transcription factors, inhibitory proteins, or phosphatases).

Most genes that have been studied so far appear to have multiple transcription factor–binding elements (Karin and Smeals, 1992; Karns et al., 1995). The transcription element and its potency dictates the synthesis of variant isoforms of proteins and, hence, changes in the expressed phenotype.

D. Summary and Implications

Although not as well described as signal diversification, signal integration is the necessary counterpart to repeated signal branching. Such integrative points can be found governing neuroendocrine hormone release and even at various amplifier steps in signal transduction pathways within the myocyte (see Figure 2A). Integrating processes are very intriguing from a therapeutic standpoint. Identification of dominant positive and negative signals to such nexuses could allow full control of the output, regardless of the presence of other confounding inputs (see Figure 2D). Moreover, emphasizing selectivity in receptor agonists might be a moot issue, since the second messenger cascades triggered by receptors nearly always include diverging pathways. Indeed, synergistic amplification arises because of the interactions between several messengers and the integration performed at certain steps. Thus, cell-type specific outcomes could be elicited by introducing subthreshold doses of relatively nonselective signals (e.g., hormone agonists). Cells lacking the capacity to utilize the synergy would ignore the agonist combination while cells possessing the specific synergy would show dramatic response due to the synergy. Lastly, it should be appreciated that the opposed enzymatic processes confer excellent regulatory control, especially for timing a response and allowing flexible threshold effects (Cohen, 1992; Alberts et al., 1996; Ferrell, 1996).

V. ADAPTATION

Stress-induced cardioadaptation is an exceptional demonstration of context-specific signaling. In such a multihumoral scenario involving several cell types, the net production of extracellular messengers depends exquisitely on the metabolic state of the organism, its age, and its species (e.g., rats are more tolerant to endotoxin than rabbits). Further, the sensing systems and even the type and efficacy of the adaptive response are constrained by evolutionary and developmental considerations. Natural stress stimuli also differ in virulence (consider hemorrhagic shock vs. septic shock). Therefore, this adaptation, too, could be particularly sensitive to combinations of messengers and may interpret each admixture accordingly.

Cardioadaptation against IR injury appears to involve *proactive* change triggered by dedicated signaling mechanisms. However, the expression of passive or proactive change is limited by the phenotype of the fully differentiated cell within evolutionary constraints. The heart is required to generate load-dependent pressure continuously, and also efficiently. Additionally, the embryological development of the organ places constraints on the design solutions that can be achieved. Thus, the immature myocardium with low output and demand utilizes numerous proteins specifically. These are replaced with adult isoforms during growth to maturity when the output requirements are greater (Katz, 1992; Opie, 1992).

A. Proactive Conservation

Proactive changes against ischemia–reperfusion are ultimately bound by the rapid energy turnover design of the heart. Myocardial ATP production is virtually dependent on mitochondrial respiration, with glycogen and glucose providing a minimal, anaerobic supply. (Katz, 1992; Opie, 1992). Reimer and Jennings (1992) evaluated the hypothesis that stress might suppress futile contractions and conserve ATP during early ischemia. It was expected that with decreased contractions and, hence, lowered ATP expenditure, cardiac pH would be preserved, which would lessen the threat of ionic dyshomeostasis. However, while contractile dysfunction persisted, protection waned within hours (Reimer and Jennings, 1992). We have re-evaluated this hypothesis with receptor signals and found that decreased contractions, ATP preservation during ischemia, or prioritized expenditure on ionic and membrane maintenance, are probably all viable mechanisms of proactive adaptation (Rehring et al., 1997). Proactive regulatory mechanisms that prevent ion pumps from conducting lethal ionic overload have been overlooked, barring a few exceptions (attenuated ischemic acidosis by inhibition of glycolysis (Steenbergen et al., 1993), hyperpolarization by adenosine triphosphate–sensitive potassium (K_{ATP}) channel opening (Gross et al., 1996). Other intriguing possibilities, such as blockade of the Na/Ca exchanger to prevent excessive calcium influx, remain to be explored.

High-energy signaling compounds (triphosphates, phospholipids, phosphoproteins) are also consumed during signal dispersion and amplification (Alberts et al., 1996). In the interest of signal *termination* (and hence improved response), cells have developed special enzymes to accelerate the degradation processes (namely, phosphatases, proteases, phospholipases, etc.) (Alberts et al., 1996). It is possible that these are proactively down-regulated by phosphorylation (during acute adaptation) or replaced by less active, energy-conserving isoforms (delayed adaptation). Indeed, broad-spectrum protein-phosphatase inhibitors appear to protect against ischemia (Armstrong and Ganote, 1992). In this sense, preconditioning stimuli resemble hypothermic cardiac arrest in reducing aspects of energy consumption. However, the preconditioned heart is more flexible in that it continues to function quite normally before ischemia.

B. Proactive Transition to Alternative Metabolic Efficiency States

The second law of thermodynamics requires that cardiac efficiency at converting substrates to energy, and thence to work, to be less than 100% (Katz, 1992; Opie, 1992). It is the nature of such *dissipative* systems (i.e., a far-from equilibrium energy-to-work-converting engine, where some energy must be lost as heat) to display *several* optima or alternate solutions (Stucki, 1982). These optima are recognizable as steady states, each with a characteristic combination of force production, power output, efficiency, and other persistent indices. At very high force-flow combinations, *efficiency* must be sacrificed (Stucki, 1982).

The signal transduction pathways must continually assess the external conditions, prioritize the output, and negotiate transitions between different optimum states. In a crisis, the relatively nonessential functions can be turned off. At the end of the crisis, the system should return to the highest efficiency feasible, possibly with an improved readiness for a subsequent crisis. Preconditioning demonstrates that the myocardium can adjust its metabolic optima and survive extreme de-energization. The adult myocardium can also reprogram gene expression, cellular structure, and function according to the changes in its environment (Karns et al., 1995). Hence, signaling might allow temporary trade-offs in efficiency and output, in order to survive low-flow conditions, or to divert energy towards remodeling and repair.

A crucial feature of the preconditioned heart is its ability to resume function after prolonged ischemia and to replenish its energy stores simultaneously (Reimer and Jennings, 1992; Banerjee et al., 1993; Steenbergen et al., 1993). Since cellular organization is energy dependent, turning off the energy supply has widespread effects on its ability to reassemble its high-energy structures. How might a prior signal affect the restoration of the cellular machinery? There are several possibilities. Perhaps, much like executing a "Save" command prior to shutting off a computer, this signal may store the steady-state parameters of the heart, such that during reperfusion normal function is resumed. Alternatively, the preconditioning signal may start up anotherwise unused power source (analogous to a battery) that provides just enough energy to survive prolonged ischemia. Lastly, degradation of high-energy aggregates (cytoskeletal polymers, multiprotein complexes) may be attenuated.

It is unknown whether the heart can remember or store the "information" describing the output characteristics of its normally perfused optimum. While this could be useful for restoring normal function upon reperfusion, it is clear that since ischemic preconditioning also protects against infarction, the preconditioning signal must protect cell viability during the profound de-energization during ischemia. Reimer and Jennings (1992) examined "back-up" ATP generation during ischemia from glycolysis and found that it was surprisingly down-regulated. Indeed, it was the decreased glycolysis and slower ATP depletion during prolonged ischemia in cyclic ischemic preconditioned hearts, that led Reimer and Jennings (1992) to their original proposal that the ischemic preconditioned heart both produced and utilized less energy. However, other organisms that are adapted to periodic hypoxia, such as diving mammals, may have developed metabolic shunts such as mitochondrial amino acid oxidation and thus produce ATP and GTP in the absence of oxygen (Katz, 1992; Opie, 1992). Signal-driven adjustment in such shunts could be an intriguing mechanism for reducing the impact of de-energization on the cellular machinery. Such signaling might require an investment of energy for phosphorylation or GTP hydrolysis, but the requisite high-energy sources are available in the heart during the well-perfused window prior to ischemia.

If the adapted state conferred by cardioprotective stimuli is superior in its resistance to ischemia, why has this not evolved to be the basal state? One likely possibility is that the adaptation against ischemia comes at a cost. Preconditioned states may represent optima where some aspects of normal function are compromised, or where normal responsiveness to external conditions is diminished. There has been a recent suggestion that anti-ischemic adaptation induced by hypothermic stress compromised functional accommodation to increased afterload or to lowered perfusate Ca^{2+} (Cornelussen et al., 1996).

Until now the therapeutic use of the signaling pathways to alter optimum set points has focused on the agonist and its effect on a single index (such as cardiac output or viability). The compromises that such a system must make in its global optimum (because of prioritization conflicts among input, efficiency, and maximum output) have been largely disregarded. Continuous signals (e.g., inotropic stimulation that overrides the intrinsic efficiency and capacity in order to support a low-output diseased state) are eventually incorporated into a new steady state, provided that the system is not driven into catastrophe. Eventually the signaling system adapts to ignore the signal input in favor of optimizing the energy economy (resulting in tachyphyllaxis, desensitization, and down-regulation).

Currently, expanding the available metabolic optima, while increasing stability to wide fluctuations, can only be achieved by cardiovascular exercise training, provided aging and disease do not compromise the processes. Acute preconditioning of the myocardium suggests that signaling can rapidly modify the system (presumably requiring energy) such that either the preconditioned state is itself resistant to an otherwise catastrophic perturbation (e.g., sustained ischemia) or that its transition to an alternate state is facilitated. Delayed preconditioning changes the cellular protein profile such that the manifold of optimum solutions now includes metabolic states that were previously unavailable to normal differentiated adult myocardium. This novel emphasis on transitions between the different possible output states of the myocardium itself (in contrast with the previous emphasis on the stimulated response) suggests that better and sustained matching (adaptation) could be tailored using the endogenous signaling pathways to reconfigure the system toward growth, or even repair.

Ultimately, understanding the mechanisms of cardiac adaptation to ischemia may yield therapeutic strategies by altering cardiac phenotype. Moreover, the ability to communicate fluently with the metabolic and genomic machinery of the myocyte may extend to our ability to communicate with other cells.

VI. SUMMARY

Progress in cardiac preconditioning has rekindled interest in exploiting cardioadaptation as a therapeutic adjunct. The phenomenon appears to involve stress and activation of several receptor signaling pathways. An adaptive phenotype may result from the integration of several signals. Cardiac adaptation is often assessed as protection against ischemia, arrhythmias, or infarction. Protection can be elicited by a variety of stress factors, present over minutes to days. However, the cellular adaptive mechanisms for preconditioning remain unclear and a full grasp of the scope of adaptation is still distant. Advances in regulation of signal transduction indicate that enzymatic kinetics in signal transduction are quite different from other known enzyme cascades involved in metabolism. Certainly, the boundaries between specificity and diversity are more melded and blurred. A better understanding of the signaling pathways through the elucidation of molecular mechanisms of cardiac adaptation could lead to the development of new targeted therapy. Most importantly, it is likely to provide useful physiological insights as to how organisms can respond to a conglomeration of extracellular stimuli in such a flexible, precise, and cohesive manner.

REFERENCES

Alberts, B., Bray, D., Lewis, J., Raff, M., and Watson, J.D. (1996). Cell signaling. In: Molecular Biology of the Cell. Garland Publishing, New York/London.

Armstrong, S., and Ganote, C.E. (1992). Effects of the protein phosphatase inhibitors okadaic acid and calyculin A on metabolically inhibited and ischaemic isolated myocytes. J. Mol. Cell. Cardiol. 24, 869-884.

Asaoka, Y., Nakamura, S., Yoshida, K., and Nishizuka, Y. (1992). Protein kinase C, calcium and phospholipid degradation. Trends Biochem. Sci. 17, 414-417.

Asimakis, G.K., Inners-McBride, K., Conti, V.R., and Yang, C.J. (1994). Transient beta adrenergic stimulation can precondition the rat heart against postischaemic contractile dysfunction. Cardiovasc. Res. 28, 1726-1734.

Banerjee, A., Gamboni-Robertson, F., Mitchell, M.B., Rehring, T.E., Butler, K., Cleveland, J.C., Meldrum, D.R., Shapiro, J.I., and Meng, X. (1996). Stress induced cardioadaptation reveals a code linking hormone receptors and spatial redistribution of PKC isoforms. Ann. N.Y. Acad. Sci. 793, 226-239.

Banerjee, A., Locke-Winter, C., Rogers, K.B., Mitchell, M.B., Bensard, D.D., Brew, E.C., Cairns, C.B., and Harken, A.H. (1993). Preconditioning against myocardial dysfunction after ischemia and reperfusion by an alpha-1 adrenergic mechanism. Circ. Res. 73, 656-670.

Bolli, R. (1996). The early and late phases of preconditioning against myocardial stunning and the essential role of oxyradicals in the late phase: an overview. [Review]. Bas. Res. Cardiol. 91, 57-63.

Brew, E.B., Mitchell, M.B., Rehring, T.E., Gamboni-Robertson, F., Mcintyre, R.C., Harken, A.C., and Banerjee, A. (1995). The role of bradykinin in cardiac functional protection after global ischemia–reperfusion in the rat heart. Am. J. Physiol. 269, H1370-H1378.

Brown, J.M., Anderson, B.O., Repine, J., Shanley, P., White, C., Grosso, M.A., Banerjee, A., Bensard, D., and Harken, A.H.. (1992). Neutrophils contribute to TNF induced myocardial tolerance to ischemia. J. Mol. Cell. Cardiol. 24, 485-495.

Brown, J.M., Grosso, M.A., Terada, L.S., Whitman, G.J., Banerjee, A., White, C.W., Harken, A.H., and Repine, J.E. (1989). Endotoxin pretreatment increases endogenous myocardial catalase activity and decreases ischemia-reperfusion injury of isolated rat hearts. Proc. Nat. Acad. Sci. 86, 2516-2520.

Brown, J.M., White, C.W., Terada, L.S., Grosso, M.A., Shanley, P.F., Mulvin, D.W., Banerjee, A., Whitman, J., Harken, A.H., and Repine, J.E.. (1990). Interleukin-1 pretreatment decreases ischemia/reperfusion injury. Proc. Nat. Acad. Sci. 87, 5026-5030.

Cairns, C.B., Bensard, D.B., Winter, C.B., Harken, A.H., and Banerjee, A. (1992). A rapid effect of endotoxin on the myocardium. Surg. Forum 43, 22-24.

Cano, E., and Mahadevan, L.C. (1995). Parallel signal processing among mammalian MAPKs. Trends Biochem. Sci. 20, 117-122.

Clark, E.A., and Brugge, J.S. (1995). Integrins and signal transduction pathways: the road taken. Science 268, 233-239.

Cleveland, J.C., Meldrum, D.R., Rowland, T., Banerjee, A., and Harken, A.H. (1997). Adenosine preconditioning of human myocardium is dependent upon the ATP-sensitive potassium channel. J Mol. Cell. Cardiol. 29, 175-182.

Cleveland, J.C., Wolmering, M., Meldrum, D.R., Rowland, R.T., Rehring, T.E., Sheridan, B.C., Harken, A.H., and Banerjee, A. (1996b). Ischemic preconditioning in human and rat ventricle. Am. J. Physiol. 271, H1786-H1794.

Cohen, M.V., and Downey, J.M. (1996). Myocardial preconditioning promises to be a novel approach to the treatment of ischemic heart disease. [Review]. Ann. Rev. Med. 47, 21-29.

Cohen, P. (1992). Signal integration at the level of protein kinases, protein phosphatases and their substrates. Trends Biochem. Sci. 17, 408-413.

Cornelussen, R., Vusse, G.J.v.d., Reneman, R.S., and Snoeckx, L.H.E.H. (1996). Inability of the heat shocked heart to adjust its preischemic and post ischemic performance to variable loading conditions. J. Mol. Cell. Cardiol. 28, 291-298.

Das, D.K., Moraru, I.I., Maulik, N., and Engelman, R.M. (1994). Gene expression during myocardial adaptation to ischemia and reperfusion. Ann. N.Y. Acad. Sci. 723, 292-307.

Daum, G., Eisenmann-Tappe, I., Fries, H.W., Troppmair, J., and Rapp, U.R. (1994). The ins and outs of Raf kinases. TIBS 19, 474-480.

Dinarello, C. (1996). Cytokines as mediators in the pathogenesis of septic shock. [Review]. Curr. Top. Microb. Immunol. 216, 133-165.

Disatnik, M.H., Buraggi, G., and Mochly-Rosen, D. (1994). Localization of PKC isozymes in cardiac myocytes. Exp. Cell. Res. 210, 287-297.

Downey, J.M., Cohen, M.V., Ytrehus, K., and Liu, Y. (1994). Cellular mechanisms in ischemic preconditioning: the role of adenosine and PKC. Ann. N.Y. Acad. Sci. 723, 82-98.

Ferrell, J.E. (1996). Tripping the switch fantastic: How a protein kinase cascade can convert graded inputs into switch like outputs. Trends Biochem. Sci. 21, 460-466.

Force, T., Pombo, C.M., Avruch, J.A., Bonventre, J.V., and Kyriakis, J.M. (1996). Stress activated protein kinases in cardiovascular disease. Circ. Res. 78, 947-953.

Goto, M., Liu, Y., Yang, X-M., Ardell, J.L., Cohen, M.V., and Downey, J.M. (1995). Role of bradykinin in protection of ischemic preconditioning in rabbit hearts. Circ. Res. 77, 611-621.

Gross, G.J., Mei, D.A., Schultz, J.J., and Mizumura, T. (1996). Criteria for a mediator or effector of myocardial preconditioning: do KATP channels meet the requirements?. [Review]. Bas. Res. Cardiol. 91, 31-34.

Hepler, J.R., and Gilman, A.G. (1992). G proteins. Trends Biochem. Sci. 17, 383-387.

Houslay, M.D. (1991). Review—Crosstalk: a pivotal role for protein kinase C in modulating relationships between signal transduction pathways. Eur. J. Biochem. 195, 9-27.

Hu, K., and Nattel, S. (1995). Mechanisms of ischemic preconditioning in rat hearts. Involvement of a_{1b} adrenoreceptors, pertussis sensitive Gproteins, and protein kinase C. Circulation 92, 2259-2265.

Inagaki, 7., Ito, M., Nakano, T., and Inagaki, M. (1994). Spatiotemporal distribution of protein kinase and phosphatase activities. Trends Biochem. Sci. 19, 439-518.

Iyengar, R. (1996). Gating by cyclic AMP: expanded role for an old signaling pathway. Science 271, 461-463.

Iyengar, R. (1997). There are GAPs and there are GAPs. Science 275, 42-43.

Jones, S.B.. and Yelich, M.R. (1987). Simultaneous elevation of plasma insulin and catecholamines during endotoxicosis in the conscious and anesthetized rat. Life Sci. 41, 1935-1943.

Karin, M., and Smeal, T. (1992). Control of transcription factors by signal transduction pathways: the beginning of the end. Trends Biochem. Sci. 17, 418-422.

Karns, L.R., Kariya, K., and Simpson, P.C. (1995). M-CAT, CArG, and Sp1 elements are required for a_1-adrenergic induction of the skeletal α actin promoter during cardiac myocyte hypertrophy. J. Biol. Chem. 270, 410-417.

Katz, A. M. (1992). Physiology of the Heart. 2nd ed. Raven Press, New York.

Klug, D., Robert, V., and Swynghedauw, B. (1993). Role of mechanical and hormonal factors in cardiac remodeling and the biologic limits of myocardial adaptation. [Review]. Am. J. Cardiol. 71, 46A-54A.

Liscovitch, M. (1992). Crosstalk among multiple signal-activated phospholipases. Trends Biochem. Sci. 17, 393-398.

Marber, M.S., Mestril, R., Chi, S.H., Sayen, M.R., Yellon, D.M., and Dillmann, W.M. (1995). Overexpression of the rat inducible 70-kD heat stress protein in a transgenic mouse increases the resistance of the heart to ischemic injury. J. Clin. Invest. 95, 446-456.

Maulik, N., Watanabe, M., Engelman, D., Engelman, R.M., Kagan, V., Kisin, E.E., Tyurin, V., Cordis, G.A., and Das, D.K. (1995). Myocardial adaptation to ischemia by oxidative stress induced by endotoxin. Am. J. Phys. 269, C907-C916.

Meldrum, D.R., Cleveland, J.C., Mitchell, M.B., Rowland, R.T., Banerjee, A., and Harken, A.H. (1996a). Constructive priming of myocardium against ischemia-reperfusion injury. Shock 6, 238-242.

Meldrum, D.R., Cleveland, J.C., Mitchell, M.B., Sheridan, B.C., Robertson, F., Harken, A.H., and Banerjee, A. (1996b). Protein kinase C mediates Ca++ induced cardioadaptation to ischemia–reperfusion injury. Am. J. Physiol. 271, R1718-R1726.

Meldrum, D.R., Cleveland, J.C., Jr., Rowland, R.T., Banerjee, A., Harken, A.H., and Meng, X. (1997). Early and delayed preconditioning: Differential mechanisms and additive protection. Am. J. Physiol. 273, H725-H733.

Meldrum, D.R., Cleveland, J.C., Sheridan, B.C., Rowland, R.T., Banerjee, A., and Harken, A. (1996c). Cardiac preconditioning with calcium: clinically accessible myocardial protection. J. Thorac. Cardiovasc. Surg. 112, 778-786.

Meldrum, D.R., Rowland, R.T., Cleveland, J.C., Harken, A.H., Banerjee, A., and Meng, X. (1997). Acute and delayed preconditioning: differential mechanisms and additive effects. Am. J. Phys. (in revison).

Meldrum, D.R., Sheridan, B.C., Cleveland, J.C., Fullerton, D.A., Banerjee, A., and Harken, A. (1996d). Neutrophil induction is required for endotoxin mediated myocardial cross-tolerance to ischemia-reperfusion injury. Arch. Surg. 131, 1203-1208.

Meng, X., Brown, J.M., Ao, L., Banerjee, A., and Harken, A.H. (1996a). Norepinephrine induces cardiac heat shock protein and delayed cardioprotection in the rat through αα1-adrenoceptors. Cardiovasc. Res. 32, 374-383.

Meng, X., Brown, J.M., Ao, L., Nordeen, S.K., Franklin, W., Harken, A.H., and Banerjee, A. (1996b). Endotoxin induces HSP70 and resistance to endotoxemic myocardial depression in the rat. Am. J. Physiol. 271, C1316-C1324.

Meng, X., Brown, J.M., Harken, A.H., and Banerjee, A. (1996c). Differential expression of myosin heavy chain and actin isogenes in rat hearts cross-resistant to ischemia and endotoxin. J. Mol. Cell. Cardiol. 28, A189.

Meng, X., Cleveland, J.C., Rowland, R.T., Mitchell, M.B., Brown, J.M., Banerjee, A., and Harken, A.H.. (1996d). Norepinephrine-induced sustained myocardial adaptation to ischemia is dependent on α1-adrenoceptors and protein synthesis. J. Mol. Cell. Cardiol. 28, 2017-2025.

Mitchell, M.B., Meng, X., Parker, C.G., Brown, J., Harken, A.H., and Banerjee, A. (1995). Preconditioning of isolated rat heart is mediated by protein kinase C. Circ. Res. 76, 73-81.

Mochly-Rosen, D. (1995). Localization of protein kinases by anchoring proteins: a theme in signal transduction. Science 268, 247-251.

Murry, C.E., Jennings, R.B., and Reimer, K.A. (1986). Preconditioning with ischemia: a delay of lethal cell injury in ischemic myocardium. Circulation 74, 1124-1136.

Nelson, D.W., Brown, J.M., Banerjee, A., Bensard, D.D., Rogers, K.B., Locke-Winter, C.R., Anderson, B.O., and Harken, A.H. (1991). Pretreatment with a nontoxic derivative of endotoxin (MPL) induces functional protection against cardiac ischemia reperfusion injury. Surgery 110, 365-369.

Nishizuka, Y. (1992). Intracellular signaling by hydrolysis of phospholipids and activation of protein kinase C. Science 258, 607-614.

Olsson, R.A., and Pearson, J.D. (1990). Cardiovascular purinoreceptors. Physiol. Rev. 70, 761.

Opie, L.H. (1992). The Heart. Physiology and Metabolism. 2nd ed. Raven Press, New York.

Przyklenk, K., and Kloner, R.A. (1995). Preconditioning: a balanced perspective [editorial]. Br. Heart. J. 74, 575-577.

Ray, K., Hansen, C.A., and Robishaw, J.D. (1996). Gβγ-mediated signaling in the heart: implications of β and γ subunit heterogeneity. Trends Cardiovasc. Med. 6, 115-121.

Rehring, T.F., Bender, P.R., Joo, K., Cairns, C.B., Friese, R.S., Shapiro, J.I., Cleveland J.C. Jr., and Banerjee, A. (1997). Different preconditioning stimuli invoke disparate electromechanical and energetic responses to global ischemia. Can. J. Physiol. Pharmacol. 75, 335-342.

Reimer, K.A., and Jennings, R.B. (1992). Preconditioning. Definitions proposed mechanisms, and implications for myocardial protection in ischemia and reperfusion. In: *Myocardial Protection: The Pathophysiology of Reperfusion and Reperfusion Injury*. Raven Press, New York.

Rowland, R.T., Meng, X., Cleveland, J.C., Jr., Meldrum, D., Harken, A., and Brown, J. (1997). Lipopolysaccharide-induced delayed myocardial adaptation enhances acute ischemic preconditioning to optimize postischemic cardiac function. Am. J. Physiol. 272, H2708-H2715.

Steenbergen, C., Perlman, M.E., London, R.E., and Murphy, E. (1993). Mechanism of preconditioning: ionic alterations. Circ. Res. 72 (1), 112-125.

Stucki, J.W. (1982). Thermodynamic organizing principles in mitochondrial energy conversion. In: Metabolic Compartmentation. Academic Press, London.

Taniguchi, T. (1995). Cytokine signaling through nonreceptor protein tyrosine kinases. Science 268, 251-255.

Ulevitch, R.J., and Tobias, P.S. (1995). Receptor dependent mechanisms of cell stimulation by bacterial endotoxin. Ann. Rev. Immunol. 13, 437-457.

Watson, S., and Girdlestone, D. (1996). 1996 receptor and ion channel nomenclature supplement. Trends Pharmacol. Sci. 7th ed.

Westfall, D., Shinozuka, K., Forsyth, K., and Fjur, R. (1990). Presynaptic purine receptors. Ann. N.Y. Acad. Sci. Reno.

Winter, C.B., Cleveland, J.C., Butler, K.L., Bensard, D.B., Mitchell, M.B., Harken, A.H., and Banerjee, A. (1997). Facilitative interactions between noradrenergic and purinergic signalling during preconditioning of the rat heart. J. Mol. Cell. Cardiol. 29, 163-173.

Yao, Z., and Gross, G.J. (1993). Role of nitric oxide, muscarinic receptors, and the ATP-sensitive K^+ channel in mediating the effects of acetylcholine to mimic preconditioning in dogs. Circ. Res. 73, 1193-1201.

Yellon, D.M., and Baxter, G.F. (1995). A "second window of protection" or delayed preconditioning phenomenon: future horizons for myocardial protection? [Review]. J. Mol. Cell. Cardiol. 27, 1023-1034.

Zhou, X., Zhai, X., and Ashraf, M. (1996). Preconditioning of bovine endothelial cells. The protective effect is mediated by an adenosine A2 receptor through a protein kinase C signaling pathway. Circ. Res. 78, 73-81.

THE ROLE OF ATP-SENSITIVE POTASSIUM CHANNELS IN MYOCARDIAL ISCHEMIC STRESS

Arpad Tosaki and Dipak K. Das

Advances in Organ Biology
Volume 6, pages 181-195.

ISBN: 0-7623-0391-3

I. INTRODUCTION

In the search for mechanisms of stress, including an ischemic event, which might be amenable to manipulation, numerous potential candidates have been identified and have been the subject of many reviews. Free radicals, calcium overload, potassium loss, gene expression, energy depletion, and the loss of key intermediates or co-factors are most frequently cited as contributing to the onset of stress or ischemia-induced functional changes in the myocardium. In reality, none of these alone is likely to be responsible, and it is probable that many and varied factors are involved. If this is the case, then it would seem highly unlikely that any single drug or intervention could do much to combat the many consequences of stress or ischemia-induced functional changes in the myocardium.

Potassium channels are diverse and ubiquitous molecular entities (Rudy, 1988). They are present in virtually every tissue in both excitable and nonexcitable cells, and exist in as many as 25 or more types, distinguishable by pharmacological, biophysical, and structural criteria. This review will focus on the pharmacological effects of the adenosine triphosphate (ATP)–sensitive class of potassium channel blockers and openers on the function of the ischemic–reperfused myocardium. Although we do note ask that readers accept what we have written here, we encourgage them to expand their knowledge of this area. The importance of ATP-sensitive potassium channel (K^+_{ATP}) blockers and openers is recognized in the regulation of stress-induced, functional changes in the myocardium (e.g., by ischemia, reperfusion, and preconditioning) and researchers must appreciate the fact that this recognition, whether we like it or not, is having a major impact on the regulation of cardiac function in ischemic–reperfused hearts.

Although it may be debatable whether K^+_{ATP} blockers or openers will ever fulfill their clinical expectations, the hope is that a new therapeutic modality could give great impetus to the search for these agents in the treatment of myocardial ischemia–reperfusion-induced injury. The use of K^+_{ATP} blockers or openers is not always beneficial and may have some detrimental effects, especially in the ischemic–reperfused myocardium. It is not the intention of this review to denigrate in any way the current high level of interest or activity in the field of K^+_{ATP} blockers and openers. Indeed, the authors have published studies in the area and continue to do so, our hope being that the current euphoria for the wonders of K^+_{ATP} blockers and openers would not lead to a loss of credibility in a field that promises to teach us much about the mechanisms underlying ischemia and reperfusion-induced injury.

II. K^+_{ATP} CHANNEL BLOCKERS

The properties of K_{ATP} channels vary according to the type of tissue, and at least three major classes are currently recognized (Ashcroft and Ashcroft, 1990). The type I channels, blocked by micromolar concentrations of intracellular ATP, are present in cardiac (Noma, 1983), skeletal (Spruce et al., 1985), smooth muscle (Standen et al., 1989), and pancreatic β cells (Cook and Hales, 1984). The pancreatic β cell is the more sensitive to ATP, and these cells are sensitive to voltage and calcium ions. The type II channel, originally described in neurons (Ashford et al., 1989), is inhibited by millimolar concentrations of ATP (Inoue et al., 1991). The type III channel, found in tracheal smooth muscle and some epithelial cells, is similar to the type II in its sensitivity to ATP, but differs in being activated by micromolar concentrations of calcium (Groschner et al., 1991).

Discovery of the hypoglycemic activity of the sulfonylurea class of agents was a consequence of the original investigation of the antibacterial activity of sulfonamides (Loubatieres, 1957). Initial evaluation by Loubatieres (1957) demonstrated that the sulfonamide group was necessary for the hypoglycemic activity of these molecules; this finding led to the development of a new treatment for noninsulin-dependent diabetes mellitus. It is now well established that sulfonylureas are specific blockers of the K^+_{ATP} channels in a variety of tissues, including pancreatic cells, insulin secretory cell lines, neuronal tissues, and skeletal and cardiac muscles (Schmid-Antomarchi et al., 1987; Miller, 1990). Sulfonylureas do not directly interact at other potassium channel types, including those activated by voltage or calcium, but do interfere with activator induced responses in smooth and cardiac muscle, albeit at relatively high concentrations (Edwards and Weston, 1990).

That the K^+_{ATP} channel can play a role in ischemia and reperfusion-induced injury is no new idea. Since the experiments of Harris et al. (1954), cellular potassium loss and extracellular potassium ($[K^+]_o$) accumulation have been considered to be major determinants of the electrophysiological changes that underlie the early malignant phase of ventricular tachyarrhythmias that follow acute myocardial ischemia (Harris, 1966; Wilde, 1996). It has been suggested that $[K^+]_o$ accumulation during the early phase of ischemia may result in part from the activation of K^+_{ATP} channels (Kakei et al., 1985). Furthermore, has been shown that K^+ accumulation results primarily from an increase in K^+ efflux rather than a decrease in K^+ influx, suggesting that Na^+–K^+ pump dysfunction is not the primary mechanism (Weiss and Shine, 1982). The two major mechanisms that have been hypothesized to explain the increase in K^+ efflux are: (1) increased membrane K^+ conductance via activation of K^+_{ATP} channels (Noma, 1983), and (2) other classes of potassium channels (Kim and Clampham, 1989). Sulfonylurea drugs, which are selective blockers of K^+_{ATP} channels, have been used to estimate the contribution of K^+_{ATP} channels to cellular K^+ loss and action potential duration (APD) shortening during myocardial ischemia and hypoxia, and to evaluate their potential antiarrhythmic usefulness (Wolleben et al., 1989; Tosaki et al., 1992; Billman et al., 1993).

In high concentrations relative to those required to suppress K^+_{ATP} channels in the myocardium, the sulfonylurea glibenclamide only modestly reduced the ischemia-induced $[K^+]_o$ accumulation in most (Kantor et al., 1990; Wilde et al., 1990) but not all (Gassner and Vaughan-Jones, 1990) studies. Other studies, although they do not differentiate the routes involved in ischemic and postischemic ionic accumulation or loss, also show that glibenclamide significantly reduced the loss of myocardial K^+ content, leading to the reduction of life-threatening ventricular fibrillation (Tosaki et al., 1993; Tosaki and Hellegouarch, 1994, Tosaki et al., 1995). Pretreatment with glibenclamide prevented APD shortening during hypoxia (Wilde et al., 1990), but glibenclamide did not reverse APD shortening if administered during hypoxia (Carmeliet et al., 1990).

A. Mechanisms of K^+_{ATP} Blockers (Sulfonylurea)

It has been previously shown that in the presence of adenosine monophosphate (ADP), K^+_{ATP} channels become less sensitive to block by ATP, with the K^+_d for ATP increasing almost fourfold in the presence of 100 μM ADP (Lederer and Nicchols, 1989). ADP is also able to increase the current through K^+_{ATP} channels in the absence of ATP (Venkatesh et al., 1991). A two-binding site model has been proposed to account for the ADP–ATP interaction on K^+_{ATP} channels (Lederer and Nicchols, 1989; Tung and Kurachi, 1991), and a similar mechanism would be consistent with the sulfonylurea–ADP interaction as well. A simple one-binding site model, in which ADP directly interferes with the ability of sulfonylureas to block the channel by competing for the same binding site, seems inconsistent with the complex effect of the ADP–sulfonylurea interaction (Venkatesh et al., 1991). Figure 1 shows the essential features of a two-binding site model. The first binding site (S1), which may be the same as the ATP binding site (and if so, located intracellularly), promotes channel closure when occupied by a ligand and has a high affinity for sulfonylureas and a low affinity for ADP. The second binding site (S2), which is analogous to the intracellular nucleotide diphosphate binding site (Tung and Kurachi, 1991), promotes channel opening when occupied by a ligand and has a high affinity for ADP and a low affinity for sulfonylureas. When ligands occupy both

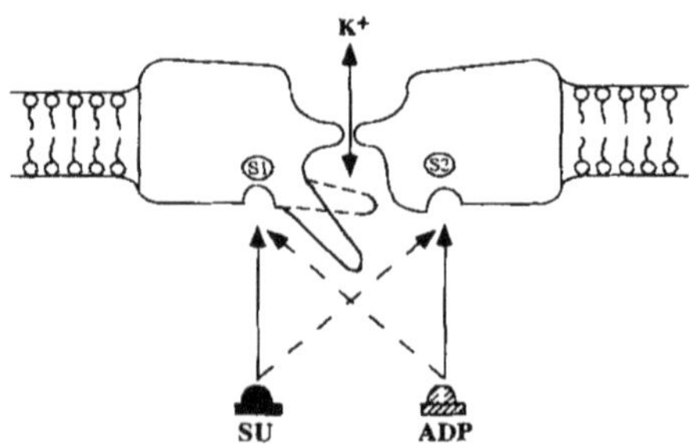

Figure 1. The two-binding site model proposed for the action mechanism of sulfonylurea (K^+_{ATP} channel blockers). ADP, adenosine monophosphate; SU, sulfonylurea; S1, binding site promoting ATP-sensitive K^+ channel closure; S2, binding site promoting the channel opening; S1 and S2 are assumed to be located on or near the cytoplasmic portion of the channel.

binding sites simultaneously, channel opening is favored. In the absence of ADP and ATP, glibenclamide binds to S1 with high affinity resulting in half-maximal channel closure at 0.5 μM. The Hill coefficient of 0.5 suggests some form of negative cooperativity, which could result from low-affinity binding of glibenclamide to S2. At the present time, however, detailed features of the model remain speculative.

B. Implication of K^+_{ATP} Blockers

Pharmacological modification of cellular K^+ loss during early myocardial ischemia is a promising antiarrhythmic strategy in view of the key role that $[K^+]_o$ accumulation plays in facilitating arrhythmias and functional failure, the major cause of mortality from coronary artery disease (Goldman et al., 1982). The moderate reduction in $[K^+]_o$ accumulation achieved with glibenclamide during early ischemia or reperfusion has been shown to be associated with a reduction in cardiac arrhythmias and improvement of cardiac function (Wolleben et al., 1989; Kantor et al., 1990; Billman et al., 1993; Tosaki and Hellegouarch 1994; Tosaki et al., 1995) and conduction slowing in the ischemic zone (Bekheit et al., 1990). However, other studies have suggested that sulfonylureas may have deleterious consequences during ischemia or reperfusion (Grover et al., 1989; Auchampach et al., 1992). For these reasons, extensive additional studies in clinically relevant models will be necessary to weigh the potential advantages and disadvantages of K^+_{ATP} channel blockade during myocardial ischemia and reperfusion. The findings of our studies suggest the potential benefit of K^+_{ATP} channel blockers in ischemic–reperfused myocardium (Tosaki et al., 1992; Tosaki et al., 1993; Tosaki and Hellegouarch 1994; Tosaki et al., 1995).

III. K^+_{ATP} CHANNEL OPENERS

The three major classes of K^+_{ATP} channel openers are chemically distinct and consist of benzopyrans, cyanoguanidines, and nitronicotinamide derivatives, represented by cromakalim, pinacidil, and nicorandil, respectively (Robertson and Steinberg, 1990). These chemical structures share the common property of producing smooth muscle relaxation and were originally developed as antihypertensive or antianginal agents (Longman and Hamilton, 1992). Electrophysiological and efflux studies have demonstrated the opening of K^+_{ATP} channels by these agents, although nicorandil is also a nitrovasodilator acting through the cyclic guanosine monophosphate (cGMP)–guanylate cyclase pathway (Hamilton and Weston, 1989; Robertson and Steinberg, 1990). Although the three primary structural groups of K^+_{ATP} channel openers may be thought of as distinct, it has been suggested that they may have common pharmacological effects (Atwal et al., 1992). Those searching for the mechanisms of action of potassium channel opening have commonly employed cromakalim and pinacidil as prototypes.

Numerous studies, especially those of Drs. Gross (Auchampach et al., 1992a, 1992b; Yao and Gross, 1994; Auchampach et al., 1994; Mizumura et al., 1995a, 1995b; Gross, 1995; Schultz et al., 1996) and Grover (Grover et al., 1989; D'Alonzo et al., 1994; Grover and Sleph, 1995; Grover et al., 1995a, 1995b; Grover et al., 1996a, 1996b) have shown that improvements in postischemic recovery can be obtained with all agents in this class. Furthermore, the extent of the protection appears to be remarkably similar, despite the use of structurally dissimilar agents (some of which, such as nicorandil with its nitrate moiety, have important secondary properties), given at different concentrations, sometimes by different routes in different species. Thus, for example, Gross et al. (1987a, 1987b, 1989), using either anesthetized or conscious dogs with 10, 15, or 30 minutes of regional ischemia and 3 to 6 hours of reperfusion, reported that intravenous nicorandil, administered 10, 15, or 30 minutes prior to and throughout the ischemic period (and being either not present during the reperfusion period or present only for the first 30 minutes of reperfusion), improved the postischemic recovery of segment shortening in every instance. Despite the multitude of protocol variations, the protection was similar in all cases, with segmental shortening at the end of reperfusion being improved from less than 20% to greater than 75% of its preischemic control value. Gross et al. (1992) also studied the effects of bimakalim given as a bolus to anesthetized dogs 15 minutes before occlusion and observed a very good protection in postischemic function. Auchampach et al. (1992a), using a near identical protocol to that of Gross et al. (1992), reported that nicorandil conferred a similar improvement of postischemic segment shortening. Auchampach et al. (1992b), using the dog, again with a similar protocol, reported that intravenous aprikalim improved recovery from less than 10% to greater than 80%. Grover et al. (1990), in addition to studying cromakalim, also investigated intravenous nicorandil, administered 15 minutes prior to and throughout 15 minutes of ischemia and 180 minutes of reperfusion; they found that it improved the postischemic recovery of segment shortening from less than 5% to greater than 40%.

It is important to note that, in all of the studies cited thus far, the drugs were administered before the onset of ischemia. It is also important to recognize that the duration of drug treatment varied greatly between studies such that in some instances treatment was carried on throughout the experiment (i.e., until the end of reperfusion), in some others the treatment was discontinued part way through the reperfusion period, while, in still others, the treatment was confined to the ischemic period or the period leading up to the induction of ischemia. Furthermore, in all studies cited in the present review and published by the laboratories of Drs. Gross and Grover, in their hands, all K^{+}_{ATP} channel openers showed very good cardiac protection in all respects of myocardial ischemia–reperfusion-induced injury independently of the administration of the drug, the method, or the species used.

On the other hand, K^{+}_{ATP} channel openers may have clinical utility as antihypertensive agents; thus, it would be useful to know if these compounds really mitigate the consequences of myocardial ischemia and reperfusion in humans. Since the population of hypertensive patients is also at risk for heart disease, this information is extremely important. It would also be useful to learn if K^+_{ATP} channel activators may have potential as anti- ischemic or antiarrhythmic agents in their own right. Although K^+_{ATP} channel activators may have the ability to increase coronary flow or reduce cardiac workload secondary to afterload reduction, it is always difficult to determine the relative importance of direct cytoprotective effects from those arising from changes in hemodynamic status. However, clinical evaluation of K^+_{ATP} channel openers in patients with essential hypertension suggests therapeutic efficacy of these agents with an incidence of dose-related side effects of edema formation, palpitation, and ventricular tachycardia (Ahnfelt-Ronne, 1988; Goldberg et al., 1988). The increased incidence of ventricular tachycardia, after the administration of a K^+_{ATP} channel opener has been reported by Fox et al. (1991) in healthy volunteers. In experimental studies, the proarrhythmic action of K^+_{ATP} channel openers has also been reported (Wolleben et al., 1989; Chi et al., 1990; Friedrichs, 1993; Tosaki and Hellegouarch, 1994; Tosaki et al., 1995). These proarrhythmic effects of K^+_{ATP} channel openers may be explained by the loss of myocardial potassium. These experimental studies do not attempt to address the problem and question of specific ionic currents. However, it has been suggested that proarrhythmic effects of K^+_{ATP} channel openers may relate to the same receptor site in which glibenclamide, a K^+_{ATP} channel blocker, may reflect a specific blockade of these channels. If this is so, the use of K^+_{ATP} channel openers may be of particular concern in that population of postinfarction patients who are known to be at high risk of sudden coronary death.

A. Mechanisms of K^+_{ATP} Channel Openers

Noma (1983) suggested that the K^+_{ATP} channel serves an endogenous cardioprotective role in the myocardium by a number of mechanisms schematically diagrammed in Figure 2. The primary electrophysiological effect of opening the K^+_{ATP} channels during ischemia or hypoxia is the shortening of the cardiac action potential due to accelerated phase III repolarization leading to a reduced calcium overload. In addition, the membrane potential may also be maintained at a more negative level by K^+_{ATP} channel opening, and this effect would result in additional lowered calcium influx via voltage-gated L-type calcium channels and sodium–calcium exchange. The intensity of K^+_{ATP} channel openers may be a function of intracellular ATP concentration, Spinelli et al. (1991) found that K^+_{ATP} channel openers produced, dose dependently, moderate to marked shortening of the action potential without modifying resting potential or action potential amplitude. The effect of K^+_{ATP} channel openers on action potential duration was promptly reversed on washout of the drugs, and repeated exposures to the same concentrations caused changes of similar magnitude. The decrease in action potential duration was accompanied by a dose-dependent increase in outward current but no consistent change in inward current in the steady-state phase I–V relationship. Both the change in action poten-

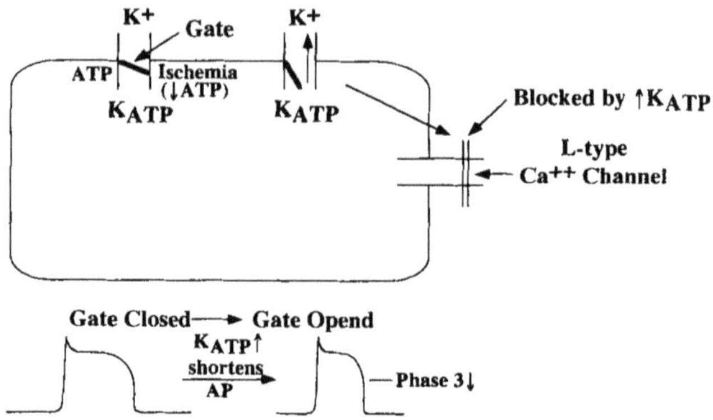

Figure 2. Mechanisms of K^+_{ATP} channel openers producing a cardioprotective effect. The opening of K^+_{ATP} channels increases the K^+ efflux, reduces the action potential (AP) duration, and blocks the L-type Ca^{2+} channels, thus preventing intracellular calcium gain. ATP, adenosine triposphate.

tial duration and the augmented outward current were prevented or reversed by glibenclamide, and so it seemed most likely that the effects observed were due to drug-induced changes in inward current (Spinelli et al., 1991).

Although K^+_{ATP} channel openers brought about the desired changes in transmembrane action potentials in cells demonstrating the early afterdepolarizations (EADs) or delayed afterdepolarizations (DADs), these agents clearly are not prototypes of new antiarrhythmic drugs. There are many reasons why K^+_{ATP} channel openers might not be good antiarrhythmic agents. Among other problems (e.g., the increase of cellular potassium loss [Tosaki et al., 1993, 1995]), they will cause relaxation of vascular smooth muscle and decrease systemic arterial pressure (Toda et al., 1985). Furthermore, since the K^+_{ATP} channel–induced current is time independent and the intensity of action of K^+_{ATP} channel openers may depend on the intracellular ATP concentration (Tseng and Hoffman, 1990), there always will be a possibility of excessive shortening of the action potential and consequent weakening of contraction. In addition to the negative inotropic effect, an excessive shortening of action potential duration will cause a proportional decrease of refractoriness, and this action may increase the likelihood of certain types of arrhythmias (Steinberg et al., 1988; Spinelli et al., 1990).

B. Implication of K^+_{ATP} Channel Openers

A number of studies report data to suggest that K^+_{ATP} channel blockers can improve cardiac function and that K^+_{ATP} channel openers can, under certain conditions, be harmful to the ischemic and reperfused myocardium. However, these potentially deleterious effects may be species dependent, model dependent, and dose dependent, and their clinical risk may have been overestimated. The ability of K^+_{ATP} channel blockers to block the ability of K^+_{ATP} channel openers provides compelling evidence to support the notion that the drugs act through a specific K^+_{ATP} channel–dependent mechanism. The nature of this mechanism, which appears to operate primarily during the ischemic period, remains to be resolved as does the balance of benefit derived from the actions of these drugs on the myocytes and the vascular smooth muscle. This applies particularly to nicorandil with its unique nitrate property, which is capable of promoting vasodilation by a direct nitric oxide–mediated mechanism that is independent of the effect of the drug on K^+_{ATP} channel–mediated vasodilation.

In conclusion, K^+_{ATP} channel opening drugs represent a new class of drugs that may offer exciting new therapeutic opportunities under certain conditions. Possible clinical applications include various forms of angina, myocardial infarction with early thrombolysis, and cardiac surgery. In all of these applications, prognosis may be affected by the consequences of the ischemia–reperfusion-induced stress mechanism. Clearly, this family of drugs is worthy of future investigation aimed at resolving a number of important questions, but the time of their clinical application has not come yet.

IV. PRECONDITIONING AND K^+_{ATP} CHANNELS

Is the involvement of K^+_{ATP} channel openers in preconditioning a clinical dilemma, laboratory artifact, or much ado about nothing? Although reperfusion is an absolute prerequisite for the survival of the ischemic myocardium, preconditioning of the heart and its potential relevance have resulted in some remarkable claims, including the suggestion that it is the most powerful form of cardioprotection known to date and that once its mechanism(s) is known it might be bottled and provide a unique approach to the management of ischemic cardiac disease. The three forms of preconditioning (ischemia induced, hypoxia induced, and chemically induced) are still a laboratory-based phenomenon and have not been conclusively documented in human subjects. In numerous studies of preconditioning and its protection, the first as well as the second window of opportunity have been demonstrated and documented very well in hearts obtained from intact animals (Thornton et al., 1993; Przynklenk et al., 1995; Tang et al., 1996; Yang et al., 1996). However, very few attempts were made to study the preconditioning-induced cardioprotection in diseased myocardium (Szilvassy et al., 1995; Tosaki et al., 1996a). There are several possible clinical conditions—such as angina pectoris, angioplasty, and some other types of cardiac surgery—in which preconditioning may occur that are amenable to experimental studies. Although preconditioning-induced protection in intact animals has been well documented in certain aspects, preconditioning in humans does not always result in a phenomenal cardiac protection (Abete et al., 1996), as was emphasized by Kloner et al. (1995).

A number of studies have been carried out to determine the cellular mechanisms of preconditioning, without consensus being reached. Murry and colleagues (1990) suggested that a reduction in ATP depletion or an increased washout of cellular metabolites might be partially responsible for the preconditioning-induced protection. Other studies have shown that the translocation of protein kinase C (Przynklenk et al., 1995; Tosaki et al., 1996b), activation of adenosine receptors (Liu et al., 1991), and induction of heat shock proteins (Marber et al., 1993) could be potential mechanisms of preconditioning in intact myocardium. The cellular mechanism by which stimulation of adenosine receptors produces this cardioprotective effect is still unknown, but Kirsch et al. (1990) have demonstrated that adenosine-1 (A_1) receptors are coupled to K^+_{ATP} channels in the myocardium. Thus, K^+_{ATP} channels have also been implicated as a potential mechanism of preconditioning (Gross and Auchampach, 1992).

In the absence of any agreed single mechanism, it becomes important to examine carefully the experimental conditions chosen for the various studies and to understand their limitations. With regard to the K^+_{ATP} channel as a potential mechanism of preconditioning, we need to clarify that: (1) K^+_{ATP} channel openers are—and blockers are not—protective in the ischemic–reperfused heart, (2) K^+_{ATP} channels have a major impact on the preconditioned myocardium, and (3) preconditioning is the most powerful cardiac protection in not only the intact but also the diseased myocardium.

In conclusion, preconditioning and K^+_{ATP} channel openers are not always beneficial, and may have some detrimental effects, in the ischemic–reperfused myocardium. Thus, there should be due care in extrapolating from experiments to any possible clinical application.

V. SUMMARY

In many studies, activators of the K^+_{ATP} channel possess anti-ischemic properties that allow them to slow the rate of development of ischemic injury such that, upon reperfusion after a fixed ischemic insult, infarct size is smaller and/or postischemic functional recovery is improved.

A number of studies report data to suggest that K^+_{ATP} channel blockers can exert antiarrhythmic effects and that K^+_{ATP} channel openers may, under certain conditions be proarrhythmic. However, these potentially deleterious effects may be species dependent, model dependent, and dose dependent and their clinical risk may have been overestimated. The ability of K^+_{ATP} channel blockers to prevent the effects of K^+_{ATP} channel openers provides compelling evidence to support the notion that the drugs act through a specific K^+_{ATP} channel–dependent mechanism. The nature of this mechanism, which appears to operate primarily during the ischemic period, remains to be resolved as does the balance of benefit derived from the actions of these drugs on the myocyte and the vascular smooth muscle. This applies particularly to nicorandil with its unique nitrate property, which is capable of promoting vasodilatation by a direct nitric oxide–mediated mechanism that is independent of the effect of the drug on K^+_{ATP} channel–mediated vasodilation.

Our studies showed that K^+_{ATP} channel openers aggravated, and K^+_{ATP} channel blockers improved the recovery of cardiac function in isolated rat and guinea pig hearts. Because the effects of K^+_{ATP} channel openers and blockers are controversial in ischemic–reperfused myocardium, the implication of K^+_{ATP} channel openers as a potential mechanism in the preconditioning phenomenon may not be appropriate. It is not the intention of this chapter to denigrate in any way the current high level of interest or activity in the field of K^+_{ATP} channel openers. Indeed, both authors have published research in the area and continue to do so. It is, however, our hope that the current euphoria for the wonders of K^+_{ATP} channel openers and blockers will not lead to a loss of credibility in a field that promises to teach us much about the mechanisms underlying ischemia–reperfusion-induced injury.

VI. CONCLUSION

In conclusion, K^+_{ATP} channel–opening drugs represent a new class of drugs that may offer exciting new therapeutic opportunities under certain conditions in which a reduction in the severity and consequences of an episode of ischemia is desirable. Possible clinical applications may include various forms of angina pectoris, myocardial infarction with early thrombolysis, and cardiac surgery. Clearly, this family of drugs is worthy of future investigations aimed at resolving a number of important, but as yet unanswered, questions. We think, however, that the time for application of K^+_{ATP} channel openers in clinical practice has not yet come.

REFERENCES

Abete, P., Ferrara, N., Cioppa, A., Ferrara, P., Bianco, S., Calabrese, C., Cacciatore, F., Longobardi, G., and Rengo, F. (1996). Preconditioning does not prevent postischemic dysfunction in aging heart. J. Am. Coll. Cardiol. 27, 1777-1786.

Ahnfelt-Ronne, I. (1988). Pinacidil: history, basic pharmacology, and therapeutic implications. J. Cardiovasc. Pharmacol. 12 (suppl. 2), S1-S4.

Ashcroft, S.J.H., and Ashcroft, F.M. (1990). Properties and function of ATP-sensitive K^+ channels. Cell Signal. 2, 197-214.

Ashford, M.L.J., Boden, P.R., and Treherne, M.J. (1989). Glucose-induced excitation of cat hypothalamic neurons in vitro is mediated by ATP- sensitive K^+ channels. Pflugers Arch. 415, P31 [Abstr.].

Atwal, K.S., Moreland, S., McCullough, J.R., O'Reilly, B.C, Ahmed, S.Z., Normandin, and D.E. (1992). Aryl cyanoguanidine potassium channel openers. Biorg. Med. Chem. Lett. 2, 83-86.

Auchampach, J.A., Cavero, I., and Gross, G.J. (1992a). Nicorandil attenuates myocardial dysfunction associated with transient ischemia by opening ATP-dependent potassium channels. J. Cardiovasc. Pharmacol. 20, 765-771.

Auchampach, J.A., Maruyama, M., Cavero, I., and Gross, G.J. (1992b). Pharmacological evidence for a role of ATP-dependent potassium channels in myocardial stunning. Circulation 86, 311-319.

Auchampach, J.A, Maruyama, M., and Gross, G.J. (1994). Cardioprotective actions of potassium channel openers. Eur. Heart J. 15 (suppl. C), 89-94.

Billman, G.E., Avendano, C.E., Halliwill, J.R., and Burroughs, J.M. (1993). The effects of the ATP-dependent potassium channel antagonist, glyburide, on coronary blood flow and susceptibility to ventricular fibrillation in unanesthetized dogs. J. Cardiovasc. Pharmacol. 21, 197-204.

Carmeliet, E.E., Storms, L., and Vereecke, J. (1990). ATP-dependent K-channel and metabolic inhibition. In: Cardiac Electrophysiology from Cell to Bedside (Zipes, D.P., and Jalife, J., Eds.), pp. 103-108. W.B. Saunders, Philadelpia.

Chi, L., Uprichard, C.G., and Lucchesi, B.R. (1990). Profibrillatory actions of pinacidil in conscious canine model of sudden coronary death. J. Cardiovasc. Pharmacol. 15, 452-464.

Cook, D.L., and Hales, C.N. (1984). Intracellular ATP directly blocks K^+ channels in pancreatic beta-cells. Nature 311, 271-273.

D'Alonzo, A.J., Darbenzio, R.B., Hess, T.A., Sewter, J.C., Sleph, P.G., and Grover, G.J. (1994). Effect potassium on the action of the K_{ATP} modulators cromakalim, pinacidil, or glibenclamide on arrhythmias in isolated perfused rat heart subjected to regional ischemia. Cardiovasc. Res. 28, 881-887.

Edwards, G., and Weston, A.H. (1990). Potassium channel openers and vascular smooth muscle relaxation. Pharmacol. Ther. 48, 237-258.

Fox, J.S., Whitehead E.M., and Shanks, R.G. (1991). Cardiovascular effects of cromakalim (BRL 34915) in healthy volunteers. Br. J. Clin. Pharmacol. 32, 45-49.

Friedrichs, G.S., Chi, L., Black, S.C., Manley, P.J., Oh, J.Y., and Lucchesi, B.R. (1993). Antifibrillatory effects of ibutilide in the rabbit isolated heart: mediation via ATP-dependent potassium channels. J. Pharmacol. Exp. Ther. 266, 1348-1354.

Gassner, R.N.A., and Vaughan-Jones, R.D. (1989). Mechanisms of potassium efflux and action potential shortening during ischemia in isolated mammalian cardiac muscle. J. Physiol. (London) 431, 713-741.

Goldberg, M.R., Sushak, C.S., Rockhold, F.W., and Thompson, W.L. (1988). Vasodilator monotherapy in the treatment of hypertension: comparative efficacy and safety of pinacidil, a potassium channel opener, and prazosin. Clin. Pharmacol. Ther. 44, 78-92.

Goldman, L., Cook, F., Hashimoto, B., Stone, P., Muller, J., and Loscalzo, A. (1982). Evidence that inhospital care for acute myocardial infarction has not contributed to the decline in coronary mortality between 1973-1974 and 1978-1979. Circulation 65, 936-942.

Groschner, K., Silberberg, S.D., Gelband, C.H., and van Breemen, C. (1991). Ca^{2+}-activated K^+ channels in airway smooth muscle are inhibited by cytoplasmic adenosine triphosphate. Pflugers Arch. 417, 517-522.

Gross, G.J. (1995) Do ATP-sensitive potassium channels play a role in myocardial stunning? Basic Res. Cardiol. 90, 266-268.

Gross, G.J., and Auchampach, J.A. (1992). Blockade of ATP-sensitive potassium channels prevents myocardial preconditioning in dogs. Circ. Res. 70, 223- 233.

Gross, G.J., Auchampach, J.A., and Maruyama, M. (1992). Cardioprotective effects of nicorandil. J. Cardiovasc. Pharmacol. 20 (suppl. 3), S22-S28.

Gross, G.J., Pieper, G., and Farber, N.E. (1989). Effects of nicorandil on coronary circulation and myocardial ischemia. Am. J. Cardiol. 63, J11-J17.

Gross, G.J., Pieper, G.M., and Warltier, D.C. (1987a). Comparative effects of nicorandil, nitroglycerin, nicotinic acid, and SG-86 on the metabolic status and functional recovery of the ischemic-reperfused myocardium. J. Cardiovasc. Pharmacol. 10 (suppl. 8), S76-S87.

Gross, G.J., Warltier, D.C., and Hardman, H.F. (1987b). Comparative effects of nicorandil, a nicotinamide nitrate derivative, and nifedipine on myocardial reperfusion injury in dogs. J. Cardiovasc. Pharmacol. 10, 535-542.

Grover, G.J., Baird, A.J., and Sleph, P.G. (1996a). Lack of pharmacologic interaction between ATP-sensitive potassium channels and adenosine A_1 receptors in ischemic rat hearts. Cardiovasc. Res. 31, 511-517.

Grover, G.J., D'Alonzo, A.J., Parham, C.S., and Darbenzio, R.B. (1995a). Cardioprotection with K_{ATP} opener cromakalim is not correlated with ischemic myocardial action potential duration. J. Cardiovasc. Pharmacol. 26, 145- 152.

Grover, G.J., McCullough, J.R., D'Alonzo, A.J., Sargent, C.A., and Atwal, K.S. (1995b). Cardioprotective profile of the cardiac-selective ATP-sensitive potassium channel opener BMS-180448. J. Cardiovasc. Pharmacol. 25, 40-50.

Grover, G.J., McCullough, J.R., Henry, D.E., Conder, M.L., and Sleph, P.G. (1989). Anti-ischemic effects of the potassium channel activators pinacidil and cromakalim and the reversal of these effects with the potassium channel blocker glyburide. J. Pharmacol. Exp. Ther. 251, 98-104.

Grover, G.J., Parham, C.S., Whigan, D.B., and Mitroka, J.G. (1996b). BMS-180448, a novel glyburide-reversible cardioprotective agent, enhances postischemic recovery of contractile function in dogs. J. Pharmacol. Exp. Ther. 276, 380-387.

Grover, G.J., and Sleph, P.G. (1995). Protective effect of K_{ATP} openers in ischemic rat hearts treated with potassium cardioplegic solution. J. Cardiovasc. Pharmacol. 26, 698-706.

Grover, G.J., Sleph, P.G., and Parham, C.S. (1990). Nicorandil improves postischemic contractile function independently of direct myocardial effects. J. Cardiovasc. Pharmacol. 15, 698-705.

Hamilton, T.C., and Weston, A.H. (1989). Cromakalim, nicorandil and pinacidil: novel drugs which open potassium channels in smooth muscle. Gen. Pharmacol. 20, 1-9.

Harris, A.S. (1966) Potassium and experimental coronary occlusion. Am. Heart J. 71, 791-802.

Harris, A.S., Bestini, A., Russel, R.A., Brigham, I.C., and Firestone, J.E. (1954). Excitatory factors in ventricular tachycardia resulting from myocardial ischemia. Potassium a major excitant. Science 119, 200-203.

Inoue, I., Nagase, H., Kishi, K., and Higuti, T. (1991). ATP-sensitive K^+ channel in the mitochondrial inner membrane. Nature 352, 244-247.

Kakei, M., Noma, A., and Shibasaki, T. (1985). Properties of adenosine triphosphate regulated potassium. J. Physiol. (London) 363, 441-462.

Kantor, P.R., Coetzee, W.A., Carmeliet, E.E., Dennis, S.C., and Opie, L.H. (1990). Reduction of ischemic K^+ loss and arrhythmias in rat hearts: effect of glibenclamide, a sulfonylurea. Circ. Res. 66, 478-485.

Kim, D., and Clampham, D.E. (1989). Potassium channel in cardiac cells activated by arachidonic acid and phospholipids. Science 244, 1174-1176.

Kirsch, G.E., Codina, J., Birnbaumer, L., and Brown, A.M. (1990). Coupling of ATP sensitive K^+ channels to A_1 receptors by G proteins in rat ventricular myocytes. Am. J. Physiol. 259, H820-H826.

Kloner, R.A., Shook, T., Przyklenk, K., Davis, V.G., Junio, L., Matthews, R.V., Burstein, S., Gibson, C.M., Poole, W.K., Cannon, C.P., McCabe, C.H., and Braunwald, E., for the TIMI 4 Investigators. (1995). Previous angina alters in-hospital outcome in TIMI 4: a clinical correlate to preconditioning? Circulation 91, 37-47.

Lederer, W.J., and Nicchols, C.G. (1989). Nucleotide modulation of the activity of rat heart (K^+[ATP]) channels in isolated membrane patches. J. Physiol. (London) 419, 193-211.

Liu, G.S., Thornton, J., Van Winkle, D.M., Stanley, A.W.H., Olsson, R.A., and Downey, J.M. (1991). Protection against infarction afforded by preconditioning is mediated by A_1 adenosine receptors in rabbit heart. Circulation 84, 350-356.

Longman, S.D., and Hamilton, T.C. (1992). Potassium channel activator drugs: mechanism of action, pharmacological properties, and therapeutic potential. Med. Res. Rev. 12, 73-148.

Loubatieres, A. (1957) The hypoglycemic sulfonamides: history and development of the problem from 1942 to 1955. Ann. NY. Acad. Sci. 71, 4-11.

Marber, M.S., Latchman, D.S., Walker, J.M., and Yellon, D.M. (1993). Cardiac stress protein elevation 24 hours after brief ischemia or heat stress is associated with resistance to myocardial infarction. Circulation 88, 1264- 1272.

Miller, R.J. (1990) Glucose-regulated potassium channels are sweet news for neurobiologists. Trends Neurosci. 13, 197-199.

Mizumura, T., Nithipatikom, K., and Gross, G.J. (1995a). Effects of nicorandil and glyceryl trinitrate on infarct size, adenosine release, and neutrophil infiltration in the dog. Cardiovasc. Res. 29, 482-489.

Mizimura, T., Nithipatikom, K., and Gross, G.J. (1995b). Bimakalim, an ATP- sensitive potassium channel opener, mimics the effects of ischemic preconditioning to reduce infarct size, adenosine release, and neutrophil function in dogs. Circulation 92, 1236-1245.

Murry, C.E., Richard, V.J., Reimer, K.A., and Jennings, R.B. (1990). Ischemic preconditioning slows energy metabolism and delays ultrastructural damage during a sustained ischemic episode. Circ. Res. 66, 913-931.

Noma, A. (1983) ATP-regulated K^+ channels in cardiac muscle. Nature 305, 147-148.

Przyklenk, K., Sussman, M.A., Simkovich, B.Z, and Kloner, R.A. (1995). Does ischemic preconditioning trigger translocation of protein kinase C in the canine model? Circulation, 92 1546-1557.

Reimer, K.A., Murray, C.E., and Jennings, R.B. (1990). Cardiac adaptation to ischemia: ischemic preconditioning increases myocardial tolerance to subsequent ischemic episodes. Circulation, 82 2266-2268.

Robertson, D.W., and Steinberg, M.I. (1990). Potassium channel modulators: scientific applications and therapeutic promise. J. Med. Chem. 33, 1529- 1541.

Rudy, B. (1980). Diversity and ubiquity of K^+ channels. Neuroscience 25, 729-749.

Schmid-Antomarchi, H., de Weille, J., Fosset, M., and Lazdunski, M. (1987). The antidiabetic sulfonylurea glibenclamide is a potent blocker of the ATP- sensitive K^+ channels in insulin secreting cells. Biochem. Biophys. Res. Commun. 146, 21-25.

Schultz, J.E.J., Hsu, A.K., and Gross, G.J. (1996). Morphine mimics the cardioprotective effect of ischemic preconditioning via a glibenclamide- sensitive mechanism in the rat heart. Circ. Res. 78, 1100-1104.

Spinelli, W., Follmer, C.H., Parsons, R., and Colatsky, T.J. (1990). Effects of cromakalim, pinacidil, and nicorandil on cardiac refractoriness and arterial pressure in open-chest dogs. Eur. J. Pharmacol. 179, 243-252.

Spinelli, W., Sorota, S., Siegal, M., Hoffman, B.F. (1991) Antiarrhythmic actions of the ATP-regulated K^+ current activated by pinacidil. Circ. Res. 68, 1127-1137.

Spruce, A.E., Standen, N.B., and Stanfield, P.R. (1985). Voltage-dependent ATP-sensitive potassium channels of skeletal muscle membrane. Nature 316, 736- 738.

Standen, N.B., Quayle, J.M., Davies, N.W., Brayden, J.E., Huang, Y., and Nelson, M.T. (1989). Hyperpolarizing vasodilators activate ATP-sensitive K^+ channels in arterial smooth muscle. Science 245, 177-180.

Steinberg, M.I., Ertel, P., Smallwood, J.K., Wyss, V., and Zimmerman, K. (1988). The relation between vascular relaxant and cardiac electrophysiological effects of pinacidil. J. Cardiovasc. Pharmacol. 12 (suppl. 2), S30-S40.

Szilvassy, Z., Ferdinandy, P., Szilvassy, J., Nagy, I., Karcsu, S., Lonovics, J., Dux, L., and Koltai, M. (1995). The loss of pacing-induced preconditioning in atherosclerotic rabbits: role of hypercholesterolaemia. J. Mol. Cell. Cardiol. 27, 2559-2569.

Tang, X.L., Qiu, Y, Park, S.W., Sun, J.Z., Kalya, A., and Bolli, R. (1996). Time course of late preconditioning against stunning in conscious pigs. Circ. Res. 79, 424-434.

Thornton, J.D., Daly, J.F., Cohen, M.V., Yang, X.M., and Downey, J.M. (1993). Catecholamines can induce adenosine receptor-mediated protection of the myocardium but do not participate in ischemic preconditioning in the rabbit. Circ. Res. 73, 649-655.

Toda, N., Nakajima, S., Miyazaki, M., and Ueda, M. (1985). Vasodilation induced by pinacidil in dogs: comparisons with hydralazine and nifedipine. J. Cardiovasc. Pharmacol. 7, 1118-1126.

Tosaki, A., Engelman, D.T., Engelman, R.M., and Das, D.K. (1995). Diabetes and ATP-sensitive potassium channel openers and blockers in isolated ischemic/reperfused hearts. J. Pharmacol. Exp. Ther. 275, 1115-1123.

Tosaki, A., Engelman, D.T., Engelman, R.M., and Das, D.K. (1996a). The evolution of diabetic response to ischemia/reperfusion and preconditioning in isolated working rat hearts. Cardiovasc. Res. 31, 526-536.

Tosaki, A., and Hellegouarch, A. (1994). Adenosine-triphosphate-sensitive potassium channel blocking agent ameliorates, but the opening agent aggravates, ischemia/reperfusion-induced injury: heart function studies in nonfibrillating isolated hearts. J. Am. Coll. Cardiol. 23, 487-496.

Tosaki, A., Maulik, N., Engelman, D.T., and Engelman, R.M. (1996b). The role of protein kinase C in ischemic/reperfused preconditioned isolated rat hearts. J. Cardiovasc. Pharmacol. 28, 723-731.

Tosaki, A., Szerdahelyi, P., and Das, D.K. (1992). Reperfusion-induced arrhythmias and myocardial ion shifts: pharmacologic interaction between pinacidil and cicletanine in isolated rat hearts. Basic. Res. Cardiol. 87, 366-384.

Tosaki, A., Szerdahelyi, P., Engelman, R.M., and Das, D.K. (1993). Potassium channel openers and blockers: do they possess proarrhythmic or antiarrhythmic activity in ischemic and reperfused rat hearts. J. Pharmacol. Exp. Ther. 267, 1355-1362.

Tseng, G.N., and Hoffman, B.F. (1990). Actions of pinacidil on membrane currents in canine ventricular myocytes and their modulation by intracellular ATP and cAMP. Pflugers Arch. 415, 414-424.

Tung, R.T., and Kurachi, Y. (1991). On the mechanism of nucleotide diphosphate activation of the ATP-sensitive K^+ channel in ventricular cell of guinea pig. J. Physiol. (London) 437, 239-256.

Venkatesh, N., Lamp, S.T., and Weiss, J.N. (1991). Sulfonylureas, ATP-sensitive K^+ channels, and cellular K^+ loss during hypoxia, ischemia, and metabolic inhibition in mammalian ventricle. Circ. Res. 69, 623-637.

Weiss, J., and Shine, K.I. (1982). Extracellular K^+ accumulation during myocardial ischemia in isolated rabbit heart. Am. J. Physiol. 242, H619- H628.

Wilde, A.A.M. (1996). ATP-sensitive potassium channels, transmural ischemia and the ECG implications for the non-insulin dependent diabetic patient? Cardiovasc. Res. 31, 688-690.

Wilde, A.A.M., Escande, D., Schumacher, C.A., Thuringer, D., Mestre, M., Fiolet, J.W.T., and Janse, M.J. (1990). Potassium accumulation in the globally ischemic mammalian heart: a role for the ATP-sensitive K^+ channel. Circ. Res. 67, 835-843.

Wolleben, C.D., Sanguinetti, M.C., and Siegl, P.K.S. (1989) Influence of ATP-sensitive potassium modulators on ischemia-induced fibrillation in isolated rat hearts. J. Mol. Cell. Cardiol. 21, 783-788.

Yang, X.M., Baxter, G.F., Heads, R.J., Yellon, D.M., Downey, J.M., and Cohen, M.V. (1996). Infarct limitation of the second window of protection in a conscious rabbit model. Cardiovasc. Res. 31, 777-783.

Yao, Z., and Gross, G.J. (1994). Effects of the K_{ATP} channel opener bimakalim on coronary blood flow, monophasic action potential duration and infarct size in dogs. Circulation, 89, 1769-1775.

DELAYED PRECONDITIONING: MECHANISMS OF ENDOGENOUS AND PHARMACOLOGIC INDUCTION OF THIS ADAPTIVE RESPONSE TO ISCHEMIA

Gary T. Elliott and Patricia A. Weber

Advances in Organ Biology
Volume 6, pages 197-217.

ISBN: 0-7623-0391-3

I. INTRODUCTION

Ischemia with or without reperfusion unquestionably represents one of the primary processes responsible for myocardial injury. Such injury manifests itself in a number of forms, including dysrhythmias, loss of contractile efficiency in viable myocytes, and frank cell death (Murry et al., 1994). Deterioration of contractility in otherwise viable cardiomyocytes can be an acute consequence of transient ischemia (myocardial stunning) (Ovize et al., 1994; Schulz and Heusch, 1994) or the result of long-term, low-flow ischemia in which profound degeneration of contractile elements within the myocyte leads to myocardial hibernation (Schulz and Heusch, 1994). Stunning and hibernation are both examples of ischemia-associated contractile dysfunction which, in contrast to contractile dysfunction associated with infarction, are reversible upon reperfusion; within hours to days in the case of stunning and over a period of weeks to months with hibernation. Dysrhythmias associated with ischemia–reperfusion take a variety of forms, including ventricular tachycardia and fibrillation, ecoptic ventricular contractures, ventricular tachycardia ,and supraventricular tachycardia, and fibrillation.

The human body responds to a variety of forms of physiological stress with adaptive processes intended to reduce the pathological consequences of such stress. For example, the athlete manifests numerous examples of such adaptive responses, including development of calluses as a response to chronic friction, skeletal muscle hypertrophy as a response to repetitive exercise, and alteration in oxygen–hemoglobin dissociation constants as a consequence of physical exertion at high altitudes.

II. PRECONDITIONING

A. Ischemic Preconditioning (Acute)

As another example of an adaptive response to physiological stress, the cardiomyocyte (Armstrong and Ganote, 1995) and vascular endothelial cell (Richard et al., 1994; De Fily and Chilian, 1991) appear to possess endogenous mechanisms that enhance tolerance of the cells to ischemic–hypoxic injury. A rapidly developing, adaptation-reported response to transient ischemia that minimizes infarct size was initially reported in the often-cited paper by Murry, Jennings, and Reimer; this response has come to be known as ischemic preconditioning (Murry et al., 1994; Murry et al., 1986). This adaptation develops within 5 to 10 minutes of reperfusion following transient ischemic periods (as short as 3 to 5 minutes in duration). It appears that reperfusion is required in most models for the preconditioned state to develop. The preconditioned state is fairly brief in duration, with susceptibility to ischemic injury (infarction) returning within 1 to 2 hours in most models. The most reproducible manifestation of this adaptative response is a profound reduction in susceptibility of myocardium to infarction upon exposure to a second prolonged (30- to 90-minute) ischemic stress. This tolerance to lethal cell injury can be observed in response to induction in a whole heart of a global or regional in situ ischemia (Murry et al., 1994; Murry et al., 1986), global ex vivo ischemia (Ytrehus et al., 1994; Schaefer et al., 1995), and following in vitro hypoxic stress of primary cardiomyocyts (Armstrong and Ganote, 1995). The ability to induce this tolerant state in cardiomyocytes following in vitro hypoxic stress suggests that at least a component of this adaptive response is operative at the level of the cardiomyocyte.

Investigators have also reported that ischemically preconditioned hearts display a reduced propensity for the development of ventricular arrhythmias upon reperfusion (Piacentini et al., 1993; Vegh et al., 1995; Parratt and Vegh, 1994; Lawson et al., 1993). Reduction in the incidence of ventricular fibrillation, tachycardia, and premature contracture in ischemically preconditioned myocardium has been associated with a bradykinin mediated nitric oxide–dependent mechanism (Parratt and Vegh, 1994).

B. Second Window of Preconditioning (Delayed Preconditioning)

It has also become apparent that a delayed adaptive response also occurs following ischemia or rapid cardiac pacing (Marber et al., 1993; Sun et al., 1996; Kuzuya et al., 1993; Vegh et al., 1994; Yang et al., 1996). Referred to as the "second window of preconditioning" (SWOP) when induced by ischemia, this adaptation evolves over a period of 12 to 24 hours following exposure of myocardium to transient preconditioning ischemia. This delayed form of preconditioning has been observed in both anesthetized and conscious animals (Yang et al., 1996). Consequently, following ischemic preconditioning, myocardium displays a bimodal state of subsequent ischemic tolerance, with a first window of preconditioning developing within minutes of reperfusion following transient ischemia and lasting for only a few hours. Thereafter, the myocardium appears to regain its normal sensitivity to ischemic injury until about 12 to 24 hours following the initial transient ischemic episode, at which time a state of ischemic tolerance returns. This delayed preconditioning is more durable in nature than the first window of protection, lasting 24 to 48 hours.

With rapid pacing, ischemia may ultimately be responsible for inducing protection and, consequently, may not be a novel inducer of tolerance. Heat stress may, on the other hand, be a unique form of physiological stress capable of activating delayed preconditioning (Marber et al., 1993; Liu et al., 1992; Shipley et al., 1995).

Not surprisingly, considerable attention has been focused on induction of heat shock proteins as mediators of delayed preconditioning. More will be said about potential mediators of this adaptation later in the chapter.

C. Preconditioning Reduces Manifestations of Ischemia–Reperfusion Injury

Apart from the prolonged nature of delayed preconditioning, there are some differences regarding manifestation of the first window and second window of preconditioning. Clearly, attenuation of myocardial necrosis following prolonged ischemia is a feature of tolerance shared by both windows of adaptation, although the effectiveness of protection, at least in the rabbit, elicited by ischemia appears to be superior in the first window of preconditioning. Antiarrhythmic effects of preconditioning are apparent in both forms of the adaptive response to ischemia (Piacentini et al., 1993; Vegh et al., 1995; Parratt and Vegh, 1994; Lawson et al., 1993; Vegh et al., 1994; Krause and Szekeres, 1995). Pronounced reduction in the occurrence of ectopic beats, ventricular tachycardia, ventricular fibrillation, and, consequently, death in the dog are reported following chronic rapid pacing (Vegh et al., 1994). Delayed preconditioning induced by ischemia has also been reported to reduce myocardial stunning in the pig 12 to 72 hours afterward (Sun et al., 1996; Sun et al., 1995). The ability of the second window of preconditioning to attenuate myocardial stunning represents a point of distinction regarding manifestation of cardioprotection following ischemia–reperfusion when compared with the first window of protection. Not only is the initial preconditioned state incapable of ameliorating myocardial stunning associated with periods of nonlethal ischemia, it is well appreciated that stunning is a consequence of first window ischemic preconditioning, the contractile dysfunction lasting for hours following the transient ischemic event (Murry et al., 1991). It is now clear that the stunned state is not a prerequisite for development of protection against infarction observed during the first window of protection, as protection clearly wanes hours before reversal of stunning is observed.

D. Biochemical Consequences of Preconditioning

A great deal of effort has been expended over the past eight years to elucidate how preconditioning alters myocardial biochemistry in a fashion that enhances tolerance to ischemia–reperfusion injury. Considerable work has been undertaken to dissect the signaling pathways that elicit the preconditioning response to ischemia. The vast majority of this work has been directed toward the first window of preconditioning. Little is known at this time about mechanisms that operate in the context of delayed preconditioning.

If we review what is known about the preconditioned state in the first window of protection, it becomes apparent that preconditioning alters the metabolic state of the cardiomyocyte (Volovsek et al., 1992; Ziegelhoffer et al., 1992; Ischihara et al.,

1994; Murry et al., 1990; Kaplan et al., 1994). Preconditioned myocardium has been reported by a number of investigators to maintain high-energy phosphate adenine nucleotide levels during ischemia. This energy conservation may be a consequence of enhanced anaerobic glycolytic adenosine triphosphate (ATP) formation (Janier et al., 1994) and reduced sarcolemmal ATPase-mediated catabolism (Ziegelhoffer et al., 1995). Consequently, in association with higher myocyte ATP, intracellular acidosis is reduced during ischemia. Transmembrane exchange of sodium for protons in an attempt to elevate intracardiomyocyte pH is reduced in the face of less acidosis, and hence, the obligatory intracellular calcium loading, which thereafter results as a consequence of activation of the sodium–calcium exchange pump, is attenuated (Steenbergen et al., 1993; Gottlieb et al., 1996; Bugge and Ytrehus, 1995). It is generally accepted that intracellular calcium loading observed during reperfusion may be an important common inducer of myocardial stunning and infarction (Gao et al., 1995; Marban and Gao, 1995; Hata et al., 1994; Yamamoto et al., 1991; Klein et al., 1995). Activation of calcium dependent proteases, stimulation of apoptosis, and degradation of myofilament contractile elements have been linked to intracellular calcium loading and, consequently, infarction and stunning following cardiac ischemia–reperfusion. In summary, the first window of ischemic preconditioning may reduce intracellular hypoxic acidosis and the consequent calcium loading.

It is not unreasonable to assume that these metabolic adaptations observed in the cardiomyocyte during ischemia with the first window (classic) form of preconditioning may also be important to the second window of preconditioning. There is, however, only limited data to support this hypothesis.

E. Mediators and Mechanism of Preconditioning

Although it is generally recognized that preconditioned myocardium displays ATP "preservation"—that is, reduced acidosis and reduced calcium overload in the face of ischemia–reperfusion—the mechanism by which this adaptive response develops is still being elucidated. There is considerably more information available regarding mediators and signal transduction pathways, which contribute to the development of classic preconditioning. A role has been established for the release of norepinephrine, adenosine, and bradykinin during transient ischemia–reperfusion, which leads to activation of G_i- or G_q-coupled receptors, phospholipase D, protein kinase C (PKC), and the ultimate opening of ATP-sensitive potassium (K^+_{ATP}) channels (Gato et al., 1995; Downey and Cohen, 1995; Gross and Auchampach, 1992; Yao and Gross, 1993; Hu et al., 1996; Liu et al, 1996). Elicitation of the preconditioned state by administration of adenosine, bradykinin, or norepinephrine to animals can all be blocked by inhibitors of the K^+_{ATP} channel, as can ischemic preconditioning, suggesting that activation of this ion channel must be a relatively late event in the development of this adaptive response (Yao and Gross, 1993; Liu et al., 1996; Gross et al., 1996). The K^+_{ATP} channel appears to be an important late mediator of preconditioning in a variety of species, including rat, rabbit, dog, pig, and, it appears, even humans (Speechly-Dick et al., 1995). It is beyond the scope of this work to discuss further the mechanism(s) of first window cardiac preconditioning. A good review of the subject was compiled by Parratt (1994) and Downey and Cohen (1995).

Considerably less is presently known regarding mechanism(s) for induction of delayed cardiac preconditioning. By definition, this adaptive process evolves slowly, suggesting intuitively that synthesis of protective factors within the cardiomyocyte and/or endothelial cell may be required. It has been proposed that these cardioprotective factors might be heat shock proteins (Marber et al., 1993; Liu et al., 1992; Parratt and Szekeres, 1995), nitric oxide synthase (Vegh et al., 1994; Parratt and Szekeres, 1995; Maulik et al., 1995), and/or superoxide dismutase (Yamashita et al., 1994; Nelson et al., 1995). In the context of delayed preconditioning, much of the early mechanistic work focused on induction of heat shock proteins as it was demonstrated early on that heat stress, in and of itself, could induce delayed preconditioning (Marber et al., 1993; Liu et al., 1992). Transient ischemia, which was demonstrated by Marber et al. to induce tolerance to infarction in rabbits 24 hours thereafter, was also shown to induce heat shock protein (HSP) 70 in myocardial biopsies (Marber et al., 1993). In addition, HSP 70 overexpressor mice were observed to be resistant to infarction (Marber et al., 1995). Both observations suggested that heat shock proteins might be cardioprotective factors of delayed preconditioning. More recent studies suggest that induction of heat shock proteins 12 to 24 hours following transient ischemia may be an epiphenomena rather than a mediator of delayed preconditioning (Shipley et al., 1995; Tanaka et al., 1994). At present there is no clear consensus about whether heat shock proteins play a role in eliciting delayed preconditioning.

Induction of manganese superoxide dismutase (Mn-SOD) was detected in rat cardiomyocyte cultures 24 hours following hypoxia (Yamashita et al., 1994). Tumor necrosis factor, which protects isolated rabbit hearts harvested from treated animals from contractile dysfunction following lethal ischemia, was also associated with induction of myocardial Mn-SOD (Nelson et al., 1995). These observations suggested that induction of antioxidant enzymes may be important in the development of delayed preconditioning.

Nitric oxide has been proposed to be an important cardioprotective factor in certain models of ischemia–reperfusion injury, and nitric oxide signaling may be important in both classic and delayed preconditioning. Although itself a free radical, nitric oxide displays relatively low oxidative reactivity and scavenges the more highly reactive oxygen free radicals such as superoxide (Maulik et al., 1995). Evidence exists that nitric oxide can function as an intracellular signal leading to activation of kineses, such as PKC (Maulik et al., 1995). Activation of PKC with ischemic preconditioning has been demonstrated to play a role in inducing the first and second windows of the preconditioned state, ultimately, it has been proposed, via activation of myocardial K^{+}_{ATP} channels (Downey and

Cohen, 1995; Hu et al., 1996; Speechly-Dick et al., 1995; Parratt, 1994). Rat cardiomyocytes synthesize inducible nitric oxide synthase (iNOS) (Luss et al., 1995; Shindo et al., 1995; McKenna et al., 1995) and delayed up-regulation of iNOS following transient ischemia may be important in eliciting the slowly developing second window of preconditioning (Yamashita et al., 1994; Nelson et al., 1995). Most recently, direct evidence using patch clamp technique suggests that nitric oxide may cause opening of K^{+}_{ATP} channels in rabbit cardiomyocytes (Cameron et al., 1996).

Baxter et al. (1994), have demonstrated that administration of an adenosine subtype nonspecific receptor blocker during transient ischemia was capable of blocking the subsequent evolution of the second window of preconditioning 24 hours thereafter. Therefore, as with classic preconditioning, binding of adenosine to complementary cardiomyocyte receptors as a consequence of ATP depletion during the transient ischemia may activate a signal transduction pathway of delayed preconditioning. It is known that binding of adenosine to an A_1 receptor coupled to G_q proteins leads to PKC activation/translocation in myocardium (Downey and Cohen, 1995). Phosphorylation of the gene transcription factor NFkB leads to dissociation of the inhibitory IkB subunit with resultant activation of nuclear factor (Kwon et al., 1995; Shirakawa and Mizel, 1989). Induction of iNOS and Mn-SOD are known to be regulated via NFkB activation (Xie et al., 1994; Das et al., 1995). Hence, induction of cardioprotective proteins associated with delayed preconditioning could be induced via G_q-coupled receptor agonists such as adenosine, which causes activation of phospholipase D, synthesis of diacylglycerol, PKC followed by NFkB activation, and, ultimately, protein synthesis.

F. Delayed Preconditioning in the Clinical Setting

Retrospective evaluation of clinical data in an attempt to identify variables that prognosticate outcome in patients hospitalized following acute myocardial infarction, suggests that prodromal angina is predictive of outcome (Yoshikawa et al., 1993; Kloner and Yellon, 1994; Anzai et al., 1994; Kloner et al., 1995; Ottani et al., 1995; Anzai et al., 1995). Interestingly, a history of symptomatic angina in patients ultimately presenting to the hospital with a diagnosis of myocardial infarction carries with it an improved prognosis. Evaluation of in-hospital death, heart failure, or shock; infarct size (assessed enzymatically, via electrocardiogram or echocardiography); and dysrhythmias all showed improvement in the group of patients who experienced pre-infarction angina. The various authors suggest that the beneficial prognostic value of prodromal angina may result from cardioprotection associated with delayed preconditioning, which is itself induced by the premyocardial infarction ischemic events. If, in fact, this hypothesis is correct, then induction of delayed myocardial preconditioning in the human may effectively reduce cardiac ischemia–reperfusion injury.

III. PHARMACOLOGICAL INDUCTION OF THE SECOND WINDOW OF PROTECTION

Few pharmacological agents are currently available that induce, in preclinical models, a state that mimics—at least on a temporal basis—the second window of ischemic preconditioning. Delayed cardiac preconditioning to ischemia has been observed in models of infarction, regional stunning, global left ventricular function, and dysrhythmia using adenosine receptor agonists, endotoxin, and monophosphoryl lipid A (MLA).

A. A_1 Receptor Agonists

A recent study has implicated adenosine receptor activation in the delayed phase of protection against myocardial infarction. Baxter et al. (1994) demonstrated blockade of the delayed protection afforded by preconditioning with intravenous (IV) administration of the adenosine receptor blocker 8-*p*-sulfophenyl--theophylline (SPT) during a standard preconditioning protocol in the rabbit model. In addition, they were able to mimic the protective effects of preconditioning by IV administration of the specific A_1 receptor agonist, 2-chloro-N^6-cyclopentyladenosine (CCPA), 24 hours prior to a 30-minute ischemic period followed by 120 minutes of reperfusion. The 100 µg/kg dose of CCPA reduced infarct size, expressed as a percentage of the area at risk, comparably to preconditioning 26.3 ± 5.7 and 32.9 ± 4.6 percent, respectively (Table 1). Kukreja and colleagues (Medical College of Virginia, Richmond, VA) have also used CCPA to induce a delayed preconditioned state in the working rat heart model and were subsequently able to ablate the protective effect with the potassium channel blocker glibenclamide (personal communication).

B. Endotoxin

Table 1. A_1 Agonist Study Values for Absolute Infarct Volume, Ischemic Risk Zone Volume, and Infarct-to-Risk Ratios in a Rabbit Model of Regional Myocardial Ischemia/Reperfusion

Group	*No. of Animals*	*Infarct Volume, cm³*	*Ischemic Risk Zone Volume, cm*	*Infarct-to-Risk Ratio %*
Saline	12	0.68±0.06	1.2±0.1	54.5±2.7
PC+VEH	6	0.36±0.06†	1.2±0.2	32.9±4.6*
CCPA 25ug/kg	6	0.40±0.08*	1.0±0.1	37.1±4.2†
CCPA 50ug/kg	6	0.43±0.05*	1.1±0.1	37.7±2.2†
CCPA 100ug/kg	6	0.27±0.09†	1.0±0.1	26.3±5.7†

Notes: CCPA indicates 2-chloro-N^6-cyclopenthyladenosine. Values are mean ± SEM.
* $p < .05$ or † $p < .01$ compared with saline-teated control group (ANOVA with Dunnett's test)
Data combined from tables 2 and 4 in *Circulation* 1994; 90 (No 6): 2993-3000.

Delayed myocardial protection has also been induced by pretreatment with bacterial endotoxin. Brown and co-workers (1989) demonstrated, in an isolated heart model, that rats administered a 500-μg/kg dose of lipopolysaccharide (LPS) intraperitoneally (IP) 24 hours prior to ischemia and reperfusion had elevated endogenous myocardial catalase activity and preservation of myocardial function including ventricular developed pressure and +/- dp/dt. Das and colleagues (Maulik et al., 1995) also found in the working rat heart model that a 500-μg/kg IP dose of LPS elevated antioxidant enzymes from 4 to 24 hours post-treatment and that significant cardiac functional preservation was evident by 24 hours. Measurements of left ventricular maximum dp/dt were similar to baseline in the endotoxin-treated rats, and aortic flow was significantly preserved in postischemic myocardium (Figures 1 and 2).

Song and colleagues (1996) pretreated rats with 1.5 or 2.5 mg/kg of LPS 4, 8, and 24 hours prior to a 30-minute occlusion and showed a significant reduction in ventricular fibrillation and in the number of ventricular premature beats (VPBs) during the occlusion period in an isolated heart model. The attenuation of VPBs was most pronounced at 8 hours post-LPS administration when the 2.5-mg/kg dose was used.

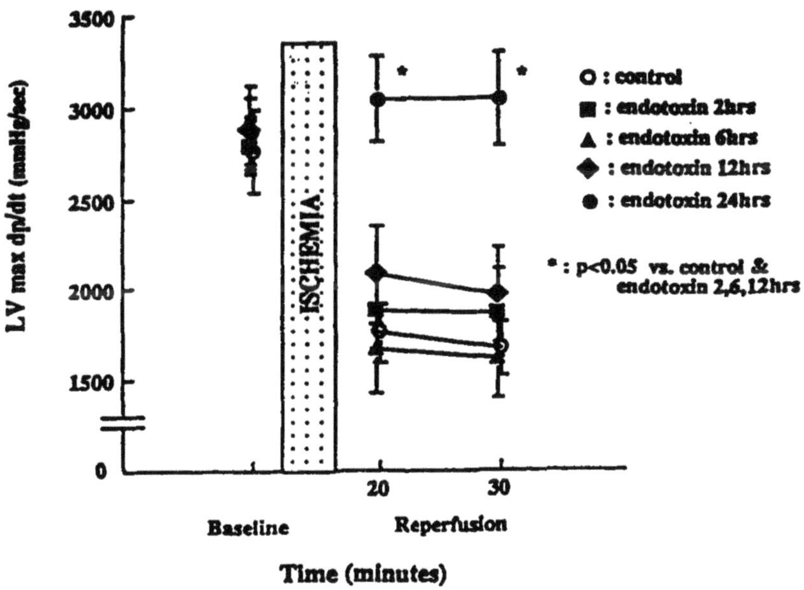

Figure 1. Effect of endotoxin in changes in left ventricular maximum dp/dt (LV max dp/dt) during ischemia and reperfusion. Rats were injected with endotoxin (0.5 mg/kg) and then killed after 24 hours to perform isolated working heart preparation. Ischemia was induced by terminating coronary flow for 30 minutes, which was followed by 30 minutes of reperfusion. LV max dp/dt was measured. Results are mean ± SE of 4 to 6 different animals per group. **P*<05 compared with control.

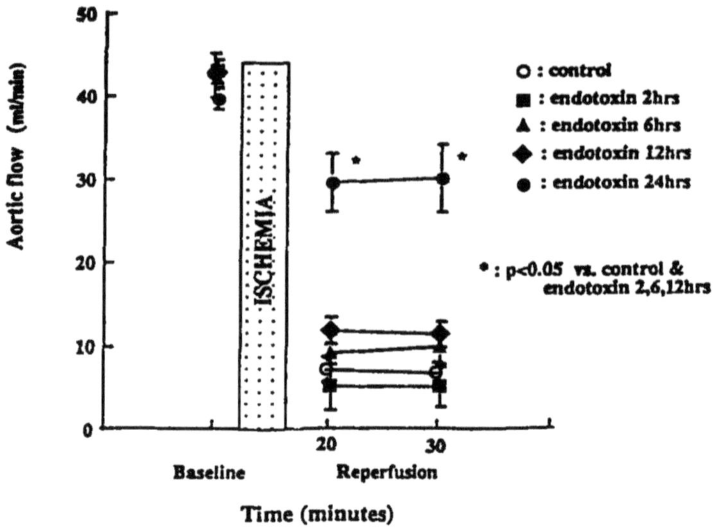

Figure 2. Effect of endotoxin on changes in aortic flow in postischemic myocardium. Rats were injected with endotoxin (0.5 mg/kg) and then killed after 24 hours to perform isolated working heart preparation. Ischemia was induced by terminating coronary flow for 30 minutes, which was followed by 30 minutes of reperfusion. Aortic flow was measured. Results are mean ± SE of 4 to 6 different animals per group. **P*<.05 compared with control.

The effect was attenuated by administration 1 hour prior to endotoxin of dexamethasone, which among other proteins and enzymes is known to block iNOS induction. These data suggest a possible role for nitric oxide in the cardioprotection induced by LPS.

Experiments by Node et al. (1995) demonstrated that 24-hour pretreatment with LPS at a dose of 60 μg/kg in dogs reduced infarct size from an average of 42% in control dogs to 19% in treated dogs. Evidence was found for the involvement of nitric oxide in the cardioprotective activity of LPS in this model. Significant increases were found in nictric oxide synthase activity (1.22 ± 0.04 vs. 0.78 ± 0.03 pmol/mg/min) and nitric oxide end products (NO^{2-} + NO^{3-}, 54 ± 5 vs. 23 ± 6) in LPS-treated versus control dogs, respectively. These investigators also found that the protective effect of LPS was blocked by administration of dexamethasone 90 minutes prior to LPS. Cardioprotection was also partially blocked by concomitant administration of N^G-nitro-L-arginine methyl ester (L-NAME), an inhibitor of both constitutive and inducible nitric oxide synthase. These data also suggest a role for nitric oxide in the cardioprotection induced by endotoxin.

More recently, it has been reported that 72-hour pretreatment with a single intravenous injection of endotoxin at doses of 5 and 10 μg/kg protected myocardium from 45 minutes of sustained occlusion in a rabbit infarct model when compared with control rabbits (22.2 ± 3.1% and 17.5 ± 1.5% vs. 45.3 ± 2.5%) (Rowland et al., 1996). The investigators found no significant increase in myocardial catalase levels but did demonstrate induction of HSP 72 within cardiomyocytes by immunohistochemistry.

C. Monophosphoryl Lipid A

There are many reports in the literature using various animal models which demonstrate that pretreatment with MLA protects myocardium from ischemia–reperfusion injury and pharmacologically mimics the second window of ischemic preconditioning. Experimental results using MLA in a rabbit infarct model consisting of 30 minutes of regional ischemia followed by 180 minutes of reperfusion found that a dose as low as 10 μg/kg given 24 hours prior to ischemia reduced infarct size by 70% (Elliott et al., 1996) (Figure 3). Studies in a dog model of ischemia–reperfusion injury have also shown that 24-hour pretreatment with doses ranging from 10 to 100 μg/kg significantly reduced infarct size when compared to vehicle treated controls (Yao et al., 1993; Przyklenk et al., 1996; Mei et al., 1996b).

In a canine model, 35 μg/kg pretreatment with MLA attenuated during the reperfusion period global left ventricular dysfunction induced by 30 minutes of normothermic ischemia associated with aortic cross-clamping and cardiopulmonary bypass (Abd-Elfattah et al., 1995). Twenty- four hour pretreatment with MLA at a dose of 35 μg/kg in a dog model of myocardial stunning associated with six 5-minute cycles of regional ischemia interspersed with 10-minute periods of reperfusion resulted in significant improvement in recovery of percent segment shortening, with control dogs displaying a 20% return of contractility by 2 hours reperfusion versus a 50% recovery in pretreated animals (Yao et al., 1995) (Figure 4).

Pretreatment with MLA at a dose of 100 μg/kg in a dog model of ischemia–reperfusion injury (25-minute coronary artery occlusion followed by reperfusion) markedly reduced the severity of ventricular arrhythmias during occlusion. The number of ventricular premature beats was reduced from 248 ± 106 in controls to 28 ± 26 in MLA-pretreated dogs. MLA pretreatment also significantly increased survival due to combined ischemia–reperfusion insult from 12% to 50% when compared to vehicle treated controls (Vegh et al., 1996).

MLA was also evaluated in a rabbit model of cardiogenic shock using prolonged regional ischemia (90 minutes) followed by reperfusion (6 hours). A 24-hour IV pretreatment with a single MLA dose of 35 μg/kg resulted in stabilization of left ventricular developed pressure, rate-pressure product, +dp/dt, –dp/dt, and mean arterial pressure during reperfusion as compared to controls where deterioration of these end points to approximately 35, 32, 48, 43, and 43% from baseline occurred, respectively (Zhao et al., 1996). Although MLA did not reduce infarct size in this

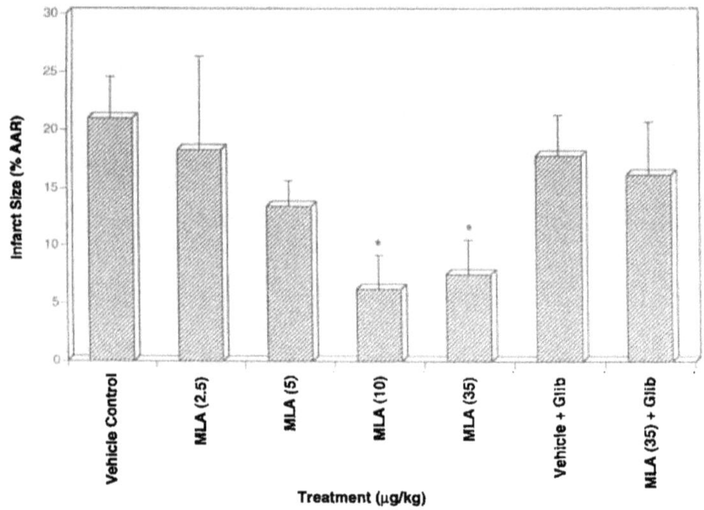

Figure 3. Dose–response effect of monophosphoryl lipid A (MLA) pretreatment, with or without pharmacologic antagonism with glibenclamide 300 µg/kg (Glib) on infarct size expressed as percentage of area at risk (AAR). MLA was given (IV) 24 hours before myocardial ischemia. Columns are means with SEM bars. Group size; vehicle control n = 9, vehicle control + Glib n = 7, MLA (5 µg/kg n = 7, 10 µg/kg n = 7, 35 µg/kg n = 7), MLA + Glib n = 7. MLA pretreatment at 10 mg/kg and 35 µg/kg significantly reduced infarct size (**P*<.05). Administration of 300 µg/kg glibenclamide at 30 minutes before occlusion blocked the cardioprotection by MLA (vehicle control versus MLA 35 µg/kg + Glib; *P* = NS, and MLA 35 µg/kg vs. MLA 35 µg/kg + Glib; *P*<.05).

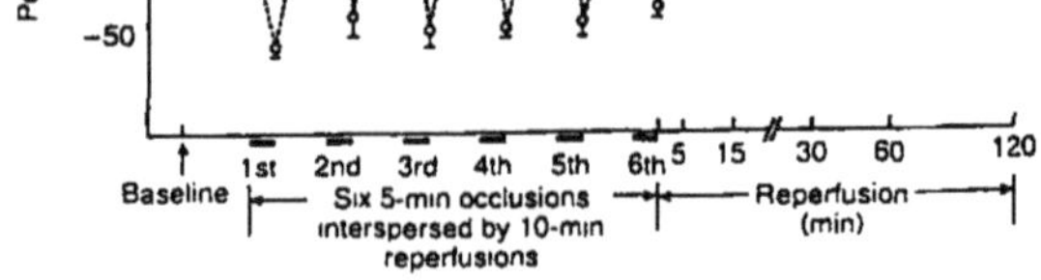

Figure 4. Pretreatment with monophosphoryl lipid A (MLA, 35 µg/kg IV) 24 hours prior to six 5-minute coronary artery occlusions followed by reperfusion markedly improved functional recovery throughout all occlusion and reperfusion periods. *Significantly different from control by *P*<.05. All values are mean ± SEM (n = 8, each group).

study, pretreatment was found to preserve ATP and adenosine diphosphate (ADP) at the end of the 90-minute ischemic period and to decrease formation of downstream adenosine catabolites including inosine, xanthine, and uric acid.

Studies designed to determine the time-course for induction of cardioprotection by MLA in myocardial infarction models have shown that more than 1 hour was necessary in the dog and full protection was evident by 12 hours following drug administration in a rabbit model (Yao et al., 1993b; Yoshida et al., 1996). Additional experiments in the rabbit infarct model have revealed that protection was not evident by 3 hours post-treatment with 35 μg/kg MLA; yet by 9 hours, protection was significant and remained apparent for 36 hours with loss of protection by 48 hours (Weber et al., unpublished data).

The time-course information suggests that protein synthesis may be necessary. Several candidate proteins have been considered, including catalase, superoxide dismutase, heat shock proteins, and nitric oxide. MLA administered at a very high dose of 5 mg/kg was found to preserve ventricular function (DP, + and – dp/dt) when given 24 hours, but not 2 hours, prior to ischemia in the isolated rat heart model. Myocardial catalase activity was significantly increased in MLA-pretreated rats as compared to controls (Nelson et al., 1991). Experimental results in the dog infarct model showed that a dose of 65 μg/kg was cardioprotective when administered 24 hours, but not 1 hour, prior to ischemia–reperfusion and, further, did not significantly elevate either myocardial catalase or superoxide dismutase levels (Yao et al., 1993). The conflicting potential importance of catalase induction in the protection induced by MLA between the rat and dog models is likely due to the extremely high dose used in the rat (5,000 μg/kg) in comparison to the doses that have been shown to be cardioprotective in the dog (10 to 100 μg/kg). The apparent lack of involvement of catalase in the delayed cardioprotection induced by MLA is supported by the previously discussed results in which low doses of endotoxin were found to be cardioprotective in the rabbit model without induction of catalase (Rowland et al., 1996). In contrast, results published in the same paper regarding the induction of HSP 72 by endotoxin in the rabbit at the 5 μg/kg dose are in conflict with what has been found with MLA. Studies from two independent laboratories have shown that the inducible 70-kDa heat shock protein (HSP 70i) is not increased by 24-hour pretreatment of rabbits with cardioprotective doses of MLA (35 μg/kg) (Yoshida et al., 1996; Baxter et al., 1996). Considering that LPS is cardioprotective 72 hours postdosing while the activity of MLA dissipates by 48 hours and, additionally, considering the disparate ability of cardioprotective doses of LPS and MLA to induce HSP 70 in the rabbit, it is likely that the mechanism of action of these two preconditioning mimetics in this model may differ in some respect.

The role of nitric oxide in the delayed cardioprotection induced by MLA was evaluated using aminoguanidine (AMG), a selective inhibitor of the inducible form of nitric oxide synthase that has far less effect on the other isoforms when administered at the appropriate in vivo concentration. The protective effect of 24-hour pretreatment with MLA at 35 μg/kg in a rabbit infarct model was completely blocked by administration of AMG 1 hour prior to the 30-minute ischemic period (Zhao et al., 1997). In addition, neutrophil infiltration into the border region of the infarcted myocardium was determined by measurement of myeloperoxidase activity. It was found that myeloperoxidase activity was significantly less in MLA-pretreated rabbits as compared to vehicle-treated controls and that this effect was partially blocked by administration of AMG. Gross and colleagues also published similiar findings regarding neutrophil infiltration in the dog infarct model (Figure 5) (Yao et al., 1993).

Evidence indicating that induction of nitric oxide synthase may be involved in delayed preconditioning in the dog was published by Vegh et al. (1994). Using a

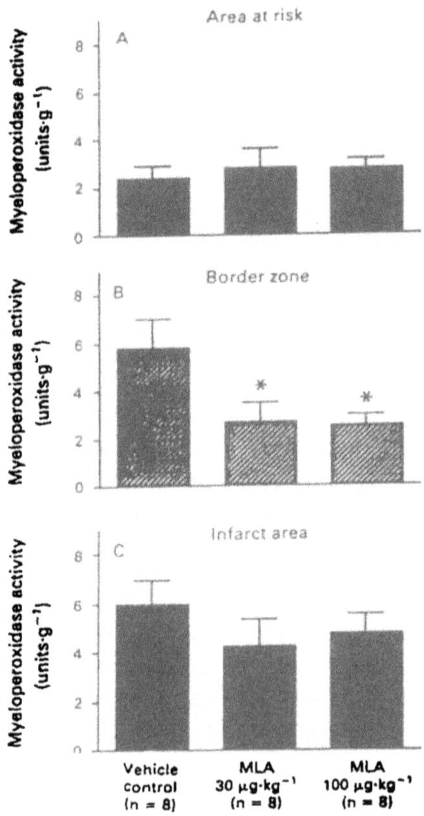

Figure 5. Effects of monophosphoryl lipid A (MLA) on myeloperoxidase activity in the area at risk distant from the infarct (**A**), the border zone immediately adjacent to the infarct (**B**), and the center of the infarct (**C**). Columns are means; bars = SEM. There was a marked ($P<.05$) reduction in myeloperoxidase activity in the border zone in both groups of MLA-treated animals. $^*P<.05$ vs. control.

model of rapid cardiac pacing that consisted of four 5-minute periods of tachycardia at a rate of 220 beats per minute to induce delayed preconditioning, it was found that the severity of ischemia-induced arrhythmias produced by a 25-minute occlusion period 20 hours after pacing was significantly reduced (e.g., reduction in ventricular fibrillation from 45% in unpaced dogs to 10% in paced dogs). The preconditioning effect was reversed by dexamethasone administered IV at a dose of 4 μg/kg, 45 minutes prior to pacing, again, suggesting a possible role for nitric oxide in delayed preconditioning.

Additional studies aimed at elaborating on the mechanism of the cardioprotection induced by MLA have been conducted in both dog and rabbit models. Gross et al. in the dog infarct and stunning models and Zhao and colleagues in the rabbit infarct model have shown that the cardioprotective effect of MLA pretreatment can be blocked by the K^+_{ATP} channel antagonist glibenclamide when it was administered IV (300 μg/kg) 15 and 30 minutes prior to the prolonged occlusion period in the dog and rabbit, respectively (Elliott et al., 1996; Mei et al., 1996b, Mei et al., 1995) (see Figure 3). Additional work in both species using the structurally unrelated K^+_{ATP} channel antagonist, 5-hydroxydecanoic acid (5-HD), has also shown blockade of cardioprotection by MLA (Mei et al., 1996b; Janin et al., 1996).

Gross and Mei have compared the effect of ischemic preconditioning with MLA pretreatment and K^+_{ATP} channel opener–induced cardioprotection on interstitial purine metabolism in dogs (Mei et al., 1996a; Mei et al., 1997). The results showed that classic preconditioning significantly elevated interstitial adenosine concentrations during each preconditioning stimulus with a progressive decrease upon each sequential stimulus. Bimakalim (a K^+_{ATP} channel opener) caused a significant decrease in baseline interstitial adenosine concentrations, while a 24-hour pretreatment with MLA resulted in an increase in adenosine levels when measured just prior to the prolonged occlusion period. Although differing effects were found prior to occlusion, all three treatments resulted in a decrease in interstitial adenosine during the prolonged occlusion period.

Data on the role of nitric oxide activation of K^+_{ATP} channels (Cameron et al., 1996), together with results regarding the involvement of iNOS in the cardioprotection induced by MLA pretreatment, support a mechanism involving nitric oxide signalling to promote the opening of K^+_{ATP} channels during ischemia. The durable cardioprotective response induced by MLA, its antiarrhythmic properties, and its low toxicity compared to endotoxin make MLA a viable pharmacological inducer of delayed preconditioning.

It is evident from the multitude of original publications regarding both classic and delayed preconditioning that there is intense interest in the pharmacological induction of a state that mimics such cardioprotective responses. Ideally, for a drug to possess clinical utility in the prevention of ischemia–reperfusion injury, it should induce a state resembling delayed preconditioning. The cardioprotective effects of classic preconditioning as induced by ischemia or pharmacologically by such compounds as adenosine are short lived, lasting no more than 1 to 2 hours. In addition, some inducers of classic preconditioning (ischemia) are associated with myocardial stunning, and certain pharmacological inducers, such as with the postassium channel openers, may be proarrhythmic. A brief window of protection is of less clinical utility in a setting of prolonged cardiopulmonary bypass and where a significant proportion of myocardial infarction from ischemia–reperfusion injury occur in the postoperative setting. The benefits of delayed preconditioning, namely, durability, lasting up to 72 hours, and antiarrhythmic and antistunning effects make it a more viable pharmaceutical target for clinical applications in the cardiac surgery arena. The ideal pharmacological agent for the induction of delayed preconditioning should possess durable effects lasting several hours to days and should protect the myocardium from dysrrhythmias and stunning as well as from infarction. Lastly, such a therapeutic agent must have an acceptable therapeutic index. Such a drug should enjoy substantial clinical use in the setting of cardiac surgery.

REFERENCES

Abd-Elfattah, A.S., Guo, J., El-Sigaby, N.R., and Elliott, G.T. (1995). Intravenous administration of monophosphoryl lipid A (MLA) 24 hours before aortic cross-clamping attenuate myocardial stunning in dogs. J. Mol. Cell. Cardiol. 27, A49.

Anzai, T., Yoshikawa, T., Asakura, Y., Abe, S., Akaishi, M., Mitamura, H., Handa, S., and Ogawa, S. (1995). Preinfarction angina as a major predictor of left ventricular function and long-term prognosis after a first Q wave myocardial infarction. J. Am. Coll. Cardiol. 26, 319-327.

Anzai, T., Yoshikawa, T., Asakura, Y., Abe, S., Meguro, T., Akaishi, M., Mitamura, H., Handa, S., and Ogawa, S. (1994). Effect on short-term prognosis and left ventricular function of angina pectoris prior to first Q-wave anterior wall acute myocardial infarction. Am. J. Cardiol. 74, 755-759.

Armstrong, S., and Ganote, C.E. (1995). In vitro ischaemic preconditioning of isolated rabbit cardiomyocytes: effects of selective adenosine receptor blockade and calphostin C. Cardiovasc. Res. 29, 647-652.

Baxter, G.F., Goodwin, R.W., Wright, M.J., Kerac, M., Heads, R.J., and Yellon, D.M. (1996). Myocardial protection after monophosphoryl lipid A: studies of delayed anti-ischaemic properties in rabbit heart. Br. J. Pharmacol. 117 (8), 1685-1692.

Baxter, G.F., Marber, M.S., Patel, V.C., and Yellon, D.M. (1994). Adenosine receptor involvement in a delayed phase of myocardial protection 24 hours after ischemic preconditioning. Circulation, 90, 2993-3000.

Brown, J.M., Grosso, M.A., Terada, L.S., Whitman, G.J., Banerjee, A., White, C.W., Harken, A.H., and Repine, J.E. (1989). Endotoxin pretreatment increases endogenous myocardial catalase activity and decreases ischemia-reperfusion injury of isolated rat hearts. Proc. Nat. Acad. Sci. U.S.A. 86, 2516-1520.

Bugge E., and Ytrehus, K. (1995). Inhibition of sodium–hydrogen exchange reduces infarct size in the isolated rat heart—a protective additive to ischaemic preconditioning. Cardiovasc. Res. 29, 269-274.

Cameron, J.S., Kibler, K.K.A., Berry, H., Barron, D.N., and Sodder, V.H. (1996). Nitric oxide activates ATP-sensitive potassium channels in hypertropied ventricular myocytes. FASEB J. 10, A65.

Das, K.C., Lewis-Molock, Y., and White, C.W. (1995). Activation of NF-kappa B and elevation of MnSOD gene expression by thiol reducing agents in lung adenocarcinoma (A549) cells. Am. J. Physiol. 269, L588-602.

De Fily, D.V., and Chilian, W.M. (1991). Preconditioning protects coronary microvascular endothelial function. Circulation, 84 (suppl. II), II434.

Downey, J.M., and Cohen, M.V. (1995). Signal transduction in ischemic preconditioning. Z. Kardiol. 84 (suppl. 4), 77-86.

Elliott, G.T., Comerford, M.L., Smith, J.R., and Zhao, L. (1996). Myocardial ischemia/reperfusion protection using monophosphoryl lipid A is abrogated by the ATP-sensitive potassium channel blocker, glibenclamide. Cardiovasc. Res. 32, 1071-1080.

Finegan, B.A., Lopaschuk, G.D., Gandhi, M., and Clanachan, A.S. (1995). Ischemic preconditioning inhibits glycolysis and proton production in isolated working rat hearts. Am. J. Physiol. 269, H1767-H1775.

Gao, W.D., Atar, D., Backx, P.H., and Marban, E. (1995). Relationship between intracellular calcium and contractile force in stunned myocardium. Direct evidence for decreased myofilament Ca2+ responsiveness and altered diastolic function in intact ventricular muscle. Circ. Res. 76, 1036-1048.

Gato, M., Lui, Y., Yang, X-M., Ardell, J.L., Cohen, M.V., and Downey, J.M. (1995). Role of bradykinin in protection of ischemic preconditioning in rabbit hearts. Circ. Res. 77, 611-621.

Gottlieb, R.A., Gruol, D.L., Zhu, J.Y., and Engler, R.L. (1996). Preconditioning rabbit cardiomyocytes: role of pH, vacuolar proton ATPase, and apoptosis. J. Clin. Invest. 97, 2391-2398.

Gross, G.J., and Auchampach, J.A. (1992). Blockade of ATP-sensitive potassium channels prevents myocardial preconditioning in dogs. Circ. Res. 70, 223-233.

Gross, G.J., Mei, D.A., Schultz, J.J., and Mizumura, T. (1996). Criteria for a mediator or effector of myocardial preconditioning: do KATP channels meet the requirements? Basic Res. Cardiol. 91, 31-34.

Hata, K., Takasago, T., Saeki, A., Nishioka, T., and Goto, Y. (1994). Stunned myocardium after rapid correction of acidosis. Increased oxygen cost of contractility and the role of the Na(+)-H+ exchange system. Circ. Res. 74, 794-805.

Hu, K., Duan, D., Li, G.R., and Nattel, S. (1996). Protein kinase C activates ATP-sensitive K+ current in human and rabbit ventricular myocytes. Circ. Res. 78, 492-498.

Ischihara, K., Miura, K., Nakai, T., and Satoh, K. (1994). Effect of preconditioning of myocardial pH during ischemia in dogs. J. Mol. Cell. Cardiol. 26, CCXIV.

Janier, M.F., Vanoverschelde, J.L., and Bergmann, S.R. (1994). Ischemic preconditioning stimulates anaerobic glycolysis in the isolated rabbit heart. Am. J. Physiol. 267, H1353-H1360.

Janin, Y., Qian, Y., Hoag, J.B., Elliott, G.T., and Kukreja, R.C. (1996). Pharmacologic preconditioning with monophosphoryl lipid A is abolished by 5-hydroxydecanoate, a specific inhibitor of the K_{ATP} channel. Cardiovasc. Res. (in press).

Kaplan, L.J., Bellows, C.F., Blum, H., Mitchell, M., and Whitman, G.J. (1994). Ischemic preconditioning preserves end-ischemic ATP, enhancing functional recovery and coronary flow during reperfusion. J. Surg. Res. 57, 179-184.

Klein, H.H., Pich, S., Lindert-Heimberg, S., Maisch, B., and Nebendahl, K. (1995). Effects of R56865, an Na(+)- and Ca(2+)-overload inhibitor, on myocardial injury in ischemic, reperfused porcine hearts. J. Cardiovasc. Pharmacol. 25, 163-167.

Kloner, R.A., Shook, T., Przyklenk, K., Davis, V.G., Junio, L., Matthews, R.V., Burstein, S., Gibson, M., Poole, W.K., and Cannon, C.P. (1995). Previous angina alters in-hospital outcome in TIMI 4. A clinical correlate to preconditioning? Circulation 91, 37-45.

Kloner, R.A., and Yellon, D. (1994). Does ischemic preconditioning occur in patients? J. Am. Coll. Cardiol. 24, 1133-1142.

Krause, E.G., and Szekeres, L. (1995). On the mechanism and possible therapeutic application of delayed adaptation of the heart to stress situations. Mol. Cell. Biochem. 147, 115-122.

Kuzuya, T., Hoshida, S., Yamashita, N., Fuji, H., Oe, H., Hori, M., Kamada, T., and Tada, M. (1993). Delayed effects of sublethal ischemia on the acquisition of tolerance to ischemia. Circ. Res. 72, 1293-1299.

Kwon, G., Corbett, J.A., Rodi, C.P., Sullivan, P., and McDaniel, M.L. (1995). Interleukin-1 beta-induced nitric oxide synthase expression by rat pancreatic beta-cells: evidence for the involvement of nuclear factor kappa B in the signaling mechanism. Endocrinology 136, 4790-4795.

Lawson, C.S., Avkiran, M., Shattock, M.J., Coltart, D.J., and Hearse, D.J. (1993). Preconditioning and reperfusion arrhythmias in the isolated rat heart: true protection or temporal shift in vulnerability? Cardiovasc. Res. 27, 2274-2281.

Liu, X., Engelman, R.M., Moraru, I.I., Rousou, J.A., Flack, J.E. 3rd, Deaton, D.W., Maulik, N., and Das, D.K. (1992). Heat shock. A new approach for myocardial preservation in cardiac surgery. Circulation 86 (suppl. 5), II358-363.

Liu, Y., Gao, W.D., O'Rourke, B., and Marban, E. (1996). Synergistic modulation of ATP-sensitive K+ currents by protein kinase C and adenosine. Implications for ischemic preconditioning. Circ. Res. 78, 443-454.

Luss, H., Watkins, S.C., Freeswick, P.D., Imro, A.K., Nussler, A.K., Billiar, T.R., Simmons, R.L., del Nido, P.J., and McGowan, F.X. Jr. (1995). Characterization of inducible nitric oxide synthase expression in endotoxemic rat cardiac myocytes in vivo and following cytokine exposure in vitro. J. Mol. Cell. Cardiol. 27, 2015-2029.

Marban, E., and Gao, W.D. (1995). Stunned myocardium: a disease of the myofilaments? Basic Res. Cardiol. 90, 269-272.

Marber, M.S., Latchman, D.S., Walker, J.M., and Yellon, D.M. (1993). Cardiac stress protein elevation 24 hours after brief ischemia or heat stress is associated with resistance to myocardial infarction. Circulation 88, 1264-1272.

Marber, M.S., Mestril, R., Chi, S.H., Sayen, M.R., Yellon, D.M., and Dillmann, W.H. (1995). Overexpression of the rat inducible 70-kD heat stress protein in a transgenic mouse increases the resistance of the heart to ischemic injury. J. Clin. Invest. 95, 1446-1456.

Maulik, N., Engelman, D.T., Watanabe, M., Engelman, R.M., and Das, D.K. (1996). Nitric oxide —a retrograde messenger for carbon monoxide signaling in ischemic heart. Mol. Cell. Biochem. 157, 75-86.

Maulik, N., Watanabe, M., Engelman, D., Engelman, R.M., Kagan, V.E., Kisin, E., Tyurin, V., Cordis, G.A., and Das, D.K. (1995a). Myocardial adaptation to ischemia by oxidative stress induced by endotoxin. Am. J. Physiol. 269, C907-916.

Maulik, N., Engelman, D.T., Watanabe, M., Engelman, R.M., Maulik, G., Cordis, G.A., and Das, D.K. (1995b). Nitric oxide signaling in ischemic heart. Cardiovasc. Res. 30, 593-601.

McKenna, T.M., Li, S., and Tao, S. (1995). PKC mediates LPS- and phorbol-induced cardiac cell nitric oxide synthase activity and hypocontractility. Am. J. Physiol. 269, H1891-H1898.

Mei, D.A., Elliott, G.T., and Gross, G.J. (1995). ATP-sensitive K^+ channels mediate the cardioprotective effect of monophosphoryl lipid A. Circulation 92, I388.

Mei, D.A., Elliott, G.T., Gross, G.J. (1996a). Comparative effect of early ischemic preconditioning (PC) and late PC produced by monophosphoryl lipid A upon myocardial infarct size and interstitial purine metabolism in dogs. Circulation 94 (suppl. I), I-185.

Mei, D.A., Elliott, G.T., and Gross, G.J. (1996b). K_{ATP} channels mediate late preconditioning against infarction produced by monophosphoryl lipid A. Am. J. Physiol. 271, H2723-H2729.

Mei, D.A., Nithipatikom, K., Lasley, R.D., and Gross, G.J. (1997). Role of adenosine in the cardioprotective effect of preconditioning produced by ischemia, hypoxia and a K-ATP channel opener. Circulation (in press).

Murry, C.E., Jennings, R.B., and Reimer, K.A. (1986). Preconditioning with ischemia: a delay of lethal cell injury in ischemic myocardium. Circulation 74, 1124-1136.

Murry, C.E., Jennings, R.B., and Reimer, K.A. (1994). What is ischemic preconditioning? In: Ischemic Preconditioning: The Concepts of Endogenous Cardioprotection (Przyklenk, K., Kloner, R.A., and Yellon, D.M., Eds.), pp. 3-17. Kluwer Academic, Boston.

Murry, C.E., Richard, V.J., Jennings, R.B., and Reimer, K.A. (1991). Myocardial protection is lost before contractile function recovers from ischemic preconditioning. Am. J. Physiol. 260, H796-H804.

Murry, C.E., Richard, V.J., Reimer, K.A., and Jennings, R.B. (1990). Ischemic preconditioning slows energy metabolism and delays ultrastructural damage during a sustained ischemic episode. Circ. Res. 66, 913-931.

Nelson, D.W., Brown, J.M., Banerjee, A., Bensard, D.D., Rogers, K.B., Locke-Winter, C.R., Anderson, B.O., and Harken, A.H. (1991). Pretreatment with a nontoxic derivative of endotoxin induces functional protection against cardiac ischemia/reperfusion injury. Surgery 110, 365-369.

Nelson, S.K., Wong, G.H., and McCord, J.M. (1995). Leukemia inhibitory factor and tumor necrosis factor induce manganese superoxide dismutase and protect rabbit hearts from reperfusion injury. J. Mol. Cell. Cardiol. 27, 223-229.

Node, K., Kitakaze, M., Komamura, K., Minamino, T., Funaya, H., Tada, M., Hori, M., and Kamada, T. (1995). Lipopolysaccharide (LPS) can induce production of adenosine and NO, and provide cardioprotection against ischemia and reperfusion injury in the canine heart. J. Am. Coll. Cardiol. 27, A512.

Ottani, F., Galvani, M., Ferrini, D., Sorbello, F., Limonetti, P., Pantoli, D., and Rusticali, F. (1995). Prodromal angina limits infarct size. A role for ischemic preconditioning. Circulation 91, 291-297.

Ovize, M., Kloner, R.A., and Przyklenk, K. (1994). Preconditioning and myocardial contractile function. In: Ischemic Preconditioning: The Concepts of Endogenous Cardioprotection (Przyklenk, K., Kloner, R.A., and Yellon, D.M., Eds.), pp. 41-60. Kluwer Academic, Boston.

Parratt, J.R. (1994). Protection of the heart by ischemic-preconditioning: mechanisms and possibilities for pharmacologic exploitation. TIPS 15, 19-25.

Parratt, J.R., and Szekeres, L. (1995). Delayed protection of the heart against ischemia. TIPS 16, 351-355.

Parratt, J., and Vegh, A. (1994). Pronounced antiarrhythmic effects of ischemic preconditioning. Cardioscience 5, 9-18.

Piacentini, L., Wainwright, C.L., and Parratt, J.R. (1993). The antiarrhythmic effect of ischemic preconditioning isolated rat heart involved a pertussis toxin sensitive mechanism. Cardiovasc. Res. 27, 674-680.

Przyklenk, K., Zhao, L., Kloner, R.A., and Elliott, G.T. (1996). Cardioprotection with ischemic preconditioning and monophosphoryl lipid A: role of adenosine regulating enzymes. Am. J. Physiol. 271, H1004-H1074.

Richard, V., Kaeffer, N., Tron, C., and Thuillez, C. (1994). Ischemic preconditioning protects against coronary endothelial dysfunction induced by ischemia and reperfusion. Circulation 89, 1254-1261.

Rowland, R.T., Cleveland, J.C., Meng, X., Ao, L., Harken, A.H., and Brown, J.M. (1996). A single endotoxin challenge induces delayed myocardial protection against infarcation. J. Surg. Res. 63, 193-198.

Schaefer, S., Carr, L.J., Prussel, E., and Ramasamy, R. (1995). Effects of glycogen depletion on ischemic injury in isolated rat hearts: insights into preconditioning. Am. J. Physiol. 268, H935-H944.

Schulz, R., and Heusch, G. (1994). Characterization of hibernating and stunned myocardium. Herz. 19, 189-203.

Shindo, T., Ikeda, U., Ohkawa, F., Kawahara, Y., Yokoyama, M., and Shimada, K. (1995). Nitric oxide synthesis in cardiac myocytes and fibroblasts by inflammatory cytokines. Cardiovasc. Res. 29, 813-819.

Shipley, J.B., Qian, Y-Z., Levasseur, J.E., and Kukreja, R.C. (1995). Expression of the stress proteins HSP-27 and HSP-72 in rat hearts does not correlate with ischemic tolerance after heat shock. Circulation 92 (suppl. 1), I654.

Shirakawa, F., and Mizel, S.B. (1989). In vitro activation and nuclear translocation of NF-kappa B catalyzed by cyclic AMP-dependent protein kinase and protein kinase C. Mol. Cell. Biol. 9, 2424-430.

Song, W., Furman, B.L., and Parratt, J.R. (1996). Delayed protection against ischaemia-induced ventricular arrhythmias and infarct size limitation by the prior administration of *Escherichia coli* endotoxin. Br. J. Pharmacol. 118, 2157-2163.

Speechly-Dick, M.E., Grover, G.J., and Yellon, D.M. (1995). Does ischemic preconditioning in the human involve protein kinase C and the ATP-dependent K+ channel? Studies of contractile function after simulated ischemia in an atrial in vitro model. Circ. Res. 77, 1030-1035.

Steenbergen, C., Perlman, M.E., London, R.E., and Murphy, E. (1993). Mechanism of preconditioning. Ionic alterations. Circ. Res. 72, 112-125.

Sun, J.Z., Tang, X.L., Knowlton, A.A., Park, S.W., Qiu, Y., and Bolli, R. (1995). Late preconditioning against myocardial stunning. An endogenous protective mechanism that confers resistance to postischemic dysfunction 24 h after brief ischemia in conscious pigs. J. Clin. Invest. 95, 388-403.

Sun, J.Z., Tang, X.L., Park, S.W., Qiu, Y., Turrens, J.F., and Bolli, R. (1996). Evidence for an essential role of reactive oxygen species in the genesis of late preconditioning against myocardial stunning in conscious pigs. J. Clin. Invest. 97, 562-576.

Tanaka, M., Fujiwara, H., Yamasaki, K., Miyamae, M., Yokota, R., Hasegawa, K., Fujiwara, T., and Sasayama, S. (1994). Ischemic preconditioning elevates cardiac stress protein but does not limit infarct size 24 or 48 h later in rabbits. Am. J. Physiol. 267, H1476-H1482.

Vegh, A., Papp, J.G., Elliott, G.T., and Parratt, J.R. (1996). Pretreatment with monophosphoryl lipid A (MPL-C) reduces ischaemia reperfusion-induced arrhythmias in dogs. J. Am. Coll. Cardiol. 28, A56.

Vegh, A., Papp, J.G., and Parratt, J.R. (1994). Prevention by dexamethasone of the marked antiarrhythmic effects of preconditioning induced 20 h after rapid cardiac pacing. Br. J. Pharmacol. 113, 1081-1082.

Vegh, A., Papp, J.G., and Parratt, J.R. (1995). Pronounced antiarrhythmic effects of preconditioning in anaesthetized dogs: is adenosine involved? J. Mol. Cell. Cardiol. 27, 349-356.

Volovsek, A., Subramanian, R., and Reboussin, D. (1992). Effects of duration of ischaemia during preconditioning on mechanical function, enzyme release and energy production in the isolated working rat heart. J. Mol. Cell. Cardiol. 24, 1011-1019.

Xie, Q.W., Kashiwabara, Y., and Nathan, C. (1994). Role of transcription factor NF-kappa B/Rel in induction of nitric oxide synthase. J. Biol. Chem. 269, 4705-4708.

Yamamoto, F., Yamamoto, H., Yoshida, S., Ichikawa, H., Takahashi, A., Tanaka, K., Kosakai, Y., Yagihara, T., and Fujita, T. (1991). The effects of several pharmacologic agents upon postischemic recovery. Cardiovasc. Drugs. Ther. 5 (suppl. 2), 301-308.

Yamashita, N., Nishida, M., Hoshida, S., Kuzuya, T., Hori, M., Taniguchi, N., Kamada, T., and Tada, M. (1994). Induction of manganese superoxide dismutase in rat cardiac myocytes increases tolerance to hypoxia 24 hours after preconditioning. J. Clin. Invest. 94, 2193-2199.

Yang, X.M., Baxter, G.F., Heads, R.J., Yellon, D.M., Downey, J.M., and Cohen, M.V. (1996). Infarct limitation of the second window of protection in a conscious rabbit model. Cardiovasc. Res. 31, 777-783.

Yao, Z., Auchampach, J.A., Pieper, G.M., and Gross, G.J. (1993a). Cardioprotective effects of monophosphoryl lipid A, a novel endotoxin analogue, in the dog. Cardiovasc. Res. 27, 832-838.

Yao, Z., Elliott, G.T., and Gross, G.J. (1995). Monophosphoryl lipid A preserves myocardial contractile function following multiple, brief periods of coronary occlusion in dogs. Pharmacology 51, 152-159.

Yao, Z., and Gross, G.J. (1993). Role of nitric oxide, muscarinic receptors, and the ATP-sensitive K^{+} channel in mediating the effects of acetylcholine to mimic preconditioning in dogs. Circ. Res. 73, 1193-1201.

Yao, Z., Rasmussen, J.L., Hirt, J.L., Mei, D.A., Pieper, G.M., and Gross, G.J. (1993b). Effects of monophosphoryl lipid A on myocardial ischemia/reperfusion injury in dogs. J. Cardiovasc. Pharmacol. 22, 653-663.

Yoshida, K., Maaieh, M.M., Shipley, J.B., Doloresco, M., Bernardo, N.L., Qian, Y.Z., Elliott, G.T., and Kukreja, R.C. (1996). Monophosphoryl lipid A induces pharmacologic "preconditioning" in rabbit hearts without concomitant expression of 70-kDa heat shock protein. Mol. Cell. Biochem. 1561, 1-8.

Yoshikawa, T., Inoue, S., Abe, S., Akaishi, M., Mitamura, H., Ogawa, S., and Handa, S. (1993). Acute myocardial infarction without warning: clinical characteristics and significance of preinfarction angina. Cardiology 82, 42-47.

Ytrehus, K., Liu, Y., and Downey, J.M. (1994). Preconditioning protects ischemic rabbit heart by protein kinase C activation. Am. J. Physiol. 266, H1145-H1152.

Zhao, L., Kirsch, C.C., Hagen, S.R., and Elliott, G.T. (1996). Preservation of global cardiac function in the rabbit following protracted ischemia/reperfusion using monophosphoryl lipid A (MLA). J. Mol. Cell. Cardiol. 28, 197-208.

Zhao, L., Weber, P.A., Comerford, M.L., Smith, J.R., Kehl, J., and Elliott, G.T. (1997). Role of inducible nitric oxide synthase (iNOS) in pharmacological "preconditioning" with monophosphoryl lipid A. Circulation 94 (suppl. I), I-422.

Ziegelhoffer, A., de Jong, J.W., Ferrari, R., Nagy, L.T. (1992). Ischemic preconditioning of the myocardium as a result of adaptation of enzymes catalyzing energy consuming process to decreased accessibility of metabolic energy. J. Am. Coll. Cardiol. 24 (suppl. I), 5150.

Ziegelhoffer, A., Vrbjar, N., Styk, J., Breier, A., Dzurba, A., and Ravingerova, T. (1995). Adaptation of the heart to ischemia by preconditioning: effects on energy equilibrium, properties of sarcolemmal ATPases and release of cardioprotective proteins. Mol. Cell. Biochem. 147, 129-137.

ADAPTATION OF CELLULAR THERMOGENIC REACTIONS

T. Ramasarma

Advances in Organ Biology
Volume 6, pages 219-239.

ISBN: 0-7623-0391-3

I. CELLULAR THERMOGENESIS

Generation of heat is basically a side product of metabolism. This heat enables the cell to maintain a temperature on which depends its metabolic rate. A majority of living cells adapt their activities as dictated by the environmental temperature, always tending to equalize with the ambient temperature by losing or absorbing heat. These cells, called poikilotherms or exotherms, are at the mercy of the environment and survive only in a suitable habitat. Interestingly, extra heat is generated in some tissues (e.g., inflorescences in some plant species and brown adipose tissue in mammals), implying that a localized need for higher temperatures can be met by tissues through specialized mechanisms. Thus, temperature regulation pervades all life forms.

This mechanism developed into a fine-tuned process of maintaining constant body temperature in warm-blooded animals (a mere 2% of species), termed homeotherms or endotherms. Homeothermy endowed such animals with autonomy and an ecological advantage and extended the thermal range of their habitat. In these animals, body temperature remains constant within a narrow range of environmental temperature, determined by the basal metabolic rate and heat-exchange processes. Beyond this zone of thermoneutrality, modulation mechanisms are invoked that enhance or curtail heat generation as an adaptation of the process of thermoregulation.

A. Heat Generation in Thermoregulation

Exposure of animals to cold increases temperature differential between the body and the environment, resulting in increased heat loss. Survival under cold stress depends on the capacity to produce extra heat to compensate for the loss. This heat is obtained initially through shivering thermogenesis, in which adenosine triphosphate (ATP) turnover due to involuntary muscular contraction provides the heat. This response is of value as immediate reaction to the stress. But long-term exposure to cold stress requires a sustained increase in heat generation by adaptation of the metabolic machinery without, however, affecting essential functions. This long-term adaptation is characterized by increased basal metabolic rate, oxygen uptake, and food consumption that is diverted exclusively for increased thermogenesis without recourse to shivering. This response is referred to as "regulatory nonshivering thermogenesis" or "chemical thermogenesis."

It is often assumed that changes in the reverse direction occur on exposure to heat stress and, indeed, food consumption, oxygen uptake, and basal metabolic rate are lowered. Differences in heat loss mechanisms also are observed among animals. Adaptation to heat stress initially seems to depend on reduction of heat-generating reactions; however, in chronic stress, regulation of mitochondrial respiratory activity and components is targeted to achieve the basic objective of preventing the ill effects of hyperthermia.

Several reviews are available summarizing research on temperature regulation. These reviews present various aspects of research and give the reader an overview of the development of this area and the changes in concepts, views, and approaches over the past four decades. Among those focused on adaptation to temperature stress are reviews by Chaffee and Roberts (1971), Jansky (1973), Hochachka (1974), Nicholls et al. (1974), Ramasarma and Susheela (1976), Himms-Hagen (1976), Cannon et al. (1981), Nicholls and Locke (1984), Ramasarma et al. (1987). Many models for cellular chemical thermogenesis have been discussed in these reviews based on the original work of scores of investigators. Explanations offered include ATP turnover, as in the uncoupled state of oxidative phosphorylation, activation of ion pumps, and wasteful futile cycles of interlocking kinases and phosphatases; shunt pathways of electron transfer that bypass ATP coupling sites (at present the most favoured theory involves delinking of a proton gradient from ATP synthesis); and multiple oxidative reactions, including oxygen-consuming microsomal lipid peroxidation.

B. Appropriate Mechanism for Thermogenesis

Adaptation of animals exposed to cold stress is characterized by increase in food intake (about two-fold) without an increase in body weight, and a large increase in oxygen consumption. The *extra* oxygen consumed is used for the combustion of the extra calories taken in. The normal functions of the animals are not effected. Therefore, the vital energy transduction process necessary for the supply of ATP and nicotinamide adenine dinucleotide (phosphate) reduced NAD(P)H, or appears to be maintained, a necessary criterion to be satisfied in any mechanism.

Of the several hormones implicated in the regulation of thermogenesis, noradrenaline and thyroxine (Jansky, 1973; Himms-Hagen, 1976) play significant roles in the adaptation process. Noradrenaline acts as a signal initiating the thermogenic reactions and also restores the lowered shivering threshold in cold exposure (Zeisberger and Bruck, 1971). Mediation of α-adrenergic receptors in noradrenaline action and in the mechanism of regulation of body temperature were reported (Zeisberger and Bruck, 1976). And, as expected, treatment of the cold-exposed animals with antagonists of α-adrenergic receptors resulted in their death (Sitaramam et al., 1977). Thyroxine treatment shows a spectacular increase in basal metabolic rate. Lack of utilization of unsaturated fatty acids (18:2, 18:3, and 20:4) in the hypothyroid condition and the increase in arachidonic acid (20:4) in the hyperthyroid condition were reported. Thyroid status appears to control the activity of Δ^5-desaturase (Chen and Hoch, 1979). The importance of unsaturated fatty acids is indicated, as changes in this enzyme system accompany adaptation process.

An appropriate mechanism for cellular thermogenesis must fit with the following features: It must (1) respond to environmental stimuli; (2) include both increase and decrease in heat generation; (3) involve noradrenaline and adrenergic receptors; (4) explain the large extra oxygen consumption without disturbing the mitochondrial ATP synthesis.

The concepts of the energy conservation mechanism of mitochondrial oxidative phosphorylation have had a stranglehold on the understanding of nonshivering thermogenesis. Studies have been limited to specific aspects of the phenomenon and have focused mostly on brown adipose tissue, considered the thermogenic tissue. An overview of the current proposals on cellular thermogenic reactions is presented in this chapter, highlighting the following areas: ATP turnover, proton gradient turnover, electron shunt pathways, hydrogen peroxide (H_2O_2) turnover, and lipid turnover.

II. INCREASED ADENOSINE TRIPHOSPHATE TURNOVER

ATP hydrolysis with a high free energy rate of about 7 kcal/mole offered itself as the most plausible, natural mechanism of thermogenesis. In the skeletal muscle heat generation occurs essentially by ATP turnover during muscular exercise and shivering. Because of the need to perform continuous work, shivering mechanism is of limited use and must be replaced by a more efficient nonshivering chemical thermogenesis to sustain a state of adaptation.

A. Na^+-K^+- ATPase

The plasma- membrane–based Na^+-K^+-ATPase (sodium pump) (Figure 1) has basically high ATP-hydrolysis activity, and is further enhanced by a known thermogenic hormone, thyroxine (Ismail-Beigi and Edelman, 1970). However, the heat generated by this pump accounted for no more than 15%. Also, no stimulation of the sodium pump was observed on catecholamine infusion and in diet-induced nonshivering thermogenesis (Himms-Hagen, 1976).

B. Futile Cycles

Turnover of ATP can be enhanced by the operation of futile cycles, which interlock a kinase and a phosphatase. Several possible examples are available: glucok-

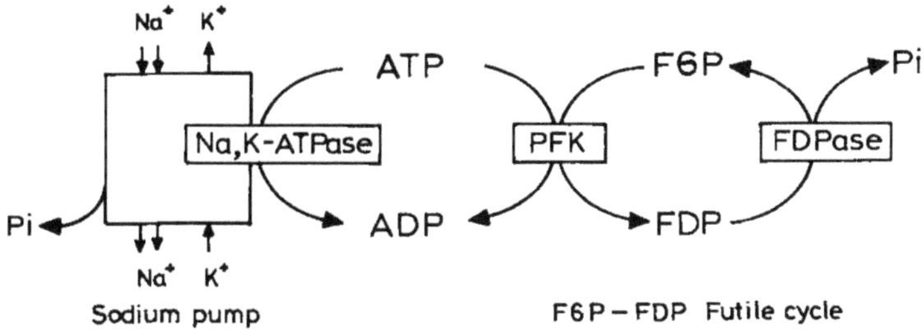

Figure 1. Adenosine triphosphate (ATP) turnover by sodium pump and F6P–FDP futile cycle. On the left is shown the utilization of ATP for exchange of Na^+ and K^+ ions by Na^+-K^+-ATPase, and on the right the wasteful hydrolysis of ATP in a futile cycle of coupling phosphofructokinase (PFK) and fructose diphosphatase (FDPase).

inse–glucose-6-phosphatase, phosphofructokinase–fructose bisphosphatase, lipase–diglyceride fattyacyl-transferase. These enzyme systems are under tight regulatory control in animals and are not allowed to operate in both directions at any time, thereby preventing short-circuiting and wastage of ATP. Such a futile cycle (see Figure 1) coupling fructose-6-phosphate and fructose diphosphate (FDP) was found in an insect, the bumble bee (Clark et al., 1973), but not in any endotherm.

A futile cycle using proton gradient instead of ATP and transporting free fatty acid across the mitochondrial inner membrane will be discussed in the next section.

C. Uncoupling of Oxidative Phosphorylation

Uncoupling of oxidation of substrates from phosphorylation of adenosine diphosphate (ADP) in mitochondria would cause dissipation of energy as heat. This view was initially supported by well-known effect of thyroxine on uncoupling of mitochondrial oxidative phosphorylation. This uncoupling byasses the need to make ATP first and then degrade it, but also has the disadvantage of depleting ATP. Indeed, "such uncoupling, were it to occur widely in the animal body, would convert all mitochondria to little furnaces; there would be plenty of heat but not enough ATP to support cellular functions," as rightly pointed out by Hochachka (1974). Mitochondria isolated from several tissues of cold-adapted animals were found to be in coupled state.

This mechanism may be used in some special cases. Mitochondria isolated from skeletal muscles of cold-adapted fur seals were found to be in a partially uncoupled state (Grav and Blix, 1979). The natural uncoupling agent is unlikely to be thyroxine, because mitochondria isolated from liver and skeletal muscle of hyperthyroid animals showed high degree of coupling (Tata et al., 1963). Another example is brown adipose tissue. Its mass increases as an adaptive response to cold stress. Isolated mitochondria from this lipid-enriched tissue show a low degree of coupling, possibly due to the abundance of fatty acids in these cells (Himms-Hagen, 1976; Cannon et al., 1981; Nicholls and Locke, 1984).

III. PROTON GRADIENT TURNOVER

A. Brown Adipose Tissue Mitochondria

It is now accepted that brown adipose tissue (BAT) is the primary site of noradrenaline-mediated nonshivering thermogenesis. This tissue is localized in cervical, thoracic, axillary, interscapular, and renal regions and is endowed with rich vascular system. It possesses a high density of mitochondria and multilocular lipid droplets that vary in content in altered thermogenic conditions. These mitochondria have large cristae content with a high concentration of respiratory components and activity. They preferentially utilize as their energy source free fatty acids released from the triglyceride-rich lipid droplets that break down under cold exposure. Conditions that demand increased nonshivering thermogenesis, such as cold exposure, a high-fat diet, arousal from hybernation, postnatal development, and infusion of noradrenaline, cause an increase in the mass and mitochondrial content of BAT. Conversely, conditions such as heat stress, fasting, and hybernation that need low heat generation decrease BAT size and lower mitochondrial population (see reviews by Himms-Hagen, 1976; Cannon et al., 1981; Nicholls and Locke, 1984).

B. Uncoupling Protein

BAT mitochondria show considerable uncoupling of oxidative phosphorylation as an index of the assigned thermogenic function (Nicholls, 1979). Addition of ADP paradoxically inhibits the oxidation, in contrast to the stimulation obtained in coupled mitochondria from other tissues. This finding led to isolation and characterization of a specific 32-kDa protein from these mitochondria, which has been variously called thermogenin, guanosine diphosphate (GDP)–binding protein, purine nucleotide–binding protein, and uncoupling protein (UCP) (19). This protein is confined to BAT-mitochondrial inner membranes, accounting for as much as 18% of its protein and mass increases during adaptation to cold (Ricquier et al., 1979).

C. Heat Generation by Discharge of Proton Gradient

Vectorial proton translocation through ATP-synthase protein is now considered the key event in the energy conservation process in oxidative phosphorylation, according to the well-entrenched Mitchell's chemiosmotic hypothesis. In order to explain diversion of energy to heat rather than ATP synthesis, it is proposed that UCP discharges the proton gradient bypassing the F_0, F_1 protein complex (Figure 2). A significant observation made by Klingenberg and Winkler (1985) was that lipo-

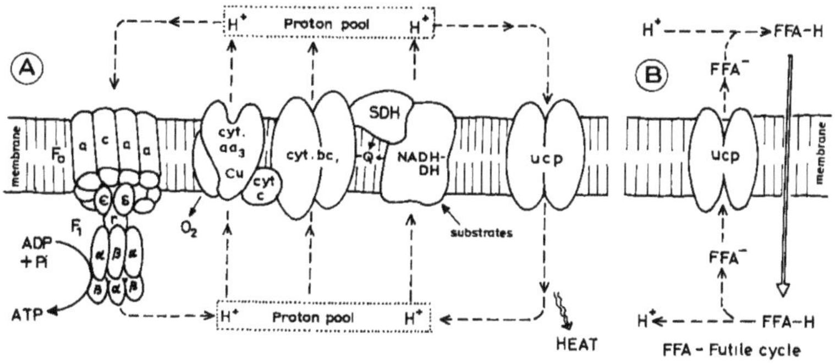

Figure 2. Heat generation by discharge of proton gradient by uncoupling protein. **A.** The electron transport system oxidizing substrates generates proton gradient. These protons translocating through F_0, F_1–ATPase will generate adenosine triphosphate (ATP), and through the uncoupling protein (UCP) will dissipate energy as heat. **B.** Fatty acid shuttle—futile cycle: protonated fatty acid (FFA-H) enters the mitochondrial matrix and is deprotonated to its anionic from (FFA^-). This is the species translocated by the uncoupling protein, which has high affinity for FFA. The FFA^- anions pick up the protons in the cytosol and the cycle continues with the release of energy as heat. ADP, adenosine diphosphate; SDH, succinate dehydrogenase.

somes enriched with purified UCP translocated protons. This activity was inhibited by purine nucleotides in the same order of GDP > GTP > ATP > ADP as their binding affinity to UCP. Although UCP is often referred as GDP-binding protein, the relative affinities of the tri- and diphosphates of both guanosine and adenosine are not very different (Klingenberg and Winkler, 1985). The presence of these nucleotides inhibits respiratory activity of BAT mitochondria, albeit at mM concentrations, and also proton translocation through UCP and permits their re-entry into the matrix through ATP-synthase and thereby ATP synthesis. Indeed, this highlights an anomaly of regulation in that a high concentration of ATP can block discharge of the proton gradient through UCP and increase ATP synthesis.

D. Fatty Acid Shuttle for Proton Translocation—A Futile Cycle

It appears that nucleotide-mediated inhibition of proton translocation by UCP is reversed by fatty acids. It is noteworthy that stimulation of respiration of BAT mitochondria by added fatty acid at an optimum concentration is indistinguishable from that obtained by noradrenaline-treatment of the animal. Thus, increased lipolysis by hormone-sensitive lipase that increases free fatty acid concentration can effectively undo the nucleotide inhibition. Fatty acyl coenzyme A (CoA) also relieves BAT-mitochondrial oxidation from inhibition by GDP, apparently by increasing proton conductance and by decreasing binding of the nucleotide to UCP. Then the question reverts to the synthesis of ATP, also necessary for the functions of the adipocyte cell.

The activation of UCP-mediated proton translocation by fatty acids is now shown to be part of the mechanism by which free fatty acid (FFA) acts as a shuttle. In the form of an anion, FFA^- picks up a H^+ from the cytosol to form protonated FFA-H. This traverses the membranes into the matrix, releases H^+, and becomes the anion. The UCP function then is to transport vectorially FFA^- into the cytosol. Evidence for this shuttle was obtained by Jezek and coworkers (1976) using photoreactive azido-fatty acid [12-(4-azido-2-nitrophenylamino) dodecanoic acid]. This behaved like a typical fatty acid in activating proton translocation and inhibition of anion transport. On photoactivation by UV light the label, now covalently linked, was prominent in the 32-kDa band of UCP, and this irreversibly inhibited the activation of proton transport while continuing to inhibit anion transport. This finding implies that fatty acid cycling is involved and that UCP acts as translocator of free fatty acid anions, similar to its kin, the ADP/ATP translocator protein, with which it shares many properties. While doing this shuttle, UCP effectively discharges the proton gradient and releases the energy as heat. In this shuttle, FFA is recycled and therefore it constitutes a futile cycle (see Figure 2B). Some questions, however, remain: What is the need and signal to transport FFA into the matrix in the first place? How does FFA in matrix lose its proton? Why this elaborate mechanism and a large amount of UCP when transmembrane proton conductance can be achieved by small molecules such as chemical uncouplers at concentrations commensurate with the respiratory chain components? If FFA is recycled in this shuttle, is the lipid in the droplets that decreases used by another reaction? Does UCP have some other unique role in BAT, being present only in this specialized tissue, especially since it is structurally related to the signal-transducing 7-transmembrane helix proteins?

E. Contribution of Brown Adipose Tissue to Total Heat Generation

The net contribution of BAT to the heat produced in nonshivering thermogenesis cannot be high, despite the attractive theory and correlative evidence. The total mass of BAT in the cold adapted rat is estimated to be about 5% of body weight, with half being located in the interscapular region. The contribution of various organs to the extra heat produced in the cold-adapted rat is estimated as follows: skeletal muscle, 50%; liver, 25%; BAT, 10%; heart, 3%; brain 4% (Jansky, 1973). Based on mitochondrial content and maximum oxidation rate, the contribution of BAT to total oxygen consumption was calculated to be 0.5% (Table 1). Furthermore, the calorigenic response to noradrenaline was unaffected by interruption of blood flow to BAT, indicating a lack of a significant role for this tissue (Foster, 1974). Based on measurement of blood flow with radioactive microspheres and differences in arterial and venous oxygen concentrations, Foster and Frydman (1978) changed their view and claimed that 60% of calorigenic response of noradrenaline was in BAT. Surprisingly, these results have been accepted to strengthen the hy-

Table 1. Contribution of Interscapular Brown Adipose Tissue to Basal Metabolic Rate

Parameter	*Cold-acclimated (0–5°, 35 days)*	*Heat Exposure of Cold-acclimated (37°, 3 hr)*
Interscapular BAT wt, mg	363	339
Mitochondrial protein isolated from BAT, mg per rat	12.3	11.5
Mitochondrial oxidation with α-glycerophosphate as substrate		
Specific activity (ng O atom/min per mg protein)	223	134
Total O_2 consumption (μg O atom/min)	2.74	1.54
Oxygen consumption in whole animal (BMR) μg O atom/min per rat	652	295
% Contribution of interscapular BAT mitochondria to BMR	0.42	0.52

Note: BAT, brown adipose tissue: BMR, basal metabolic rate.
Data represent mean values for clarity from 6 to 10 experiments. Taken from Gaikwad et al., (1990).

pothesis of the role of BAT in thermogenesis. Later, Ma and Foster (1989) came to the conclusion that contribution of BAT was less than 3% in diet-induced thermogenesis, and no attempt was made to clarify the difference.

BAT does indeed generate more heat relative to other tissues. It can provide local heat in different areas of the body to the neighboring organs, although small in size. The increased mass of BAT in cold adaptation may have a direct bearing on this function. The contribution of the small mass of BAT to the total heat produced in the animal can only be small because surgical removal or interference with vascular supplies to BAT did not suppress noradrenaline-induced thermogenesis in cold-adapted rats (Foster, 1974; Himms-Hagen et al., 1975). The marked proliferation of BAT is often assumed to correlate with increased thermogenesis. Increase in BAT mass also occurs at ambient temperatures as a response to increased food intake (Ma and Foster, 1989), and in creatine-depleted rats (Yamashita et al., 1995) without thermogenic response.

No doubt blood flow to BAT is high, but this is likely to remove and distribute to other tissues a "cold-acclimation factor" proposed by Himms-Hagen (1976). More evidence is now available that BAT may act as an endocrine tissue and as an internal, indirect sensor of environmental temperature and excessive energy stores. The adipocyte cell is, in fact, acquiring a new role besides lipid storage in providing signals that would limit food intake. A 16-kDa secretary protein, called leptin, expressed solely in adipose tissue as the product of the ob gene and absent in obese mice, is considered to be a satiety factor (Zhang et al., 1994). Brown adipocytes do possibly produce such secretory factors that influence other tissues. BAT would thereby gain a regulatory function of thermogenesis in all the tissues rather than merely being seen as producing some extra heat.

F. Change in Cytochromes as an Alternative Regulatory Mechanism

The wide acceptance of uncoupling protein mediation in thermogenesis seems to have precluded studies of other possible mechanisms. Rapid, selective change in cytochromes was identified as a mechanism of regulation of oxidation activities in BAT mitochondria under conditions of acute heat stress in cold-adapted animals, which decreased basal metabolic rate by 55%. BAT mitochondria in such animals showed a decrease in substrate-dependent oxidations and H_2O_2 generation, as well as concentration of cytochrome b, the protein entirely coded and synthesized in mitochondria and therefore amenable for rapid response to changes in energy needs (Gaikwad et al., 1990).

IV. ELECTRON SHUNT PATHWAYS

Generation of extra heat by the augmentation of "calorigenic shunts" was postulated by Potter (1958). In this mechanism, the energy of electron transfer is dissipated as heat by short-circuiting electrons through alternative nonphosphorylating pathways during oxidation of substrates by mitochondria. Response to cold can be achieved by channelling electrons to the shunt pathway. A succinate-dependent shunt measured by reduction of the dye neotetrazolium was found to increase in cold stress (Aithal and Ramasarma, 1971) and thyroxine treatment (Ramasarma and Susheela, 1974). This activity depends on concentration of ubiquinone, which increased under conditions of cold stress and treatment with thyroxine or nondrenaline (Ramasarma, 1968). Several properties of this enzyme system differ from that of respiratory chain succinate dehydrogenase measured by reduction of phenazine methosulfate (Ramasarma and Susheela, 1974). This finding prompted the proposal that it represents an alternative shunt pathway with its specially localized pool of ubiquinone acting as a switch. The regulatory molecule of ubiquinone is localized differently from the main electron transfer component and is available for reduction by the dehydrogenase and oxidation by molecular oxygen with H_2O_2, rather than H_2O, as the product.

A. Cyanide-insensitive Respiration

Respiration insensitive to cyanide and independent of cytochrome oxidase occurs to a significant extent in some plants (Rich, 1978) and also in rat liver mitochondria in presence of bypass redox compounds such as vitamin K_3 (Meera and Ramasarma, 1979). This process is called "alternate oxidase" as it diverts electrons to oxygen through ubiquinone and considered to generate H_2O_2. It is characteristically inhibited by free-radical scavengers such as salicylhydroxamate, propylgallate, and disulfiram. Spadices of Arum species have high cyanide-insensitive respiration and ubiquinone, and they exhibit a core temperature 10°C to 15°C higher than ambient (Raskin et al., 1987). The thermogenic appendages of flowers of water lilies of Victoria and Nelumbo species were found to possess cyanide-insensitive respiration inhibited by salicylhydroxamate when both the tissues and isolated mitochondria were tested (Skubatz et al., 1990).

B. Heat Generation in Inflorescences of Plants

Heat is generated in special organs of some plants. Heat-producing sterile florets are arranged in the middle of the spadix (appendix) protected by a sheath called a spathe. At the time of pollination, the spathe unfolds, exposing the florets to the environment, and the appendix starts generating heat (Raskin et al., 1987). This is part of an elegant adaptive mechanism wherein localized heat is necessary during a few hours in the afternoon to volatalize amines and indoles (Smith and Meeuse, 1966), whose putrescent odour attracts insect pollinators. Amazingly, temperatures of about 40°C are reached notwithstanding low ambient temperature (Nagy et al., 1972). During this process, exceptionally high rates of oxygen consumption were observed, as expected of the burst in heat generation. Such episodes of heat generation, enhanced oxygen consumption, and regulated high temperatures occur for extended period of 2 days during flowering (Seymour and Schultz-Motel, 1996).

An endogenous developmental regulator, termed the "calorigen," was identified as salicyclic acid, which "induced concentration-dependent and photoperiodically controlled heat and odour production in the appendix tissue of *Sauromatum guttatum*" (Raskin et al., 1987). Salicylic acid is suggested by these authors to serve as a "natural thermogenic trigger" in plants.

All of these processes seem to use alternative oxidation pathways, although the substrates oxidized remain to be clarified. One common feature is the involvement of ubiquinone, which funnels the electrons to reduce molecular oxygen to H_2O_2 in some cases, and H_2O in others. The alternative oxidase was also identified as quinol oxidase and lipoxygenase (Parrish and Leopald, 1978). These activities are the likely targets for activation by the calorigen, salicylate, and inhibition by its derivative, salicylhydroxamate.

V. TURNOVER OF H_2O_2

Generation of H_2O_2 and its degradation by catalase are exergonic reactions. In the bombardier beetle, an elegant system of generating and degrading H_2O_2 is used for thermogenesis. The reactor glands of this insect possess a mixture of H_2O_2 and hydroquinone, which is sequezed into an outer compartment containing the enzymes catalase and peroxidase. This results in oxidation of hydroquinone and degradation of H_2O_2—all thermogenic reactions. The heat thus produced vaporises water into steam which, along with the quinone formed, ejects as yellow puffs that are effectively used for defense against predators (Hochachka, 1974).

Generation of H_2O_2 in cells is now considered a necessary physiological process and is implicated in phagocytic killing of bacteria, in synthesis of vital biochemicals such as thyroid hormones and prostaglandins, as a possible intermediate in the actions of insulin and oxytocin, and in other redox-dependent metabolic modulations (Chance et al., 1979; Ramasarma, 1982). Oxidations in mitochondrial, microsomal, and peroxisomal membranes are the major sources of H_2O_2. Several flavorproteins, copper-containing enzymes, and heme and iron-sulfur proteins undergo auto-oxidation and generate H_2O_2.

A. Photorespiration is Thermogenic

The phenomenon of light-dependent reoxidation of part of photosynthetic carbon in temperate zone C_3 plants is known as photorespiration (Tolbert, 1981). This process generates large quantity of H_2O_2 as a cooperative effort of intracellular organelles in the leaf—chloroplasts, peroxisomes, and mitochondria (Figure 3). Dur-

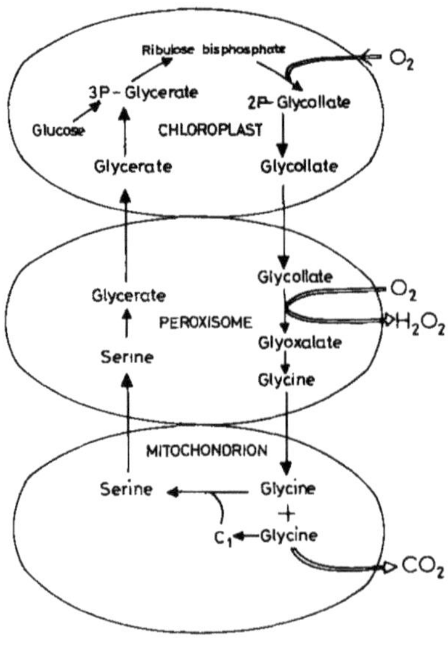

Figure 3. H_2O_2 generation during photorespiration. This is a cooperative effort of chloroplasts, peroxisomes, and mitochondria. Glycollate is produced in chloroplasts by oxygenation of ribulose bisphosphate. Transported into peroxisomes, glycollate is oxidized producing H_2O_2 and glyoxylate, which is converted to glycine. In mitochondria, two glycine molecules are converted to serine + CO_2. Serine is then converted to glycerate in peroxisomes and transferred to chloroplasts. Through the Calvin cycle, glycerate, and glucose-derived intermediates, ribulose bisphosphate is formed. No attempt is made to show the stoichiometry. The large quantity of H_2O_2 produced will be degraded in peroxisomes by catalase, thus generating heat.

ing oxidation of glycollate, H_2O_2 is formed and is degraded by catalase in peroxisomes. This thermogenic energy-dissipating pathway may be needed by these plants to maintain a minimal temperature in the cells, ensuring sufficiently high reaction rates that otherwise would not be possible in the cold environment. It is therefore not surprising that attempts to increase productivity by suppressing photorespiration led to decreased growth or even death, and that in some plants half the photosynthetic carbon is expended for photorespiration (Tolbert, 1981).

B. Mitochondrial H_2O_2 Generation Responds to Thermogenic Conditions

A small fraction (1% to 2%) of the total oxygen consumed by mitochondria in state 4 (ADP exhausted) is reduced through a pool of ubiquinone to form H_2O_2 (Boveris et al., 1976). This activity responds to changes in thermogenic conditions of the animals with a high degree of correlation.

Concentration of ubiquinone in the liver and the mitochondrial activities of ubiquinone-dependent neotetrazolium reductase and H_2O_2 generation increased in cold exposure or in animals treated with thyroxine or noradrenaline (Aithal and Ramasarma, 1971; Swaroop et al., 1983; Swaroop and Ramasarma, 1985). Phenoxybenzamine, an α-adrenergic receptor antagonist, blocked the change with noradrenaline (Swaroop et al., 1983). Decreases were observed in these activities when the animals were exposed to heat stress or thiouracil, an antithyroid compound (Swaroop and Ramasarma, 1985). Mitochondria from brown adipose tissue also showed increased H_2O_2 generation upon cold exposure of the animal, and this decreased on deacclimation and exposure to heat (Gaikwad et al., 1990). However, the amount of H_2O_2 thus generated is quantitatively small and does not explain the overall changes in thermogenesis. A role of signal is thus assigned for H_2O_2, which needs amplification by increasing another oxygen-consuming reaction (Ramasarma, 1982).

VI. LIPID TURNOVER

Cyanide-insensitive respiration is now associated with oxygen consumption by lipoxygenase reaction, especially in plants. In this pathway, peroxy radicals (ROO•) act as electron acceptors in ubiquonol oxidation, and the terminal reduction products of oxygen are likely to be ROH and H_2O. A large consumption of polyunsaturated fatty acids (PUFA) and, therefore, the need for high amounts in the diet are implicit. Respiration of heat-generating sterile florets of *Philodendron selloum* was found to depend on direct oxidation of lipids (rather than conversion to carbohydrate intermediates), stored in abundantly occurring lipid bodies that become depleted during maximum heat generation (Walker et al., 1983). The presence of starch and lipid bodies in the thermogenic staminal appendages of Nelumbo flowers, and their utilization during heat generation, has been reported (Skubatz et al., 1990).

It is more than a coincidence that similar lipid vesicles stored in brown adipose tissue also become depleted during thermogenesis. In this tissue, fatty acyl CoA is a major substrate for mitochondrial oxidation. Nearly one-third of fatty acids are of the polyunsaturated type and may be directly oxidized as in the plant tissue. In such specialized thermogenic tissues, lipid turnover appears to be a common feature. Can lipid peroxidation be the common basis of thermogenesis?

A. Lipid Peroxidation—A Thermogenic Reaction

Besides mitochondrial cytochrome oxidase, the other major cellular oxygen-consuming reaction is microsomal lipid peroxidation (Table 2). This is considered basically to be a reaction by which PUFA are oxidatively degraded and, therefore, may be responsible for some pathophysiological states. Microsomal lipid peroxidation is triggered by a small concentration of Fe^{2+}–ADP complex, is initiated by formation of lipid-peroxy radicals (LOO•), and is characterized by uptake of oxygen and oxidation of NADPH at rates higher than formation of the degradation product, malondialdehyde, with the stoichiometry of about 1:4:16 for malondialdehyde: NADPH:O_2 (Ernster and Nordenbrand, 1961). Experiments in our laboratory indicated that the ratio can be as high as 1:8:30. The variable is the proportion of lipid degraded to malondialdehyde. Liver cytosolic proteins which inhibit lipid peroxidation were found (Kamataki et al., 1974; McCay et al., 1981; Ramasarma et al., 1984). One of these seemed to spare PUFA and lipid-peroxy radicals from breakdown to malondialdehyde by the minor reaction while allowing NADPH oxidation and oxygen consumption, the major reactions (Ramasarma et al., 1984). The stoichiometry of O_2:NADPH is about 4:1. The excess oxygen consumed is used for the formation of lipid hydroperoxides and is a consequence of attack by oxygen of lipid alkyl radicals, generated during initiation by H-abstraction by reactive-iron species and also by lipid-peroxy radicals during propagation in the chain reaction (Ernster and Nordenbrand, 1961). Only about 1% to 2% of the lipid-peroxy molecules are degraded to malondialdehyde. NADPH is oxidized in generating the reactive-iron species to initiate H-abstraction.

Table 2. Comparison of Rate of Oxygen Consumption During Mitochondrial Substrate Oxidation and Microsomal Lipid Peroxidation

	nmole O_2/min per mg Protein	
Tissue	*Mitochondria*	*Microsomes*
Liver	66 ± 10	132 ± 5
Brown adipose tissue	120 ± 20	182 ± 18

Note: Mitochondria and microsomes were prepared from liver and brown adipose tissue from rats adapted to cold (0 to 5C) for 30 days. The rates of oxygen consumption were determined in an oxygraph equipped with a Clark Oxygen electrode with mitochondria oxidizing succinate and microsomes oxidizing NADPH in presence of Fe^{2+}-ADP. The values are mean ± S.D. of independent determinations of six rats. Adapted from Aithal and Ramasarma (1971), Gaikwad et al. (1990), and Sekhar et al. (1990).

This oxidation of NADPH, a quantitatively significant reaction, is thermogenic. NADPH availability increases in cold exposure as glucose-6-phosphate dehydrogenase activity increases. The only component needed to start the reaction in the cell is free iron ions, normally present at less than 1 μM concentration. Hence, controlled release of Fe^{2+} from stores within the cell ensures regulated NADPH oxidation, lipid peroxidation, and thermogenesis.

B. Lipid Peroxidation Responds to Thermogenic Conditions

Lipid peroxidation, measured by formation of malondialdehyde and by consumption of oxygen, shows adaptive changes as expected of a thermogenic reaction and fulfills many criteria required for an appropriate mechanism (Ramasarma et al., 1987), as follows:

1. The microsomal activities in the livers of rats of NADPH-dependent oxygen consumption and formation of malondialdehyde in the presence of Fe^{2+}–ADP increased in cold exposure and noradrenaline treatment and decreased in heat exposure and hypothyroid condition. The increase in activity was prevented upon treatment with an α-adrenergic receptor antagonist (Sekhar et al., 1990).
2. Rats fed a diet deficient in iron were unable to maintain their body temperature, and their survival was reduced to a few hours when simultaneously exposed to cold (Dillman et al., 1980).
3. Rats fed a diet containing vegetable oil rich in PUFA adapted to cold, but those fed diets containing hydrogenated fat or free of fat did not. Their body temperature and oxygen consumption were drastically reduced before death (Ramasarma et al., 1987). In addition, symptoms of PUFA deficiency, such as hair loss and stunted-curved tail, "precipitated" in animals on a fat-free diet upon exposure to cold for short periods, indicating rapid loss of residual PUFA (unpublished experiments).

C. Requirement of Polyunsaturated Fatty Acids in Cold Adaptation

The high turnover and conversion to peroxidized products according to the preceding proposed function in NADPH oxidation explain why PUFA are needed in large quantities in animal diets. In this context, the high content of PUFA in the diets of Eskimos living in cold climate is worthy of note. The PUFA derived from marine oils in these diets are mostly of the ω-3 class (20:5, 22:5, and 22:6 fatty acids), which are known to increase lipid peroxidation (Hammer and Wills, 1978). Therefore, their presence in arctic animals becomes meaningful as an adaptation to the extreme cold environment. Survival under these conditions appear to depend on the adaptation to the high PUFA as cell constituents, and their metabolic turnover.

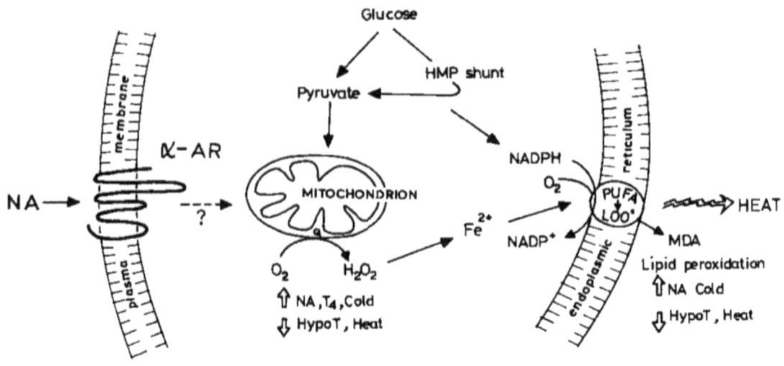

Figure 4. Cellular thermogenesis using lipid peroxidation. Generation of heat in animals depends on activation of α-adrenergic receptors (α-AR) on plasma membranes. It is increased upon cold exposure and treatment with thyroxine (T_4) and noradrenaline (NA) and decreased in heat exposure and hypothyroid (Hypo T) conditions. The mechanism proposed involves reading of the signal by altering H_2O_2 generation in mitochondria and lipid peroxidation in microsomes (endoplasmic reticulum). Hydrogen peroxide (H_2O_2) signal is amplified through release of Fe^{2+} from stores, which then triggers lipid peroxidation. This reaction involves a high rate of O_2 consumption, oxidation of NADPH (provided by HMP shunt), and peroxidative loss of polyunsaturated fatty acids (PUFA) measured as malondialdehyde (MDA), and the resulting energy is converted to heat.

Regulation of thermogenesis, both an increase and a decrease in heat generation, can easily be carried out through iron-dependent lipid peroxidation by release or sequestration of free iron. Iron loading in animals caused an increase in microsomal lipid peroxidation (Osgino and Awai, 1988). Release of free iron from cellular sources is likely to be controlled by stimuli of α-adrenergic receptors through mitochondrial generation of H_2O_2 or superoxide anions, known to release ferritin-bound iron (54). Adaptive change sensing the need of heat generation is implicit without disturbing the vital ATP synthetic machinery.

The combined actions of activation of α-adrenergic receptors by noradrenaline (the signal), increase in mitochondrial H_2O_2 generation (transduction of the signal), release of iron from endogenous stores (amplification of thermogenic signal), and stimulation of lipid peroxidation–dependent NADPH oxidation and oxygen consumption (thermogenic reaction) would explain the basic features of nonshivering chemical thermogenesis, shown schematically in Figure 4.

VII. CONCLUSION

Conceptually there are a number of ways whereby extra heat can be generated in a cell. The mechanisms discussed in this chapter are all plausible and may be used in some cases. Indeed, there is no need to invoke a single unique mechanism and one specialized tissue to undertake this activity in an animal. In such a case, removal of that tissue should result in heat insufficiency and inability of the animal to survive cold stress. In cold-adapted rats, removal of a major portion of the BAT did not decrease their ability to survive in cold stress (Foster, 1974). But deprivation of PUFA in the diet did cause their death (Ramasarma et al., 1987).

The foregoing discussion highlights a common feature of increased temperature and heat generation of specialized tissues—inflorescence in plants and BAT in animals. In both the cases, direct oxidation of the stored lipid occurs. This extra heat generation in the plant plays the role of vaporizing insect attractants, or possibly even dispersal of pollen. Does the BAT have a parallel role in exporting material or signals that need a high thermal state? The adipocyte cell may well be an endocrine tissue as it secretes a 16-kD protein product of ob gene (absent in obese animals), considered to be a satiety factor. Intriguingly this cell also has genes expressing tumor necrosis factor α (TNFα), a cytokine found in macrophages, and peroxisomal proliferator activated receptor $\gamma 2$ (PPAR$\gamma 2$), a lipid-dependent transcription factor (Flier, 1995). These new developments indicate additional roles for adipocyte cells besides storage of lipid and heat generation.

Massive evidence collected by a large number of laboratories showed BAT mitochondria with the unique uncoupling protein can effectively divert energy to heat. A succinct overview of heat-generating reactions and adrenergic control is provided by Lowell (1996) and presented schematically in Figure 5 (upper portion). This overview explains the primary features of thermogenesis: noradrenaline as a signal, β-adrenergic receptor activation, PKA-dependent release of FFA needed as an energy source and for FFA shuttle for proton translocation, and thereby heat generation. This scheme has no need for α-adrenergic receptors, for PUFA turnover, or for an oxygen-consuming reaction other than mitochondrial cytochrome oxidase, and has confined the heat-generation function to BAT.

Studies in our laboratory provide an alternative scheme, also shown in Figure 5 (lower portion) based on mitochondrial H_2O_2 generation and microsomal NADPH-dependent lipid peroxidation. This Scheme includes the following sequence of events: noradrenaline as a signal; β-adrenergic receptors for release of FFA as the energy source; α-adrenergic receptors for activation of the mitochondrial H_2O_2 generating system as well as microsomal lipid peroxidation; release of free iron from stores by H_2O_2 (or O_2^-); increased utilization of PUFA during lipid peroxidation, which consumes the extra oxygen and oxidizes NADPH; and release of energy as heat. These changes occur in the liver and BAT (possibly other tissues) and respond to alterations in thermogenic conditions of the rat. This scheme has an advantage in that it leaves the essential ATP synthesis untouched. The need for PUFA in the diet and reduced survival in its absence in cold exposure give solid support for this scheme.

BAT as a source of heat in nonshivering thermogenesis has dominated thought and research for nearly four decades. This focus yielded excellent new information

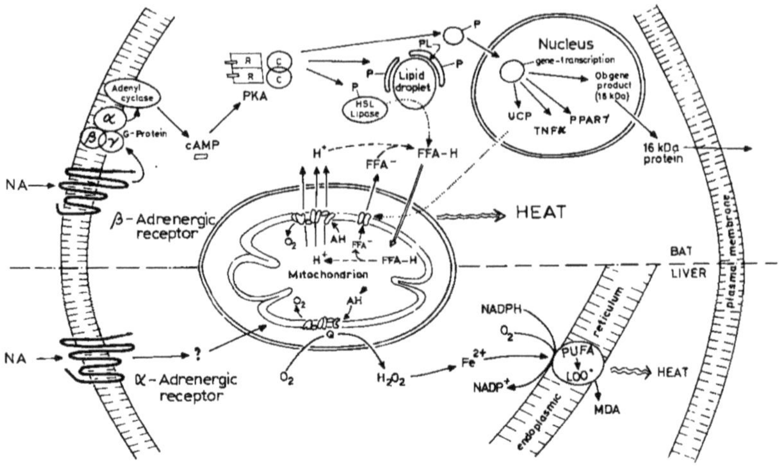

Figure 5. An integrated model of chemical thermogenesis involving proton gradient discharge in BAT mitochondria, and mitochondrial H_2O_2 generation and microsomal lipid peroxidation in liver. The top portion for BAT shows the following: noradrenaline (NA)-mediated signal transduction of β-adrenergic receptor, G protein, and adenyl cyclase; cyclic adenosine monophosphate (cAMP), which activates protein kinase A (PKA) by dissociating the catalytic subunits (C) from regulatory subunits (R); activation by protein phosphorylation of gene transcription in the nucleus including secretory ob gene product of 16-kD protein, tumor necrosis factor α (TNFα), peroxisome prolifierator activated receptor γ2 (PPARγ2), and the uncoupling protein (UCP). UCP is transferred to mitochondria where it is present exclusively. PKA also phosphorylates perilipin (PL) and hormone-sensitive lipase (HSL) and releases free fatty acids (FFA). Through a futile cycle, FFA discharges the electrogeneic proton gradient that generates heat. The lower portion for liver and BAT implicates α-adrenergic receptor, mitochondrial hydrogen peroxide (H_2O_2) generation, amplification of this signal by Fe^{2+} release that triggers peroxidation of polyunsaturated fatty acids (PUFA), and it degradation through lipid peroxyradicals to malondialdehyde (MDA) accompanied by rates of NADPH oxidation and oxygen consumption. The energy of this oxidation is released as heat.

and led to the development of attractive and plausible proposals that fitted many features. But it is known that other tissues do contribute to heat generation. Thus, regulatory role during the adaptation process of greater significance than is presently assigned is emerging for BAT.

ACKNOWLEDGMENT

The author is a senior scientist of the Indian National Science Academy, New Delhi, India, and acknowledges the contributions of many of the colleagues for the work from this laboratory: C.K.R. Kurup, H.N. Aithal, L. Susheela, V. Sitaramam, M.A.A. Namboodiri, R. Manjunath, Meera Rau, Vidya Shivaswamy, A. Swaroop, S. Sivaramakrishnan, R.S. Puranam, B.S. Sekhar, A. Gaikwad, and H.N. Ravishankar.

REFERENCES

Aithal, H.N., and Ramasarma, T. (1971). Changes in liver mitochondrial oxidation of succinate during cold-exposure. Biochem. J. 123, 677-682.

Boveris, A., Cadenas, E., and Stoppani, A.O.M. (1976). Role of ubiquinone in mitochondrial generation of hydrogen peroxide. Biochem. J. 156, 435-444.

Cannon, B., Nedergaard, J., and Sundin, U. (1981). Thermogenesis, brown fat and thermogenin. In: Survival in the Cold, Hybernation and Other Adaptations (Mussachia, X., and Jansky, L., Eds.), pp. 99-120. Elsevier/North-Holland, Amsterdam.

Chaffee, R.R.J., and Roberts, J.C. (1971). Temperature acclimation in birds and mammals. Ann. Rev. Physiol. 33, 155-202.

Chance, B., Sies, H., and Boveris, A. (1979). Hydrogen peroxide metabolism in mammalian organs. Physiol. Rev. 59, 527-605.

Chen, Y.D.I., and Hoch, F.L. (1977). Thyroid control over biomembranes. Rat liver mitochondrial inner membrane. Arch. Biochem. Biophys. 181, 470-483.

Clarke, M.G., Bloxham, D.P., Holland, P.G., and Lardy, H.A. (1973). Estimation of the fructose diphosphate-phosphofructokinase substrate cycle. Biochem. J. 134, 589-597.

Dillman, E., Gale, C., Green, W., Johnson, D.G., Mackler, B., and Finch, C. (1980). Hypothermia in iron deficiency due to altered triiodothyronine metabolism. Am. J. Physiol. 239, R 377-381.

Ernster, L., and Nordenbrand, K. (1961). Microsomal lipid peroxidation. Methods Enzymol. 10, 574-580.

Flier, J.S. (1995). The adipocyte: storage depot or node on the energy information superhighway? Cell 80, 15-18.

Foster, D.O. (1974). Evidence against a mediatory role of brown adipose tissue in the calorigenic response of cold acclimated rats to noradrenaline. Can. J. Physiol. Pharmacol. 52, 1051-1062.

Foster, D.O., and Frydman, M.L. (1978). Nonshivering thermogenesis in the rat. II. Measurements of blood flow with microspheres point to brown adipose tissue as the dominant site of the calorigenesis induced by noradrenaline. Can. J. Physiol. Pharmacol. 56, 110-122.

Gaikwad, A.S., Ramasarma, T., and Kurup, C.K.R. (1990). Decrease of oxidative activities in brown adipose tissue mitochondria of cold acclimated rats on short term exposure to heat stress. Biochem. Biophys. Acta 1017, 242-250.

Grav, H.J., and Blix, A.S. (1979). Uncoupled mitochondria of skeletal muscle in cold-adapted fur-seal. Science 204, 87-89.

Hammer, C.T., and Wills, E.D. (1978). The role of lipid components of the diet in the regulation of the fatty acid composition of the rat liver endoplasmic reticulum and lipid peroxidation. Biochem. J. 174, 585-593.

Himms-Hagen, J. (1976). Cellular thermogenesis. Ann. Rev. Physiol. 38, 315-351.

Himms-Hagen, J., Behrens, W., Muirhead, M., and Hbous, A. (1975). Adaptive changes in the calorigenic effect of catecholamines: role of changes in adenyl cyclase system and of changes in the mitochondria Mol. Cell. Biochem. 6, 15-21.

Hochachka, P.W. (1974). Regulation of heat production at the cellular level. Fed. Proc. U.S.A. 33, 2162-2167.

Ismail-Beigi, F., and Edelman, I.S. (1970). Mechanism of thyroid calorigenesis, role of sodium transport. Proc. Nat. Acad. Sci. U.S.A. 67, 1071-1078.

Jansky, L. (1973). Non-shivering thermogenesis and its thermoregulatory significance. Biol. Rev. 48, 85-132.

Jezek, P., Hanus, J., Semyad, C., and Garlid, K.D. (1996). Photoactivated azido fatty acid irreversibly inhibits anion and proton transport through the mitochondrial uncoupling protein. J. Biol. Chem. 271, 6199-6205.

Kamataki, T., Ozawa, N., Kitada, M., Kitagawa, H., and Sato, R. (1974). The occurrence of an inhibitor of lipid peroxidation in rat soluble fraction and its effect on microsomal drug oxidations. Biochem. Pharmacol. 23, 2485-2490.

Klingenberg, M., and Winkler, E. (1985). The reconstituted isolated uncoupling protein is a membrane potential driven H^+ translocator. EMBO J. 4, 3087-3092.

Lin, C.S., and Klingenberg, M. (1982). Characteristics of the isolated purine nucleotide binding protein from brown fat mitochondria. Biochemisty 21, 2950-2956.

Lowell, B.B. (1996). Slimming with leaner enzyme. Nature 382, 585-586.

Ma, S.W.Y., and Foster, D.O. (1989). Brown adipose tissue, liver and diet-induced thermogenesis in cafeteria diet-fed rats. Can. J. Physiol. Pharmacol. 67, 376-381.

McCay, P.B., Gibson, D.B., and Hornbrook, K.R. (1981). Glutathione-dependent inhibition of lipid peroxidation by a soluble, heat-labile factor not glutathione peroxidase. Fed. Proc. U.S.A. 40, 199-205.

Meera, R., and Ramasarma, T. (1979). Stimulation of cyanide-insensitive respiration in rat liver mitochondria by menadione. Indian J. Biochem. Biophys. 16, 379-383.

Nagy, K.A., Odell, D.K., and Seymour, R.S. (1972). Temperature regulation by the inflorescence of philodendron. Science 178, 1195-1197.

Nicholls, D.G. (1979). Brown adipose tissue mitochondria. Biochem. Biophys. Acta 549, 1-29.

Nicholls, D.G., Cannon, B., Grav, H.J., and Lindberg, O. (1974). Energy dissipation in non-shivering thermogenesis. In: Dynamics of Energy-transducing Membranes (Ernster, L., Estabrook, R., and Slater. E.C., Eds.), pp. 529-537. Elsevier, Amsterdam.

Nicholls, D.G., and Locke, R.M. (1984). Thermogenic mechanisms in brown fat. Physiol. Rev. 64, 1-64.

Ogino, T., and Awai, M. (1988). Lipid peroxidation and tissue injury by ferric citrate in paraquat-intoxicated mice. Biochem. Biophys. Acta 958, 388-395.

Parrish, D.J., and Leopald, A.C. (1978). Confounding of alternate respiration by lipoxygenase activity. Plant Physiol. 62, 470-472.

Potter, V. (1958). Possible biochemical mechanisms underlying adaptation to cold. Fed. Proc. U.S.A. 17, 1060-1063.

Ramasarma, T. (1968). Studies on ubiquinone. J. Sci. Industr. Res. 27, 147-164.

Ramasarma, T. (1982). Generation of H_2O_2 in biomembranes. Biochem. Biophys. Acta 694, 69-93.

Ramasarma, T., Maukkassah-Kelly, S., and Hochstein, P. (1984). Inhibition of microsomal lipid peroxidation by cytosolic protein in presence of ADP and high concentration of Fe^{2+}. Biochim. Biophys. Acta 796, 243-250.

Ramasarma, T., Sekhar, B.S., and Kurup, C.K.R. (1987). Cellular thermogenesis: a new approach. In: Bioenergetics: Structure and Function of Energy Transducing Systems (Ozawa, T., and Papa, S., Eds.), pp. 225-233. Japan Sci. Soc. Press, Tokyo.

Ramasarma, T., and Susheela, L. (1974). A mechanism of thermogenesis by modification of succinate dehydrogenase. In: Biomembranes—Architecture, Biogenesis, Bioenergetics and Differentiation (Packer, L., Ed.), pp. 261-277. Academic Press, New York.

Raskin, I., Ehmann, A., Melander, W.R., and Meeuse, B.J.D. (1987). Salicylic acid: a natural inducer of heat production in Arum lilies. Science 237, 1601-1602.

Rich, P.R. (1978). Quinol oxidation in *Arum maculatum* mitochondira and its application to the assay, solubilization and partial purification of alternate oxidase. FEBS Lett. 96, 252-256.

Ricquier, D., Mory, G., and Hemon, P. (1979). Changes induced by cold-adaptation in the brown adipose tissue from several species of rodents with special reference to mitochondrial components. Can. J. Biochem. 57, 1262-1266.

Sekhar, B.S., Kurup, C.K.R., and Ramasarma, T. (1990). Increase in hepatic microsomal lipid peroxidation mediated by α-adrenergic system under cold stress and noradrenaline treatment. Mol. Cell. Biochem. 94, 61-70.

Seymour, R.S. and Schultze-Motel, P. (1996). Thermoregulatory lotus flowers. Nature 383, 305.

Sitaramam, V., Meera, R., and Ramasarma, T. (1977). Influence of some drugs on body temperature and induction of hepatic tryptophan pyrrolase in animals exposed to cold. Indian J. Exp. Biol. 15, 98-104.

Skubatz, H., Williamson, P.S., Schneider, E.L., and Meeuse, B.J.D. (1990). Cyanide-insensitive respiration in thermogenic flowers of *Victoria* and *Nelumbo*. J. Exp. Botany 41, 1334-1390.

Smith, B.N. and Meeuse, B.J.D. (1966). Production of volatile amines and skatole at anthesis in some Arum lily species. Plant Physiol. 41, 343-347.

Swaroop, A., Patole, A.S., Puranam, R.S., and Ramasarma, T. (1983). Noradrenaline treatment of rats stimulated H_2O_2 generation in liver mitochondria. Biochem. J. 214, 745-750.

Swaroop, A. and Ramasarma, T. (1985). Heat exposure and hypothyroid conditions decrease hydrogen peroxide generation in liver mitochondria, Biochem. J. 226, 403-408.

Tata, J.R., Ernster, L., Lindberg, O., Arrhenius, E., Pederson, S., and Hedman, R. (1963). The action of thyroid hormones at the cell level. Biochem. J. 86, 408-428.

Thomas, C.E., Morehouse, L.A., and Aust, S.D. (1985). Ferritin and superoxide-dependent lipid peroxidation. J. Biol. Chem. 260, 3275-3280.

Tolbert, N.E. (1981). Metabolic pathways in peroxisomes and glyoxysomes. Ann. Rev. Biochem. 50, 133-157.

Walker, D.B., Gysi, J., Sternberg, L., and De Niro, M.J. (1983). Direct respiration of lipids during heat production in the inflorescence of *Philodendron selloum*. Science 220, 419-421.

Yamashita, H., Ohira, Y., Wakatsuki, T., Yamamato, M., Kizaki, T., Ohnishi, S., and Ohno, H. (1995). Increased growth of brown adipose tissue but its reduced thermogeneic activity in creatine-depleted rats β-guanidinopropionic acid. Biochem. Biophys. Acta 1230, 69-73.

Zeisberger, E. and Bruck, K. (1971). Central effect of noraderenaline on the control of body temperature in the guineapig. Pflugers Arch. 322, 152-166.

Zeisberger, E. and Bruck, K. (1976). The significance of central adrenergic α-receptive structures in the control of thermogenesis and in cold adaptation. Israel J. Med. Sci. 12, 1103-1106.

Zhang, Y., Proenca, R., Maffei, M., Barone, M., Leopald, M., and Friedman, J.M. (1994). Positional cloning of the mouse obese gene and its human homologue. Nature 372, 425-432.

FROM RAYNAUD'S PHENOMENON TO SYSTEMIC SCLEROSIS (SCLERODERMA): LACK OR EXHAUSTION OF ADAPTATION?

Marco Matucci Cerinic, Sergio Generini,
Alberto Pignone, and Mario Cagnoni

Advances in Organ Biology
Volume 6, pages 241-253.

ISBN: 0-7623-0391-3

I. INTRODUCTION

Systemic sclerosis (SSc) is a multisystem disease leading to fibrosis of the skin and internal organs. However, in the large majority of SSc patients before the onset of fibrosis, the disease affects the microcirculatory endothelium (arterioles and capillaries) (Matucci Cerinic et al., 1995), generating a vascular tone instability that is clinically known as Raynaud's phenomenon (RP) (Coffman, 1994). RP is considered a sentinel symptom for the early diagnosis of SSc (Matucci Cerinic et al., 1995). In fact, the disease, in particular the limited form (Ter Borg et al., 1994), starts with RP, which may last few or many years before the clinical manifestations of skin and internal organ involvement become evident. For this reason, RP and the dysregulation of vascular tone represent, from the clinical and pathogenetic point of view, a key event intimately linked to SSc onset and progression. Vascular ischemia and reperfusion are the pivotal events in the early phase of the disease. In this paper, attention will be focussed on ischemia–reperfusion and its importance to the evolution and maintenance of SSc, and a possible lack or exhaustion of adaptation of endothelial cells to reperfusion–injury will be hypothesized in order to explain the evolution of RP to SSc.

II. RAYNAUD'S PHENOMENON

RP is characterized by episodes, in particular when exposed to cold, of finger blanching (white) followed by ischemia (blue) and hyperemia (red). This phenomenon may take place many times during the day, is prevalent during the winter, and may attenuate during the summer. In SSc, the vasoconstrictive phenomenon may be divided into two phases: a first, functional phase, when the vessel is still able to accomodate and may thus close and open generating the classic triad of RP (blanching, cyanosis, and redness); and a second, organic phase, when the profound alteration of the vessel wall (hypertrophic intima and media) does not allow vessel accomodation, and the initial blanching of RP is replaced by chronic cyanosis. Between these two main phases, a series or variegation of colors, from red to blue-purple, may be observed on the fingers as an expression of the transition from the functional to the organic phase.

RP may not be limited to the digits, but may also involve the nose, ears, face, tongue, and lips in some patients, and may be more common in the primary form. In patients with RP, the spasm of the vasculature is potentially present in all internal organs, including heart (Alexander et al., 1981), lungs (Fahey et al., 1984), kidneys (Cannon et al., 1974), and retina (Salmenson et al., 1992). Several factors are involved in the pathogenesis of RP in SSc (Matucci Cerinic et al., 1995). Endothelial injury is heavily responsible for the modification of the vascular tone. A recent hypothesis is that a neurotransmitter deficiency may contribute to the functional and vascular disorder (Kahaleh and Matucci, Cerimic, 1995). However, the main causative factor is still unknown, and the cascade of events leading to the dysfunction of vascular tone is still elusive. The knowledge we have of RP mechanisms allows us only to focus attention, in the pathogenetic evolution of SSc, on the initial phase of vascular tone modification when the episodes of ischemia are continuously followed by reperfusion several times every day, for a period that varies from patient to patient but may also span many years. In this phase, in the fingers and probably in internal organs as well, a large number of microenvironmental modifications take place that may influence the onset, development, and maintenance of SSc. Vascular ischemia and reperfusion are important factors that may lead progressively not only to the development of periungueal and fingertip ulcers but, in particular, to skin fibrosis.

III. ISCHEMIA–REPERFUSION, OXYGEN FREE RADICALS, AND TISSUE INJURY

Reperfusion after ischemia has always been considered a "positive" event. In the past two decades, it has been demonstrated that reperfusion causes a rapid degeneration of endothelial cell function and of the vascular environment not only in the heart but also in the lung, gut, and kidney (Fischer et al., 1994; Paller and Greene, 1994; Kurose and Granger, 1995). Thus, ischemia and reperfusion are two critical events that may lead to: (1) impairment of β oxidation, (2) generation of oxygen free radicals and development of oxidative stress, (3) inactivation of antioxidant enzymes, and (4) breakdown of sarcolemmal phospholipids.

Oxygen free radicals are involved in several injurious and pathological processes. Mitocondrial electron transport reduces oxygen in several stages that progressively include the superoxide anions, hydrogen peroxide, hydroxyl radicals, and water. Various other processes can also generate these species from oxygen or oxygen-derived materials. Generation of free radicals has been implicated at the onset of reperfusion; it is particularly augmented by postischemic reperfusion, especially under hyperoxic conditions. During ischemia, lactic acidosis can lead to protonation of some peroxide anions that can better penetrate the membrane and initiate lipid peroxidation, while cellular metabolism becomes anaerobic, lactate increases, and pH drops. Lactic acid represents a particular problem for the cell because it is more lipophilic than smaller organic acids and has a high cell permeability. Acidosis aggravates injury by making iron more soluble. The sources of oxygen free radicals include the arachidonic acid cascade, cathecolamine oxidation, mitocondrial leak, xanthine oxidase, and oxidation of extravasated hemoglobin. Free radicals can also be derived from *neutrophils*. Activated neutrophils may adhere to damaged endothelium and amplify traumatic ischemic or ischemic–reperfusion injury in the microcirculation (Kurose and Granger, 1995). Free radicals possess a single unpaired electron, and they are highly reactive toward a variety of cellular constituents. Their entry in the inflammatory site may exacerbate the damage induced by reperfusion through the release of arachidonic acid (Curmulte et al.,

1984). Adhesion molecules are capable of attachment to activated polymorphonuclear neutrophils, allowing transendothelial migration and cytotoxic damage. Elevation of intercellular adhesion molecule 1 (ICAM 1), E-selectin, and vascular cell adhesion molecule 1 (VCAM 1) indicates activation or damage to vascular endothelium (Gearing and Newman, 1993). Their increase thus greatly contributes to magnifying the inflammatory endothelial cell response after ischemia–reperfusion injury. *Macrophages* also are important sources of reactive oxygen and may thus contribute to tissue injury.

One main target of damage is the cellular membrane. The membrane of endothelial cells may undergo peroxidation induced by oxygen free radicals and catalyzed by transferrin, iron, ferritin, or lowered pH. *Lipid peroxidation* at the membrane level is a chain reaction that alters or destroys membrane phospholipids. The injury of the cell and lipid peroxidation occurs during the early period of reperfusion. Primary reactive free radicals interact with a polyunsaturated fatty acid to initiate a complex series of reactions resulting in the appearance of a variety of degradation products such as malonaldehyde. Lipid peroxidation induced by oxygen-derived free radicals increases progressively with the duration of reperfusion. Dysruption of the phospholipid bilayer of the cellular membrane perturbs its regulatory mechanisms and causes irreversible cellular injury: the entrance of calcium into the cell activates the calcium-dependent phospholipases and protein kinases, which cleave fatty acids from phospholipids causing further changes in the chemical composition and physiological state of the cell membrane. Liberation of arachidonic acid and enhanced prostaglandin synthesis caused by phospholipid breakdown is augmented in reperfusion–injury. Free fatty acids are transformed by cyclo-oxygenase in prostaglandins and thromboxane, and new radicals species are formed during the cascade (Otani et al., 1989). Recent research into macrovascular disease has highlighted the role of oxidation of plasma lipoproteins as well as the cytotoxicity of these oxidation products. Oxidized low-density lipoproteins (LDLs) are highly immunogenic, can activate T lymphocytes, increase the release of interleukin-1β and enhance the proliferation of smooth muscle cells by increasing the expression of platelet-derived growth factor by these cells. Furthermore, mildy oxidized LDL can induce in endothelial cells the expression of genes for cellular adhesion molecules as well as for monocyte chemotactic protein 1. Oxidized LDL may also influence the metabolism of lipid in fibroblasts and, conceivably, collagen biosynthesis via endothelial cells which secrete factors that mediate matrix synthesis.

Endothelin is released only after a prolonged period of reperfusion and until significant endothelium injury occurs. The release process seems to be controlled by calcium and phospholipase but not by free radicals generated at the onset of reperfusion. The release process is likely to be controlled by the intracellular calcium influx and phospholipase activation (Maulik et al., 1992).

Reperfusion causes a rapid degeneration of endothelial functions and a consequent decreased release of *nitric oxide*. Ischemia–reperfusion is associated with decreased vasorelaxation of isolated coronary arteries (Kurose et al., 1994). Nitric oxide is an endogenous inhibitor of platelet aggregation, leukocyte chemotaxis, adherence, and activation, as well as an inactivator of superoxide free radicals. Nitric oxide synthesis can thus consistently reduce oxidative stress: in particular, it may block the formation of cytotoxic OH° by directly scavenging superoxide anion O_2^- (Win et al., 1993), and it may indeed reduce inflammatory mediators, expression of adhesion molecules, and neutrophil endothelial interaction (Kubes et al., 1991). However, nitric oxide can rapidly react with O_2^- to form highly reactive peroxynitrite radical ($ONOO^-$) (that may become the primary metabolite of nitric oxide), which is highly reactive and mutagenic and accounts for the cytotoxic activity of nitric oxide.

Prolonged ischemia and hypoxia are associated with a decrease of antioxidant enzymes, while short duration of ischemia may enhance antioxidant enzymes. This evidence focuses attention on the capacity of the cells to resist repeated ischemia–reperfusion through the induction of antioxidant enzymes and to adapt in order to survive against stress by eliminating oxidative assault.

IV. ADAPTATION

Repeated episodes of ischemia stress the cells, inducing them to adapt in order to overcome any subsequent stressful event (Duncan and Hershey, 1989). More information about this cell capacity has been derived from the study of preconditioning, in particular, in the heart. The term "preconditioning" is derived from the evidence that a system can, after repeated stunning, moderate and delay the onset of lethal ischemic injury (Murry et al., 1990). Preconditioning reduces myocardial infarct size, the number and severity of reperfusion arrhythmias (Schott et al., 1990), and the degree of autonomic denervation preserving both sympathetic and parasympathetic reflexes during ischemia (Miyazaki and Zipes, 1989).

The development of cell adaptation to oxygen-derived free radicals has been related to an increase in enzyme activities such as glucose-6-phosphate dehydrogenase, glutathione peroxidase, catalase, and superoxide dismutase. Preconditioning reduces endothelial cell dysfunction of coronary arterioles after ischemia–reperfusion (DeFily and Chilian, 1993). Endothelial cells promptly respond to oxidative stress by synthesis of several oxidative stress–inducible proteins and by stimulating antioxidative enzymes, apparently to provide protection from the toxic effects of stress, allowing the cells to recover and survive (Lu et al., 1993). Adaptation is a long term process: several changes occur during the initial phase, allowing the cells to counteract stressful conditions. Depending on the nature of the stress, the cells synthesize new proteins, modify the lipid bilayer, and modulate many enzyme activities (Hammond et al., 1982).

Cells are constitutively protected against stress by intracellular antioxidants (glutathione, α tocopherol, ascorbic acid, β carotene), and antioxidant enzymes (superoxide dismutase, catalase, glutathione peroxidase). These factors may reduce stress induced by oxygen free radicals, quenching them before vital cellular components are damaged. If this system of defense cannot counteract oxidative stress, oxygen free radicals reach nucleic acids, proteins, and lipids, causing DNA strand breaks, protein degradation, and lipid peroxidation, respectively. However, the cell may use several lypolytic and proteolytic enzymes (proteases, phospholipases, etc.) that are specialized in the detection and removal of damaged cells. A last resource of the cellular defensive system is an inducible pathway allowing the synthesis of proteins necessary for cell protection or repair of the damage. In the early phase of a stress, different genes are rapidly and transiently expressed (heat shock proteins, glucose proteins, superoxide dismutase and catalase genes, and protooncogenes) (Das et al., 1995). In endothelial cells, oxidative stress induces heat shock protein (HSP) 70 expression, which is protective against ischemic injury (Jornot et al., 1991).

V. RAYNAUD'S PHENOMENON, OXIDATIVE STRESS, AND SYSTEMIC SCLEROSIS

Since the vascular and microvascular lesions of SSc and RP occur together, a common pathogenesis based on the injury of the microcirculatory system has been proposed (Matucci Cerinic et al., 1995). Different mechanisms have been suggested to explain the vascular dysfunction: a protease enzyme, a cytotoxic factor sensitive to proteolysis by trypsin, a 5-kD protein which may be a leukotriene, and enhanced platelet activation (Matucci Cerinic et al., 1995). Altered immune processes have also been suspected, including autoantibody-mediated damage caused directly by the adhesion of lymphocytes and granulocytes to the endothelium via the increased expression of the vascular adhesion molecules E-selectin, ICAM 1, and VCAM 1 (Matucci Cerinic et al., 1995). Lymphocyte activation results in an increased released cytokines and granzymes (cytotoxic for the endothelium) (Matucci Cerinic et al., 1995). In this scenario, antiendothelial cell antibodies may activate endothelial cells, in part due to autocrine or paracrine actions of interleukin-1 (Carvalho et al., 1996).

In SSc, the most interesting potential candidate for vascular impairment seems to be the free radical–induced oxidative stress, which may occur either as a result of hypoxic/ischemic episodes and upon reperfusion of tissues, or because of altered immune processes (Murrell, 1993). The formation of free radicals and the consequent lipid peroxidation may be considered important factors in the initiation of tissue damage in RP and SSc, and may be directly or indirectly involved in most of the other pathogenetic mechanisms described earlier.

In SSc, an increase in oxidized lipoproteins (Blann et al., 1993) as well as an increased susceptibility to oxidation of LDL (Bruckerdorfer et al., 1995) and increased levels of antibodies against oxidized LDL (Simonini et al., 1998 in press) have been detected. However, although oxidized lipoproteins are abnormally increased in RP and in SSc, the lack of consistent correlations with other indicators of endothelial damage (e.g., von Willebrand factor, angiotensin-converting enzyme, endothelial cell cytotoxicity) suggested that additional mechanisms other than lipid peroxides may be responsible for the insult to the vascular tree (Blann et al., 1993). Deficient levels of the antioxidants selenium and ascorbic acid in patients with SSc and RP have been found, suggesting abnormalities of oxidation in these diseases (Herrick et al., 1994). The high rate of chromosomal instability found in RP and SSc patients was linked to the increased levels of oxygen free radicals (Housset et al., 1969; Emerit, 1976), Recently, it has been shown that scleroderma autoantigens are cleaved at highly specific sites in a reaction that requires metal binding and the generation of reactive oxygen species: these fragments may be potential initiators of the autoimmune process, and autoantibody response may be considered the immune marker of a unique protein structure induced and driven by cycles of ischemia–reperfusion (Casciola-Rosen et al., 1997).

Patients with SSc were found to have complex abnormalities of the arachidonic acid cascade and metabolism of eicosanoids. Abnormal levels of circulating prostaglandins (Belch et al., 1985), depletion of prostaglandins precursors (Horrobin, 1986; Reilly et al., 1985), and low serum phospholipase A2 activity (Pruzanski et al., 1990) have been described in SSc. In particular, increased levels of urinary F2-isoprostane, a measure of oxidative stress, provide further evidence that free radical–induced oxidative injury occurs in SSc (Stein et al., 1996).

The role of polymorphonuclear (PMN) cells in the pathophysiology of SSc has been recognized (Spisani et al., 1981), and increased PMN activity in patients with RP alone as well as in those with SSc associated with RP has been described (Lau et al., 1992b). Activated PMN cells, via the generation of leukotriene B4 and release of free radicals, may enhance oxidative stress, further tissue damage, and contribute to the reduction of blood flow by physical obstruction (Lau et al., 1992a).

Other pivotal steps in SSc pathogenesis involve an increase of circulating endothelin and a decreased release of nitric oxide. Conspicuous circulating levels of endothelin (Kahaleh, 1991; Vancheeswaran et al., 1994) contribute to disease pathogenesis not only through vasoconstriction, but also through the stimulation of smooth muscle cells and fibroblast proliferation (Kahaleh, 1991). In SSc, the capacity of the endothelium to release nitric oxide is reduced. This is demonstrated by the reduced circulating levels of nitric oxide in basal conditions and, in particular, by the lack of adjustment to cooling (Kahaleh et al., submitted for publication) and of response to substance P stimulation (Matucci Cerinic et al., 1990) with an increased release of nitric oxide. Indeed, endothelial cells, stimulated in vitro with the serum of SSc patients, reduce the production of nitric oxide (Kaheleh et al., submitted for publication). The impairment of nitric oxide in SSc is an additional factor reducing the scavenging potential and the adaptive capacity of the endothelium.

Local ischemia and reperfusion during repeated attacks of RP may contribute to the vascular damage, enhancing oxidative stress and increasing oxygen free radical release. In such circumstances, a vicious cycle of local ischemia and reperfusion, endothelial damage with imbalance of the factors responsible for vascular tone control, sensory nervous system failure, and white blood cell and platelet activation could develop, leading to further ischemia.

VI. HYPOTHESIS

The close relationship between RP and SSc and, in particular, the transformation of RP in SSc is not yet fully understood. Peripheral vasoconstriction in response to cold is physiological, and vasoconstriction sufficient to produce digital pallor or cyanosis may occur in normal individuals after a prolonged or severe cold exposure. Healthy individuals with RP, irrespective of etiology, have undue intolerance to environmental cold. A key question is whether these attacks represent an exaggeration of the normal vasoconstrictive mechanism or are due to a specific abnormality. Other key questions are whether primary and secondary RP have the same pathogenetic substrate, and which is (if there is one) the culprit event that makes RP shift toward SSc. Is SSc effectively an evolution of RP? A different response (or adaptation) to the same injurious factors? Or is RP just a sentinel symptom accompanying SSc?

New theories have been proposed to answer these questions, but the identification of the culprit lesion is still missing, or even if there is not a culprit lesion but rather an accumulation of conditions having nothing in common but few symptoms. Moreover, a vasospastic attack may result not from a single event, but from a cascade of events.

In SSc, the tenet is that an endothelial injury is the early pathogenetic event (Matucci Cerinic et al., 1995). It is now known that there is an increase of oxidative stress (free radicals, lipid peroxidation, and PMN cells) as well as a failure of the control of the vascular tone with a shift toward vasoconstriction due to an increase in endothelin and a reduction in nitric oxide. However, the precise starting point of the cascade of these events still remains elusive, in particular, whether the dysregulation of the vascular tone (i.e., RP) is the key initial factor or just a consequence of endothelial injury.

A large proportion of RP subjects (approximately 70%) will evolve to SSc, while the rest will continue to have RP without evolving to a connective tissue disease. Continuous episodes of RP and reperfusion may be considered as a stimulus to the endothelium to achieve a physiological adaptation able to counteract oxidative injury. The knowledge of the physiological processes of reperfusion–injury and preconditioning–adaptation may suggest a hypothesis when applied to RP and SSc. In fact, the main question is whether in RP patients the process of adaptation takes place and is kept active. The hypotheses are as follows (Figure 1):

- *Adaptation is exhausted:* Continuous episodes of daily RP and reperfusion exhaust and overwhelm the capacity of the endothelium to counteract oxidative stress, which leads progressively to chronic injury to the arteriolar and capillary bed.

- *Adaptation is lacking*: The endothelium is not able to provide an adequate line of defense to counteract oxidative stress. This failure may depend upon previous endothelial injury due to the causative agent of the disease, or may represent a constitutive lack of adaptative capacity to stress of the endothelium of those RP patients who will evolve to SSc.

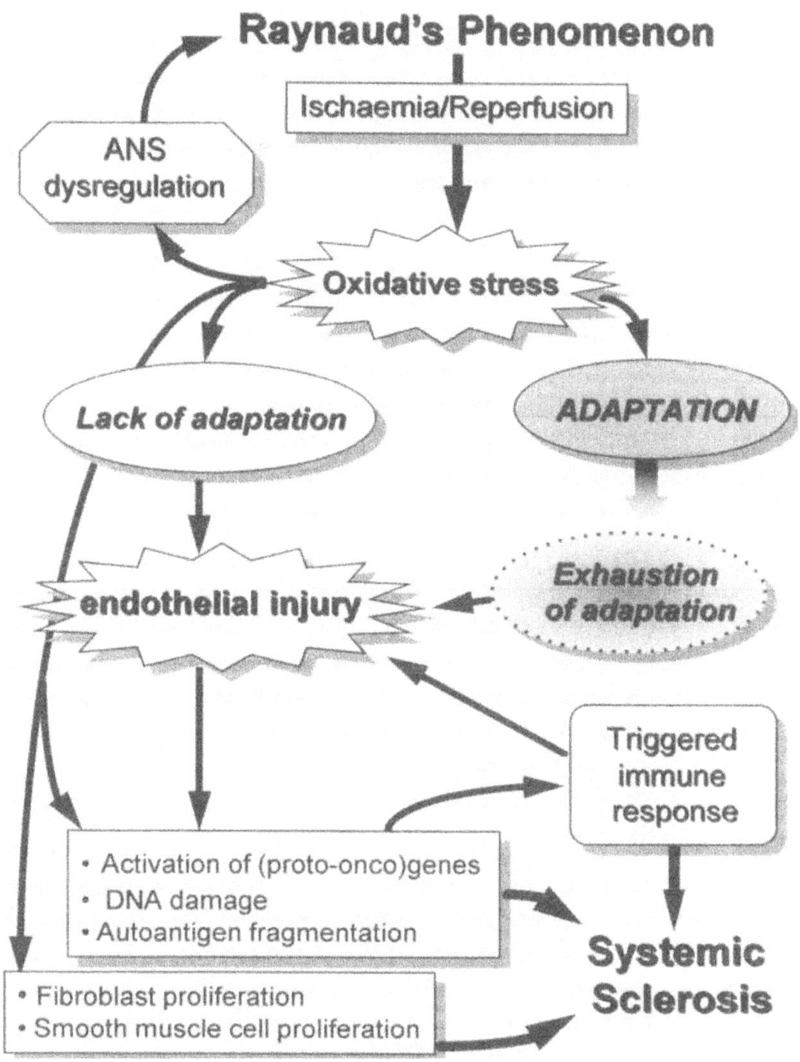

Figure 1. Two hypotheses assuming lack or exhaustion of adaptation to oxidative stress. Both pathways may lead to endothelial injury and trigger the evolution of Raynaud's phenomenon to systemic sclerosis. ANS, autonomic nervous system.

The lack or exhaustion of adaptation and the collapse of defense mechanisms thus permit a chronic injurious oxidative stress to the endothelium with DNA damage and the possibility of a late activation of new detrimental (proto-onco)genes. At the same time, oxygen free radicals stimulate proliferation of cultured human fibroblasts that are also able to release superoxide upon stimulation (O_2^-) (Murrell et al., 1990). Oxygen free radicals released by a host may thus provide a very fast, specific, and sensitive trigger for fibroblast proliferation and fibrosis, and stimulate endothelial cells to recruit monocytes that migrate through the vessel wall (Berliner et al., 1990). To this process, endothelial injury may greatly contribute via the enhanced release of endothelin, which not only potentiates fibroblast proliferation but also induces smooth muscle cell proliferation. All these factors may thus contribute to trigger the disease and transform RP in SSc (see Figure 1).

VII. CONCLUSIONS

We believe that SSc should be treated from the earliest prefibrotic phase. The therapeutic strategy should be broad and consider oxygen free radical formation and lipid peroxidation during RP as early important steps in SSc pathogenesis. Their damaging potential is detrimental and might be fought in different ways. Vasodilating treatment is very important in order to avoid continous episodes of RP and reperfusion. For this purpose, calcium antagonists, prostanoids, and nitroderivates are useful. Oxygen free radical formation and lipid peroxidation can be counteracted by scavengers inhibiting the chain reaction (vitamin E, catalase, superoxide dismutase, iron chelating agents, etc.). Agents lessening phospholipid mobility and stabilizing cellular membrane can limit the propagation of lipid peroxidation: glucocorticoids are potent antioxidants and may effectively interrupt lipid peroxidation, but their chronic use in SSc is contraindicated. Nonglucocorticoid steroids with very high antioxidant activity (Lazaroids) might be potent inhibitors of lipid peroxidation in SSc. Immunosuppressants are now used in overt disease and might become in the future a useful therapeutic choice in RP that will evolve to SSc.

We believe that a careful investigation of the modification of the tight cross-talk between the peripheral nervous system and the endothelium in RP and SSc might shed a new light on the understanding of the dysfunction of the vascular tone, the process of adaptation, and the genesis of endothelial injury.

VIII. SUMMARY

The dysregulation of vascular tone in RP represents a key event intimately linked to the onset and progression of SSc. RP is usually the earliest symptom of Ssc. RP provokes vascular ischemia followed by reperfusion: this phenomenon favors the generation of oxygen free radicals and oxidative stress, impairment of β oxidation, inactivation of antioxidant enzymes, and breakdown of sarcolemmal phospholipids. Repeated episodes of ischemia and reperfusion induce endothelial cells to adapt in order to overcome any subsequent stressful event. Endothelial cells promptly respond to oxidative stress by the synthesis of several intracellular antioxidants, oxidative stress–inducible proteins and antioxidative enzymes, as well as by the synthesis of proteins necessary for cell protection or repair of the damage. In the early phase of a stress, different genes are rapidly and transiently expressed (heat shock proteins, glucose proteins, superoxide dismutase and catalase genes, and protooncogenes). In RP and Ssc, a reduction of antioxidative potential as well as an increase of circulating free radicals has been reported.

The knowledge of the physiological processes of reperfusion–injury and preconditioning–adaptation applied to RP and Ssc encourage the hypotheses that a lack or exhaustion of adaptation, and the collapse of defense mechanisms toward injurious oxidative stress to the endothelium, may be fundamental factors contributing, at least in part, to the evolution of RP in SSc.

REFERENCES

Alexander, E.L., Firestein, G.S., and Leitl, G. (1981). Scleroderma heart disease: evidence for cold-induced abnormalities of myocardial function and perfusion. Arthritis Rheum. 24 (suppl.), 58 [Abstr.].

Belch, J.J.F., McLaren, M., Anderson, J., Lowe, G.D., Sturrock, R.D., Capell, H.A., Forbes, C.D., Mikhailidi, S.D.P., Jeremy, J.Y., and Dandona, P. (1985). Increased prostacyclin metabolites and decreased red cell deformability in patients with systemic sclerosis and Raynaud's syndrome. Prostaglandins Leukot. Essent. Fatty Acids 17, 1-9.

Berliner, J.A., Territo, M.C., Sevanian, A., Ramin, S., Ai, K.J., Bamshad, B., Esterson, M., and Fogelman, A.M. (1990). Minimally modified low density lipoprotein stimulates monocyte endothelial interactions. J. Clin. Invest. 85, 1260-1266.

Blann, A.D., Illingworth, K., and Jayson, M.I.V. (1993). Mechanisms of endothelial cell damage in systemic sclerosis and Raynaud's phenomenon. J. Rheumatol. 20, 1325-1330.

Bruckdorfer, K.R., Hillary, J.B., Bunce, T., Vancheeswaran, R., and Black, C.M. (1995). Increased susceptibility to oxidation of low density lipoproteins isolated from patients with systemic sclerosis. Arthritis Rheum. 38, 1060-1067.

Cannon, P.J., Hassar, M., Case, D.B., Casarella, W.J., Sommers, S.C., and LeRoy, E.C. (1974). The relationship of hypertension and renal failure in scleroderma (progressive systemic sclerosis) to structural and functional abnormalities of the renal cortical circulation. Medicine 53, 1-46.

Carvalho, D., Savage, C.O.S., Black, C.M., and Pearson, J.D. (1996). IgG antiendothelial cell autoantibodies from scleroderma patients induce leucocyte adhesion to human vascular endothelial cells in vitro. Induction of adhesion molecule expression and involvement of endothelium derived cytokines. J. Clin. Invest. 97, 111-119.

Casciola-Rosen, L., Wigley, F., and Rosen, A. (1997). Scleroderma autoantigens are uniquely fragmented by metal-catalyzed oxidation reactions: Implications for pathogenesis. J. Exp. Med. 185, 71-79.

Coffman , J. (1994). The diagnosis of Raynaud's phenomenon. Clin. Rheumatol. 12, 283-290.

Curnulte, J., Badwes, J.A., Robinson, J.M., Karnovsy, M.J., and Karnovsky, M.L. (1984). Studies on the mechanism of superoxide release from human neutrophils stimulated with arachidonate. J. Biol. Chem. 259, 11851-11857.

Das, D.K., Maulik, K.N., and Moraru, I.I. (1995). Gene expression in acute myocardial stress. Induction by hypoxia, ischemia, reperfusion, hypertermia and oxidative stress. J. Mol. Cell. Cardiol. 27, 181-185.

DeFily, D., and Chilian, W.M. (1993) Preconditioning protects coronary arteriolar endothelium from ischemia–reperfusion injury. Am. J. Physiol. 34, H700-H706.

Duncan, R.F., and Hershey, J.W.B. (1989). Protein synthesis and protein phosphorilation during heat stress recovery and adaptation. J. Cell. Biol. 109, 1467-1481.

Emerit, I. (1976). Chromosomal breakage in systemic sclerosis and related disorders. Dermatologica 153, 145-146.

Fahey, P.J., Utell, M.J., Condemi, J.J., Green, R., and Hyde, R.W. (1984). Raynaud's phenomenon of the lung. Am. J. Med. 76, 263-269.

Fischer, A.B., Dodia, C., Ayene, I., and Al-Mehdi, A. (1994). Ischemia reperfusion injury to the lung. Ann. N.Y. Acad. Sci. 273, 197-207.

Gearing, A.J.H., and Newman, W. (1993). Circulating adhesion molecules in disease. Immunol. Today 14, 506-512.

Hammond, S.L., Loi, Y.K., and Markert, C.L. (1982). Diverse forms of stress lead to new patterns of gene expression through a common and essential metabolic pathway. Proc. Nat. Acad. Sci. 79, 3485-3488.

Herrick, A.L., Rieley, F., Schofield, D., Hollis, S., Braganza, J.M., and Jayson, M.I.V. (1994). Micronutrient antioxidant status in patients with primary Raynaud's phenomenon and systemic sclerosis. J. Rheumatol. 21, 1477-1483.

Horrobin, D.F. (1986). Essential fatty acid and prostaglandin metabolism in Sjoegren syndrome, systemic sclerosis and rheumatoid arthritis. Scand. J. Rheumatol. 61 (suppl.), 242-245.

Housset, E., Emerit, I., and Baulon, A. (1969). Anomalies chromosomiques dans la sclerodermie generalisée: études de 10 malades. C. R. Acad. Sci. (Paris) 269, 413-420.

Jornot, L., Mirault, M.E., and Junod, A.F. (1991). Differential expression of Hsp 70 stress proteins in human endothelial cells exposed to heat shock and hydrogen peroxide. Am. J. Resp. Cell. Mol. Biol. 5, 265-275.

Kahaleh, B.M. (1991). Endothelin an endothelial dependent vasoconstrictor in scleroderma: enhanced production and profibrotic action. Arthritis Rheum. 34, 978-983.

Kahaleh, B.M., Fan, P.S., Matucci Cerinic, M., and Ignarro, L. (submitted for publiclation). Nitric oxide in scleroderma: reduced production and blunted response to cold.

Kahaleh, B.M., and Matucci Cerinic, M. (1995). Raynaud's phenomenon and scleroderma: dysregulated neuroendothelial control of vascular tone. Arthritis Rheum. 38, 1-4.

Kubes, P., Suzuli, M., and Granger, D.N. (1991). Nitric oxide: an endogenous modulator of leucocyte adhesion. Proc. Nat. Acad. Sci. 88, 4651-4655.

Kurose, I., and Granger, D.N. (1995). Evidence implicating xanthine oxidase and neutrophils in reperfusion induced microvascular dysfunction. Ann. N.Y. Acad. Sci. 723, 158-179.

Kurose, I., Wolf, R., and Granger, D.N. (1994). Modulation of ischemia reperfusion induced microvascular dysfunction by nitric oxide. Circ. Res. 74, 376-382.

Lau, C.S., Bridges, A.B., Muir, A., Scott, N., Bancroft, A., and Belch, J.J.F. (1992a). Further evidence of increased polymorphonuclear cell activity in patients with Raynaud's phenomenon. Br. J. Rheumatol. 31, 375-380.

Lau, C.S., O'Dowd, A., and Belch, J.J.F. (1992b). White blood cell activation in Raynaud's phenomenon of systemic sclerosis and vibration induced white finger syndrome. Ann. Rheum. Dis. 51, 249-259.

Lu, D., Maulik, N., Moraru, I., Kreutzer, D.L., and Das, D.K. (1993). Molecular adaptation of vascular endothelial cells to oxidative stress. Am. J. Physiol. 264, C715-C721.

Matucci Cerinic, M., Kahaleh, B.M., and Leroy, E.C. (1995). Vascular involvement: pathogenesis. In: Systemic Sclerosis (Furst, D., and Clements, P., Eds.) p. 153. New York, Lea & Febiger.

Matucci Cerinic, M., Pietrini, U., and Marabini, S. (1990). Local venomotor response to intravenous infusion of substance P and glyceryl trinitrate in systemic sclerosis. Clin. Exp. Rheumatol. 8, 561-565.

Maulik, N., Liu, X., Subramanian, R., and Das, D.K. (1992). Release of endothelin during reperfusion of ischemic myocardium. Endothelin 1 release from reperfused heart. Am. J. Cardiovasc. Pathol. 4, 133-144.

Miyazaki, T., and Zipes, D.P. (1989). Protection against autonomic denervation following acute myocardial infarction by preconditioning ischemia. Circ. Res. 64, 437-448.

Murrel, D.F. (1993). A radical proposal for the pathogenesis of scleroderma. J. Am. Acad. Dermatol. 28, 78-85.

Murrel, G.A.C., Francis, M.J.O., and Bromley, L. (1990). Modulation of fibroblast proliferation by oxygen free radicals. Biochem. J. 265, 659-665.

Murry, C.E., Richard, V.J., Reimer, K.A., and Jennings, R.B. (1990). Ischemic preconditioning slows energy metabolism and delays ultrastructural damage during a sustained ischemic episode. Circ. Res. 66, 913-931.

Otani, H., Prasad, M.R., Jones, R.M., and Das, D.K. (1989). Mechanisms of membrane phospholipid degradation in ischemic–reperfused rat hearts. Am. J. Physiol. 257, H252-H258.

Paller, M.S., and Greene, E.L. (1994). Role of calcium in reprfusion injury of the kidney. Ann. N.Y. Acad. Sci. 273, 59-70.

Pruzanski, W., Lee, P., Stefanski, E., Sternby, B., and Vadas, P. (1990). Hyphospholipasemia A2 in systemic sclerosis. J. Rheumatol. 17, 1182-1186.

Reilly, I.A.G., Roy, L., and Fitzgerald, G.A. (1986). Biosynthesis of thromboxane in patients with systemic sclerosis and Raynaud's phenomenon. Br. Med. J. 292, 1037-1039.

Salmenson, B.D., Reisman, J., Sinclar, S.H., and Burge, D. (1992). Macular capillary hemodynamic changes associated with Raynaud's phenomenon. Ophthalmology 99, 914-919.

Schott, R.J., Rohmann, S., Braun, E.R., and Schaper, W. (1990). Ischemic preconditioning reduces infarct size in swine myocardium. Circ. Res. 66, 1133-1142.

Simonini, G., Matucci Cerinic, M., Generini, S., Zoppi, M., Anichini, M., Cesaretti, A., Pignone, A., Falcini, F., Lotti, T., and Cagnoni, M. (1998). Oxidative stress in systemic sclerosis. Mol. Cell. Biochem. (In press.)

Spisani, S., Dovigo, L., and Colamussi, V. (1981). Leukocyte migration and phagocytosis in progressive systemic sclerosis. Scand. J. Rheumatol. 81, 299-300.

Stein, C.M., Tanner, S.B., Awad, J.A., Roberts, L.J. II, and Morrow, J.D. (1996). Evidence of free radical–mediated injury (isoprostane overproduction) in scleroderma. Arthritis Rheum. 39, 1146-1150.

Ter Borg, E.J., Piersma-Wichers, G., Smit, A.J., Kallemberg, C.G.M., and Wounda, A.A. (1994). Serial nailfold capillary microscopy in primary Raynaud's phenomenon and scleroderma. Semin. Arthritis Rheum. 24, 40-47.

Vancheeswaran, R., Magoulas, T., Efrat, G., Penny, R.M., and Black, C.M. (1994). Circulating endothelin 1 levels in systemic sclerosis subsets: A marker of fibrosis or vascular dysfunction? J. Rheumatol. 21, 1838-1844.

Win, K.D.A., Haubauer, I., Krishna, M.C., DeGraff, W., Ganson, J., and Mitchell, J.B. (1993). Nitric oxide protects against cellular damage and cytotoxicity from reactive oxygen species. Proc. Nat. Acad. Sci. 90, 13-17.

MOLECULAR ADAPTATION TO TOXIC CHEMICALS AND DRUGS

Prasanta K. Ray and Tanya Das

Advances in Organ Biology
Volume 6, pages 255-269.

ISBN: 0-7623-0391-3

I. ADAPTATION—IN THE LIGHT OF EVOLUTION

Evolution has witnessed cumulative changes in the genetic characteristics of populations that are related by descent. Such changes resulted from differences in the ability of the phenotypes in these populations to obtain representation in the next generation. Natural selection is the mechanism that drives evolution, and it favors those individuals who are best fitted to interact with the environment in ways that will promote their survival and reproduction (Watts, 1971). This fitness is the essence of adaptation, and natural selection brings about cumulative improvement in the adaptive characteristics of populations, which is conventionally termed "survival of the fittest." However, organisms respond to the solicitations of their surroundings, either by diversification of their genotypes or by an adjustment of the translation of the genetic code to the new conditions. Adaptation, therefore, is a biochemical continuum (Schoffeniels, 1971) of the total of living mass and its metabolic extensions, as well as its relations to the ever-changing conditions of the environment. In fact, biological systems have evolved hierarchial and distributed control mechanisms that greatly enhance their adaptability.

II. MOLECULAR ADAPTATION: THE SECRET OF THE ORGANISM–ENVIRONMENT RELATIONSHIP

"Molecular adaptation" starts at the level of the organism–environment relationship through a series of biochemical processes. In fact, nucleic acids and proteins have undergone, in the phylogenetic descent, all pervasive and continuous changes at the level of their primary structure. This continuous process of changes in the informational macromolecules, as influenced by the environmental factors, is the basis of molecular adaptation. Each organism orchestrates a series of molecular phenomena to adapt to the changing environment, of which the two most effective ones are:

- *Induced Reactions,* in which the inducer leads to the expression of inducible enzymes or to activation of the signal transduction phenomenon, ultimately potentiating gene expression and cellular functions.
- *Mutation* that occurs as a result of an exposure to mutagens, which can be both physical and chemical in nature. As a consequence, certain properties of the organism may be genomically changed.

It is obvious, therefore, that the study of evolution will permit us to understand more about the "molecular adaptation" process that holds the key to morphological and physiological adaptations. In this chapter, we will discuss the molecular adaptation phenomena of the host against toxic chemical (drug) insults.

III. WHY CHEMICALS AND DRUGS BECOME TOXIC

It is interesting to note that each organism has the built-in capacity to tolerate each chemical up to a certain concentration (i.e., "threshold dose"), which of course varies from one chemical to another and is also dependent on the physiological status of the individual. Sources of exposure to these chemicals include (1) the environment, over which we have no control, and (2) drugs that are taken as medicine.

Human interaction with the products of nature and those developed by human ingenuity does not lead normally to physiological catastrophe. Evolutionary gains have allowed humans and other higher organisms to develop a number of anatomical, physiological, biochemical, and immunological barriers that provide various means of molecular adaptation to noxious agents. Thus, following the exposure to any noxious agent, the physiological system may initiate a large number of processes, as detailed below:

- Several types of enzyme synthesis are induced to catalyze the oxidation, reduction, cleavage, rearrangement, and conjugation reactions that will process the foreign agents for their rapid elimination.
- These chemicals or their metabolites send signals to activate gene expression for the production of new biomolecules to fight back the toxic insults.
- Pre-exposure to toxic chemicals or drugs strengthens the antioxidant system, which in turn protects the host from further oxidative stress.
- The body's immune system is alerted to destroy the foreign agent and/or eliminate it from the system.
- Cell growth and DNA repair mechanisms are potentiated to replenish the damage caused by noxious chemicals and/or other agents.

IV. ENZYMATIC BASIS OF MOLECULAR ADAPTATION

The higher organisms appear to have developed a class of genes encoding various proteins and enzymes designed to meet the need for detoxication of compounds that are without physiological value. Such detoxication reactions may be classified very broadly into four types, occurring in two distinct phases: phase I—biotransformation, and phase II—detoxification. In principle, the products of metabolism are, in general, more water soluble and thus more easily eliminated than their precursors.

A. Phase I Biotransformation Enzymes

These enzymes are broadly classified into three groups: (1) mixed function mono-oxygenase systems; (2) oxidation reduction systems; and (3) hydrolytic enzyme systems.

Table 1. Other Phase I and Phase II Enzymes

Enzyme	*Activity*
Phase I	
Alcohol dehydrogenase	Alcohol oxidation
L-gulonate dehydrogenase, aldose reductase, glycerol dehydrogenase	Reductive biotransformation of aldehydes and ketones
Aldehyde dehydrogenases	Aldehyde oxidation
Ketone reductase	Reduction of carbonyl compounds
Xanthine oxidase	Oxidative hydroxylation of nitrogen-containing heterocyclic compounds and aldehydes
Superoxide dismutase	Dismutation of superoxide radical
Glutathione peroxidase	Detoxification of hydroperoxide
Monoamine oxidase	Oxidative deamination of amines
Microsomal epoxide hydrolase, carboxyesterases, and amidases	Transformation of toxic drugs, insecticides, and antibiotics
Phase II	
Glucuronidase	Conversion of toxic chemicals and drugs to polar water-soluble glucuronides
Methyltransferases	Biological *N*- and *O*- methylation
Glutathione-S-transferases	Catalyze the reaction of glutathione and epoxide

1. *Mixed Function Mono-oxygenase Systems*

Many of the detoxification reactions require the participation of the mixed function oxygenase systems, such as cytochrome P-450, which is the central enzyme of the mixed function oxygenase system. Treatment of animals with various inducers results in the appearance of different types of cytochrome P-450. This enzyme brings about chemical changes both in physiologically important substrates, such as fatty acids, steroids, and prostaglandins, and in a host of foreign substances, such as petroleum products, drugs, pesticides, anesthetics, various chemical carcinogens, and miscellaneous organic substances (Coon and Pearson, 1980).

The cytochrome P-450 enzyme system is induced as an adaptive response of the body to toxic chemicals and drugs. It is apparent that a selection mechanism must exist for the elevation of specific forms of cytochrome P-450 by specific inducers.

The requirement of NADPH–cytochrome C reductase in cytochrome P-450 mediated reactions has also been demonstrated. This enzyme may also function in other physiologically important reactions, not involving cytochrome P-450, for example, the heme oxygenase reaction (Yoshida and Kikuchi, 1978) which involves NADPH–cytochrome C (P-450) reductase and an apoprotein.

2. *Microsomal Flavin-containing Mono-oxygenase*

A microsomal flavin–containing cytochrome-free mono-oxygenase catalyzes oxygenation of nucleophiplic organic nitrogen compounds (Gorrod, 1978) and is a major route for the metabolism and disposition of trimethylamine, different tertiary and secondary amines, aryl or alkyl thiols, thiones, or thioamides.

B. Phase II Detoxification Enzymes

Among other phase I enzymes and phase II conjugation enzymes are those listed in Table 1. Induced synthesis of glutathione and many of the enzymes listed in Table 1 occurs as an adaptive response of the body. These responses are well illustrated in any textbook of pharmacology and metabolism.

V. SIGNALING EVENTS INVOLVED IN MOLECULAR ADAPTATION

The interaction of a ligand with its cognate receptor not only activates signal transduction pathways, but also determines adaptive responses of the cell through gene expression and synthesis of new proteins. The regulatory mechanisms involved can be divided according to two distinct time frames, acute and chronic.

Short-term regulation involves post-translational mechanisms that alter the functional status of the elements without changing steady state levels or gene expression. Protein phosphorylation plays a prominent role in these acute adaptive responses. Long-term regulation involves transcriptional (gene expression), post-transcriptional (mRNA stability), and post-translational (protein phosphorylation) regulation of the turnover of the elements in the information (signal) transduction pathway. Signaling systems thus enable organisms to sense their environment and mount an appropriate adaptive response. Signal transduction and gene regulation are the two major global regulatory networks of molecular adaptation.

Recent findings show that neurotransmitters and drugs that affect neurotransmission and signal transduction have important influences on gene expression. Among the commonly used drugs, ethanol produces specific changes in several signal transduction cascades. It stimulates phospholipase C, increases cytosolic calcium levels, and activates the expression of several genes. The product of one of these genes has extensive sequence homology to phosducin, a phosphoprotein that modulates trimeric G protein function by binding to G protein β subunits (Miles et al., 1993). Such alterations in signaling are thought to be a crucial aspect of the central nervous system's adaptive response, which occurs with chronic exposure to ethanol (Miles et al., 1993). Moreover, there is a depressed unsaturation: polyunsaturation index (Ellingson et al., 1991) during chronic ethanol administration, which can modify the membrane function. Alprazolam, busprione, benzodiazepines, and morphine may also evoke the adaptive response involving Ca^{2+} channels, second messengers, and protein phosphorylation (Brennan and Littleton, 1991).

Chronic benzodiazepine treatment elicits adaptive responses in the central nervous system that involve down-regulation of benzodiazepine receptors, β-adrenergic receptor, α_2-adrenoreceptor, and postsynaptic serotonergic receptor (Bourin and Bradwejn, 1991). The antidepressants imipramine hydrochloride and phenelzine sulfate both downregulate adrenal peripheral-type benzodiazepine receptors and upregulate hepatic peripheral-type benzodiazepine receptors as the adaptive responses to drug-induced hormonal and cellular responses (Weizman et al., 1993). Antidepressants also downregulate the mRNA level and the expression of the tyrosine hydroxylase gene (Nestler et al., 1990).

The antidepressant 1-sulpiride downregulates β receptor–linked adenylate cyclase activity in the frontal cortex and blocks dopamine D_2 receptors, thus controlling norepinephrine release as an adaptive response (Sigala et al., 1991). On the other hand, chronic exposure of the brain to nicotine induces upregulation of $\alpha_4\beta_2$-receptor protein, inducing a conformational change of the receptor that is removed from the surface and degraded slowly (Peng et al., 1994).

Studies of the mechanisms of opiate drug-induced changes in proenkephalin gene expression demonstrated that AP-1 complex differentially regulates proenkephalin transcription at Ca^{2+}-/calmodulin-dependent responsive element (CRE)-2. cJun constitutively activates transcription but JunD activation is dependent on cyclic adenosine monophosphate (cAMP)–dependent protein kinase (Comb et al., 1992).

Peroxisome proliferators induce transforming growth factor (TGF)-β1 and insulinlike growth factor (IGF)-II/mannose-6-phosphate receptor expression, indicating an adaptive response to limit the initial hyperplastic effects of such compounds (Rumsby et al., 1994). On the other hand, phosphoinositide hydrolysis and activation of protein kinase C are the reactions involved in the adaptive response of many cells to different chemicals and drugs (Wojcikiewicz and Nahorski, 1991). The induction of an adaptive response by hydrogen peroxide (H_2O_2) in human lymphocytes involves free Ca^{2+} but not inositol triphosphate (IP_3)–dependent Ca^{2+} release (Wojewodzka et al., 1994). Moreover, H_2O_2 can also act as a signal transducer in oxidative stress by oxidizing a dithiol protein to disulfide form, which then activates transcription of stress-inducible genes as a part of the adaptive response (Crapo and Tierney, 1974).

Adrenal corticosteroids and stress proteins play important roles in regulating cell survival in response to toxic chemicals or drugs. Corticosteriods can produce a significant reduction in the apoptotic death of L929 mice fibroblasts produced by tumor necrosis factor (TNF)-α (Pagliacci et al., 1993) by inducing synthesis of a hippocampal cytosolic protein having characteristics of glycerol phosphate dehydrogenase.

Lipopolysaccharide (LPS), the *E. coli* endotoxin, causes a time-dependent increase in glucose-6-phosphate dehydrogenase (G6PD) mRNA expression in Kupffer and endothelial cells (Spolarics and Navarro, 1994). Increased G6PD expression is important for the support of up-regulated NADPH-dependent pathways, such as superoxide anion and nitric oxide production, macromolecular synthesis, or maintenance of cellular glutathione status. On the other hand, reduction in receptor-linked nicric oxide release, as occurs during diabetes, may be counteracted by sodium nitroprusside–stimulated cyclic guanosine monophosphate (cGMP) and prostagladin E1-stimulated cAMP production (Miller et al., 1994). Adaptive changes in endogenous nitric oxide are known to play a critical role, among others, in sodium and blood pressure homeostasis. Methyl mercury hydroxide (MMH) induces glutathione synthesizing rate-limiting enzyme—glutamyl-cysteine synthetase—and such increase in gluthatione protects renal cells from oxidative tissue damage during MMH exposure (Woods et al., 1992). These reports and numerous others clearly demonstrate how molecular adaptation is governed by purposeful, but regulated, signal transduction mechanisms.

VI. PRECONDITIONING BY CHEMICALS REDUCING OXIDATIVE STRESS

Oxidative stress is a major threat to life processes. Oxygen-derived free radicals are probable mediators of such toxicity. Antioxidants thus can protect cultured bovine lung endothelial cells from endotoxin mediated cellular injury (Brigham et al., 1987). Protein A, a *Staphylococcus aureus* Cowan I cell wall protein, can ameliorate cyclophosphamide and carbon tetrachloride–induced toxicity, probably due to its antioxidant property (Agarwal and Ray, 1991). Hydrogen peroxide at a low dose renders protection to the host from further oxidative injury (Collinson and Dawes, 1992). Recent studies revealed that pre-exposure to LPS, the *E. coli* endotoxin, can induce an oxidative stress that is subsequently translated into the stimulation of antioxidant enzymes and enhanced antioxidant reserve, ultimately resulting in myocardial adaptation to ischemia (Maulik et al., 1995). Continuous infusion of deprenyl for 1 week showed significant increase in superoxide dismutase and calatase activities, the enzymes of the antioxidant system (Carrillo et al., 1992). All of these studies indicate that preconditioning of the system by chemicals can reduce oxidative stress.

VII. MUTATIONS AS ANOTHER TOOL OF MOLECULAR ADAPTATION

In the modern theory of molecular adaptation, mutations have been ascribed to have an important role to play. Mutations may be induced by irradiation, chemicals, drugs, and so forth. If the mutations that occur, provide some advantage to their possessors, natural selection favors them. During mutations, substitution or deletion of one nucleotide in the DNA molecule can result in the production of altered protein(s). Proteins serving as enzymes catalyze all living processes. Thus, changes in genes by mutation result in the production of enzymes which, in turn, results in changes in metabolism and developmental processes. Hence, these genetic mutations become a part of the molecular adaptation process. For example, during the adaptive response toward the oxidative stress of human skin fibroblasts by UV radiation, an increase in levels of the heme catabolizing enzyme, heme oxygenase 1, and the iron storage protein, ferritin, was observed (Vile et al., 1994) which protects human skin fibroblasts from damage by further irradiation. Studies of tumor suppressor gene p53 demonstrate that mutation alters the protein in such a way that it gains function, and subsequently gains some of the characteristics of an activated oncogene (Lane and Benchimol, 1990). It is interesting to note that overexpression of p53 in tumors can be taken to indicate the presence of mutant forms of p53. Thus alteration of one gene may have far-reaching effects upon developmental and metabolic processes and hence upon survival.

Mutagenic agents alter the DNA molecule, either directly or probably more commonly by altering chemical structures in the cellular environment of the DNA molecule. For example, the so-called "junk DNAs," which during normal conditions are not expressed, may be caused to respond after mutation, becoming activated to synthesize new proteins as the adaptive response toward those mutagens. Nature may have preserved them as weapons in reserve to deal with the various onslaughts it has to suffer from environmental agents.

VIII. ADAPTIVE PROTECTION TOWARD DNA DAMAGE AND HARMFUL MUTATIONS

Most of the mutations we study in our laboratories are harmful or not beneficial. The multitudinous mutations are responsible for almost all of the abnormalities and malformations (e.g., reduction in the viability or fertility) of the possessors. To overcome these lethal mutational effects, the body's adaptive response induces repair activity that acts on lesions in DNA through the synthesis of new or altered proteins. The repair of UV-induced postreplication DNA gaps has been observed in *E. coli* cells adapted to methylmethane sulfonate and ethylmethane sulfonate (Zhestianikov and Savel'eva, 1994).

A low dose of γ-ray or H_2O_2 pre-exposure leads to a decreased susceptibility to gene deletions and rearrangements after high-dose irradiation as the adaptive response. Actinomycin-D, bleomycin sulfate, or interferon might make bone marrow cells resistant to the clastogenic effects of radiation (Mozdarani and Saberi, 1994). Thus, using these drugs in combination with radiation may decrease the adverse biological effects of radiation. Interferon inducers such as protein A and 6 micelial fraction acetone also protect normal cells and tissues from toxic damage caused by cyclophosphamide (Zaidi et al., 1990), carbon tetrachloride (Singh, et al., 1990), benzene (Shankar et al., 1993), azidothymidine (AZT) (Subbulakshmi et al., 1998), aflatoxin (Raisuddin et al., 1994), and dimethyl benzanthracene (Shukla et al., 1996), among others. The effect of such molecular adaptations may be selective for specific mutagens.

Finally, pre-exposure to different alkylating agents can induce protection against UV radiation by promoting *ada* and *alkA* gene expression in the adaptive process (Saget and Walker, 1994).

IX. ADAPTIVE RESPONSE OF THE IMMUNE SYSTEM

To reproduce and to protect the "self" from environmental odds are the two intrinsic urges of all mammals. The immune system is the body's major defense mechanism to fight against the insults of foreign substances, including toxic chemicals and drugs.

Upon first contact with an antigen, the system is altered to leave behind some memory that would enable the response to any subsequent encounter to be faster and greater in magnitude. The basic principle of immunization is the preconditioning of the system with required antigen. Thus, utilizing the specificity and memory of the acquired immune response, the foundation of molecular adaptation is laid. Molecular adaptation of lymphocytes to different chemicals, drugs, or ionizing radiation requires new and/or altered synthesis of proteins (Shadley, 1994). A dose-dependent stimulation of immunoglobulin synthesis by B lymphocytes in the presence of adriamycin or ethidium bromide has been observed, which involves free radical processes (Meliksetian et al., 1993). Studies of the induction of adaptive response in bone marrow cells to ethyl methane sulfonate, a monofunctional alkylating agent, indicate that a low dose of this agent offers resistance to these cells against further clastogenic effects of any challenge dose of this chemical (Mahmmod and Vasudev, 1993).

Exposure of mouse bone marrow–derived macrophages to β-1,3-glucan, a particulate inflammatory stimulus, up-regulates the expression of acid hydrolase, β-glucuronidase, and platelet-derived growth factor β, and to polyinosinate polycytidylate, a stimulus of macrophage cytocidal activation, and also stimulates the expression of complement component (Laszlo et al., 1993), indicating the molecular adaptation of macrophages to potentiate host defence.

In another example of molecular adaptation of the immune system, immunoglobulins form networks of immune complexes with foreign antigenic compounds and remove them from the circulatory pool. This process reduces the residency time of foreign substances in the system, as well as their harmful actions.

It is known that immunocytes (e.g., macrophages, T cells, and B cells), produce various lymphokines, which are important biomodulators and regulate growth, differentiation, and proliferation of immunocytes and other cell types. These important molecules take part in a molecular adaptation process either directly or by modulating an immune system response and taking part in replenishing the damage caused to cells and tissues. Thus, if the host is competent immunologically, it should be able to adapt itself better to the adverse environmental conditions.

X. FAILURE OF THE SYSTEM TO ADAPT AGAINST A TOXIC INSULT

As the preceding discussion makes clear, the better an organism can adapt to the environmental odds, the greater its chance of survival. The human or animal body thus maintains different intrinsic regulatory programs to ensure proper functioning of its adaptive machineries to fight against the adverse effects of toxic chemicals and drugs—but only up to a limit.

Many toxic chemicals depress phase I and phase II biotransformation enzymes, which otherwise are required to detoxify and eliminate the unwanted metabolic products from the body. Undue mutations may also become harmful and fatal for the body. Accumulated toxic chemicals can interfere with the signaling events and hamper molecular adaptation. Many toxicants and carcinogens are known as potent immunosuppressors. Thus, continuous exposure of the body to various toxic agents may jeopardize the adaptive response of the host, ultimately causing anergy. As a result of such a failure, an organism or animal may lose its fitness for survival.

XI. HOW TO POTENTIATE ADAPTIVE MECHANISMS: A NEW CONCEPT

Molecular adaptation is the result of multifunctional processes in which various metabolic, genomic, and immunologic reactivities are included. If the host is competent in all these areas of its functioning, it should be able to develop a proper molecular adaptation mechanism to keep itself fit in the otherwise hostile environment. Our hypothesis is that by using suitable biological response modifiers, one may be able to reconstitute the intrinsic capacity of a host's molecular adaptation mechanisms, enabling the host to fight off toxic, carcinogenic insults (Ray's hypothesis).

We have constructed this hypothesis using protein A (PA) of *Staphylococcus aureus* Cowan I as a probe. PA has a vast array of biological response modifying activities and thus may play a crucial role in regulating some of the body's adaptive systems. PA is an anticarcinogen, antitoxic molecule (Shukla et al., 1996), having immunostimulatory properties (Mishra et al., 1992; Singh et al., 1992; Prasad and Ray, 1991). PA can modulate toxicity induced by various toxic chemicals such as cyclophosphamide, carbon tetrachloride, aflatoxin, and benzene (Zaidi et al., 1990; Shankar et al., 1993; Raisuddin et al., 1994). The carcinogenic potential of dimethyl benzanthracene (Shukla et al., 1996) could also be decreased substantially by PA. It has the unique ability to revert the depressed enzyme activites of both the phase I and phase II biotransformation and detoxifying systems (Dwivedi et al., 1989). Since continuous metabolic transformation and elimination of toxic intermediates may help in molecular adaptation, accelerated repletion and recovery of those crucial enzymes by PA reconstitute the host's adaptive system.

The unique ability of PA to replete the depleted hematopoietic cells is also of major significance in repairing the cellular damage and/or replenishing the damage.

PA has been found to be a potent immunostimulator. It shows increased macrophage phagocytic activity (Mishra et al., 1992) and increased elicitation of interleukin-1 (Prasad and Ray, 1991). It stimulates specific clones of T cells to show an increased release of interleukin-2 (Singh et al., 1992) TNF-α (Paul et al., 1993) and γ-interferon (Ray et al., 1997). Lymphokine-activated killer cell activity and antibody-dependent cell-mediated cytotoxicity reactions are also stimulated by PA (Singh et al., 1992; Catalona et al., 1981). PA also helps the host to effect an accelerated removal of metabolites from AZT-treated hosts (Subbulakshmi et al., 1998).

Very recently, α_1 signaling and signal transduction pathways have been implicated in the preconditioning phenomenon during ischemia (Das, 1993). Recent observation from our laboratory substantiated the view of PA as a signal inducer in B cell proliferation, following a cascade of reactions involving tyrosine kinase, phospholipase C, IP_3, Ca^{2+}, protein kinase C (PKC), and mitogen-activated protein kinase, as signal mediators (Ray et al., 1997). These important signaling molecules can themselves add up to the adaptive power of the host, or the antibodies produced as a result of increased B cell proliferation may potentiate the immunological pathways of molecular adaptation. PA also induces the production of different lymphokines via a PKC-dependent pathway (unpublished data). These lymphokines may, in turn, initiate cascades of molecular reactions, increasing the interactions among various compartments of metabolic, genetic, and immune systems.

Because PA has been shown to induce immunostimulation, antioxidant reactivities, and activation of detoxication enzymes, as well as to provoke signal transduction pathways, any one or more of these phenomena may be instrumental in prompting the host to regain its potential for molecular adaptation.

As suggested in case of lipopolysaccharide preconditioning, such adaptation may involve gene expression and protein biosynthesis, and indeed, PA-treated immunocytes showed an increase in the protein levels of some important signaling molecules and oncoproteins, including c-fos, c-myc, and PKC (Das et al., 1995). Since the oncoproteins (e.g., c-myc) are thought to convert biochemical signals into changes in gene expression, the role of PA in gene expression is also implicated in its overall functions. These multifactorial effects of PA support our hypothesis (see Figure 1).

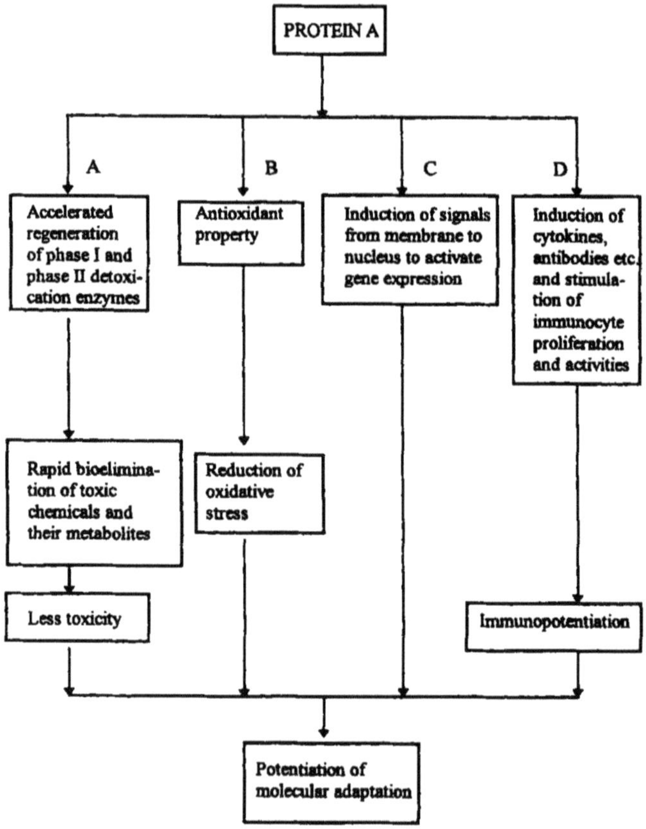

Figure 1 Ray's hypothesis to potentiate molecular adaptation. A. Adaptation through detoxication enzymes. B. Adaptation through antioxidant property. C. Adaptation through signaling events. D. Adaptation through immunopotentiation.

CONCLUSIONS

If the metabolic, genetic, and immunologic profiles of the host can be maintained in a functionally operative state from the point of view of biotransformation, mutation repair, signaling, detoxification, and immunopotentiation, it would provide the host with the potency to adapt and withstand larger-than-normal dosages of toxic chemicals or drugs. Ray's hypothesis thus may add a new dimension to the knowledge of molecular adaptation and, ultimately, biological evolution.

In the course of our further understanding, such strategies of induction of specific molecular adaptation phenomena might help in strengthening the host to withstand an increasing order of stressor-induced phenomena, either during toxic drug therapy or as a preventive measure to tune up the molecular adaptation mechanisms in high-risk individuals who work in an environment containing toxic and/or carcinogenic chemicals.

ACKNOWLEDGMENT

The authors acknowledge V. Subbulakshmi and S. Goenka for general assistance and R. Das for typing the manuscript.

REFERENCES

Agarwal, D.K., and Ray, P.K. (1991). *In vitro* antioxidant property of protein A of *Staphylococcus aureus*. Indian J. Exp. Biol. 29, 1130-1133.

Bourin, M., and Bradwejn, J. (1991). Clinical and neurobiological aspects of long-term administration of psychotropic drugs. Can. J. Psychiatry 36, 260-264.

Brennan, C.H., and Littleton, J.M. (1991). Chronic exposure to anxiolytic drugs, working by different mechanisms causes upregulation of dihydropyridine binding sites on cultured bovine adrenal chromaffin cells. Neuropharmacology 30, 199-205.

Brigham, K.L., Meyrick, L.C., Berry, L.C. Jr., and Repine, J.E. (1987). Antioxidants protect cultured bovine lung endothelial cells from injury by endotoxin. J. Appl. Physiol. 63, 840-850.

Carrillo, M.C., Kanai, S., Sato, Y., Ivy, G.O., and Kitani, K. (1992). Sequential changes in activities of superoxide dismutase and catalase in brain regions and liver during (–) diprenyl infusion in male rats. Biochem. Pharmacol. 44, 185-189.

Catalona, W.J., Ratcleff, T.L., and McCool, R.E. (1981). Interferron induced by *S. aureus* protein A augments natural killing & ADCC. Nature 291, 77-79.

Collinson, L.P., and Dawes, I.W. (1992). Inducibility of the response of yeast cells to peroxide stress. J. Gen. Microbiol. 138, 329-335.

Comb, M.J., Kobierski, L., Chu, H.M., Tan, Y., Borsook, D., Herrup, K., and Hyman, S.E. (1992). Regulation of opioid gene expression: a model to understand neural plasticity. NIDA Res. Monogr. 126, 92-112.

Coon, M.J., and Persson, A.V. (1980). Microsomal cytochrome P-450: a central catalyst in detoxication reactions. In: Enzymatic Basis of Detoxication (Jakoby, W.B., Ed.), pp. 117-134. Academic Press, New York.

Crapo, J.D., and Tierney, D.L. (1974). Superoxide dismutase and pulmonary oxygen toxicity. Am. J. Physiol. 226, 1401-1407.

Das, D.K. (1993). Ischemic preconditioning and myocardial adaptation to ischemia. Cardiovasc. Res. 27, 2077-2079.

Das, T., Sa, G., Subbulakshmi, V., Sen, P., Biswas, S. and Ray, P.K. (1995). Unique properties of protein A of *Staphylococcus aureus*. IV. Tracking the signaling pathway for protein A from membrane to nucleus [Abstr.]. International Symposium on Trends in Microbiology, December 5-8, Calcutta.

Dwivedi, P.D., Verma, A.S., Mishra, A., Singh, K.P., Prasad, A.K., Saxena, A.K., Dutta, K.K., Mathur, N., and Ray, P.K. (1989). Protein A protects mice from depletion of biotransformation enzymes and mortality induced by *Salmonella typhimurium* endotoxin. Toxicol. Lett., 49, 1-13.

Ellingson, J.S., Janes, N., Taraschi, T.F., and Rubin, E. (1991). The effect of chronic ethanol consumption on the fatty acid composition of phosphatidylinositol in rat liver microsomes as determined by gas chromatography and 1H-NMR. Biochim. Biophys Acta 1062, 199-205.

Gorrod, J.W. (1978). On the multiplicity of microsomal N-oxidase systems. In: Mechanisms of Oxidizing Enzymes (Singer, T.P., and Ondarza, R.N., Eds.), pp. 189-197. Elsevier, Amsterdam.
Lane, D.P., and Benchimol, S. (1990). p53: oncogene or anti-oncogene Genes Dev. 4, 1-8.
Laszlo, D.J., Henson, P.M., Remigio, L.K., Weinstein, L., Sable, C., Noble, P.W., and Riches, D.W. (1993). Development of functional diversity in mouse macrophages. Mutual exclusion of two phenotypic states. Am. J. Pathol. 143, 587-597.
Mahmood, R. and Vasudev, V. (1993). Inducible protective processes in animal systems: IV. Adaptation of mouse bone marrow cells to a low dose of ethyl methane sulfonate. Mutagenesis 87, 83-86.
Maulik, N., Watanabe, M., Engelman, D., Engelman, R.M., Kagan, V.E., Kishin, E., Tyurin, V., Cordis, G.A., and Das, D.K. (1995). Myocardial adaptation to ischemia by oxidative stress induced by endotoxin. Am. J. Physiol. 269, 907-916.
Meliksetian, M.B., Davtian, T.K., Ivanova, I.V., Aleksanian, I.T., and Ignatova, T.N. (1993). The adaptive response of B cells to the action of cytotoxic preparations in culture III. The possible mechanism of the action of adriamycin on immunoglobin synthesis by the cells of murine myeloma. Isitologiia 35, 98-104.
Miles, M.F., Barhite, S., Sganga, M. and Elliott, M. (1993). Phosducin like protein: an ethanol responsive potential modulator of guanine nucleotide binding protein function. Proc. Nat. Acad. Sci. U.S.A. 96, 10831-10835.
Miller, M.A., Morgan, R.J., Thompson, C.S., Mikhailidis, D.P., and Jeremy, J.Y. (1994). Adenylate and guanylate cyclase activity in the penis and aorta of the diabetic rat: an *in vitro* study. Br. J. Urol. 74, 106-111.
Mishra, A., Dwivedi, P.D., Verma, A.S., and Ray, P.K. (1992). Mechanism of enhanced phagocytic response in protein A treated rat macrophages. Immunol. Lett. 34, 289-296.
Mozdarani, H., and Saberi, A.H. (1994). Induction of cytogenetic adaptive response of mouse bone marrow cells to radiation by therapeutic doses of bleomycin sulfate and actinomycin D as assayed by the micronucleus test. Cancer Lett. 78, 141-150.
Nestler, E.J., McMohon, A., Sabban, E.L., Tallman, J.F., and Duman, R.S. (1990). Chronic antidepressant administration decreases the expression of tyrosine hydroxylase in the rat locus coeruleus. Proc. Nat. Acad. Sci. U.S.A. 87, 7522-7526.
Pagliacci, M.C., Migliorati, G., Smacchia, M., Grignani, F., Riccardi, C., and Nicoletti, I. (1993). Cellular stress and glucocorticoid hormones protect L929 mouse fibroblasts from tumor necrosis factor alpha cytotoxicity. J. Endocrinol. Invest. 16, 591-599.
Paul, B.N., Saxena, A.K., and Ray, P.K. (1993). *In vivo* induction of tumor necrosis factor alpha by soluble protein A from *Staphylococcus aureus*. Immunol. Infect. Dis. 3, 295-298.
Peng, X., Gerzanich, V., Anand, R., Whiting, P.J., and Lindstrom, J. (1994). Nicotine-induced increase in neuronal nicotinic receptors results from a decrease in the rate of receptor turnover. Mol. Pharmacol. 46, 523-530.
Prasad, A.K., and Ray, P.K. (1991). Protein A induced elicitation of interleukin-1 by activated peritoneal macrophages. Int. J. Toxicol. Environ. Health 1, 101-106.
Raisuddin, S., Singh, K.P., Zaidi, S.A., and Ray, P.K. (1994). Immunostimulating effect of protein A in immunosuppressed aflatoxin intoxicated rats. Int. J. Immunopharmacol. 16, 977-984.
Ray, P.K., Datta, P.K., Das, T., Srivastava, M., and Subbulakshmi, V. (1997). Structural analysis of protein A to localize immuno-modulatory motifs. Immunomodulation 101-110.
Rumsby, P.C., Davies, M.J., Price, R.J., and Lake, B.G. (1994). Effect of some peroxisome proliferators on transforming growth factor beta 1 gene expression and insulin like growth factor II mannose-6-phosphate receptor gene expression in rat liver. Carcinogenesis 15, 419-421.
Saget, B.M., and Walker, G.C. (1994). The Ada protein acts as both a positive and a negative modulator of *Escherichia coli*'s response to methylating agents. Proc. Nat. Acad. Sci. U.S.A. 91, 9730-9734.
Schoffeniels, E. (1971). Adaptation at the molecular scale. In: Biochemical Evolution and the Origin of Life (Schoffeniels, E., Ed.), pp. 314-335. North-Holland Publishing Company, Amsterdam/London.

Shadley, J.D. (1994). Chromosomal adaptive response in human lymphocytes. Radiat. Res. 138, 9-12.

Shankar, U., Kumar, A., Rao, G.S., Dwivedi, P.D., Pandya, K.P., and Ray, P.K. (1993). Modulation of benzene induced toxicity by protein A. Biochem. Pharmacol. 46, 517-529.

Shukla, Y., Verma, A.S., Mehrotra, N.K., and Ray, P.K. (1996). Antitumor activity of protein A in a mouse skin model of two stage carcinogenesis. Cancer Lett. 103, 41-47.

Sigala, S., Rizzonelli, P., Zanelli, E., Forgione, A., Missale, C., and Spano, P. (1991). Low doses of 1-sulpiride down regulate striatal and cortical dopamine receptors and beta adrenoceptors. Eur. J. Pharmacol. 199, 247-253.

Singh, K.P., Shau, M., Gupta, R.K., Kopald, K., and Ray, P.K. (1992). Protein A potentiates lymphokine activated killer cell induction in normal and melanoma patient lymphocytes. Immunopharmacol. Immunotoxicol. 14, 79-103.

Singh, K.P., Zaidi, S.I., Raisuddin, S., Saxena, A.K., Dwivedi, P.D., Seth, P.K., and Ray, P.K. (1990). Protection against carbon-tetrachloride induced lymphoid organotoxicity in rats by protein A. Toxicol. Lett. 51, 339-351.

Spolarics, Z., and Navarro, L. (1994). Endotoxin stimulates the expression of glucose-6-phosphate dehydrogenase in Kupffer and hepatic endothelial cells. J. Leukoc. Biol. 56, 453-457.

Subbulakshmi, V., Ghosh, A.K., Das, T., and Ray, P.K. (1998). Mechanism of protein A induced amelioration of toxicity of anti-AIDS drug, zidovudine. Biochem. Biophys. Res. Commun. (In press).

Vile, G.F., Basu-Modak, S., Waltner, C., and Tyrrell, R.M. (1994). Heme oxygenase I mediates an adaptive response to oxidative stress in human skin fibroblasts Proc. Nat. Acad. Sci U.S.A. 91, 2607-2610.

Watts, R.L. (1971). Genes, chromosomes and molecular evolution. In: Biochemical Evolution and the Origin of Life (Schoffeniels, E., Ed.), pp. 14-42. North-Holland Publishing Company, Amsterdam/London.

Weizman, R., Burgin, R., and Gavish, M. (1993). Modulatory effect of antidepressants on peripheral type benzodiazepine receptors. Eur. J. Pharmacol. 250, 289-294.

Wojcikiewicz, R.J., and Nahorski, S.R. (1991). Chronic muscarinic stimulation of SH-SY5Y neuroblastoma cells suppresses inositol 1,4,5-triphosphate action. Parallel inhibition of inositol 1,4,5 triphosphate-induced Ca^{2+} mobilization and inositol 1,4,5 triphosphate binding. J. Biol. Chem. 266, 22234-22241.

Wojewodzka, M., Wallicka, M., Sochanowicz, B., and Szumiel, I. (1994). Calcium antagonist, TMB-8, prevents the induction of adaptive response by hydrogen peroxide or x-rays in human lymphocytes. Int. J. Radiat. Biol. 66, 99-109.

Woods, J.S., Davis, H.A., and Baer, R.P. (1992). Enhancement of gamma-glutamylcysteine synthetase mRNA in rat kidney by methyl mercury. Arch. Biochem. Biophys. 296, 350-353.

Yoshida, T., and Kikuchi, G. (1978). Features of the reaction of heme degradation catalyzed by the reconstituted microsomal heme oxygenase system. J. Biol. Chem. 253, 4230-4236.

Zaidi, S.I., Singh, K.P., Raisuddin, S., Saxena, A.K., and Ray, P.K. (1990). Protein A-induced abrogation of cyclophosphamide toxicity is associated with concomitant potentiation of immune function of host. Immunopharmacol. Immunotoxicol. 12, 479-512.

Zhestianikov, V.D., and Savel'eva, G.E. (1994). The repair of UV-induced postreplication DNA gaps in *Escherichia coli* cells adapted to methylmethane sulfonate and ethylmethane sulfonate. Isitologiia 36, 194-199.

INDEX

www.ingramcontent.com/pod-product-compliance
Ingram Content Group UK Ltd.
Pitfield, Milton Keynes, MK11 3LW, UK
UKHW020246290926
15470UKWH00002B/771

9 780762 3039